Protein
Functionality
in
Food Systems

ift Basic Symposium Series

Edited by
INSTITUTE OF FOOD TECHNOLOGISTS
221 N. LaSalle St.
Chicago, Illinois

1. Foodborne Microorganisms and Their Toxins: Developing Methodology, *edited by Merle D. Pierson and Norman J. Stern*
2. Water Activity: Theory and Applications to Food, *edited by Louis B. Rockland and Larry Beuchat*
3. Nutrient Interactions, *edited by C. E. Bodwell and John W. Erdman*
4. Food Toxicology: A Perspective on the Relative Risks, *edited by Steven L. Taylor and Richard A. Scanlan*
5. Biotechnology and Food Process Engineering, *edited by Henry G. Schwartzberg and M. A. Rao*
6. Sensory Science Theory and Applications in Foods, *edited by Harry T. Lawless and Barbara P. Klein*
7. Physical Chemistry of Foods, *edited by Henry G. Schwartzberg and Richard W. Hartel*
8. Flavor Measurement, *edited by Chi-Tang Ho and Charles H. Manley*
9. Protein Functionality in Food Systems, *edited by Navam S. Hettiarachchy and Gregory R. Ziegler*

Protein
Functionality
in
Food Systems

edited by

Navam S. Hettiarachchy
Department of Food Science
University of Arkansas
Fayetteville, Arkansas

Gregory R. Ziegler
Department of Food Science
Pennsylvania State University
University Park, Pennsylvania

CRC Press
Taylor & Francis Group
Boca Raton London New York

CRC Press is an imprint of the
Taylor & Francis Group, an **informa** business

CRC Press
Taylor & Francis Group
6000 Broken Sound Parkway NW, Suite 300
Boca Raton, FL 33487-2742

First issued in paperback 2019

ISBN-13: 978-0-8247-9197-1 (hbk)
ISBN-13: 978-0-367-40205-1 (pbk)

Library of Congress Cataloging-in-Publication Data

Protein functionality in food systems / edited by Navam S.
  Hettiarachchy, Gregory R. Ziegler.
      p.   cm.—(IFT basic symposium series ; 9)
    Includes bibliographical references and index.
    ISBN 0-8247-9197-5
    1. Food—Protein content—Congresses.   2. Proteins in human
  nutrition—Congresses.   3. Proteins—Analysis—Congresses.
  I. Hettiarachchy, Navam S.   II. Ziegler, Gregory R.
  III. Series.
  TX553.P7P758   1994
  664—dc20
                                                              94-4799
                                                              CIP

Visit the Taylor & Francis Web site at
http://www.taylorandfrancis.com

and the CRC Press Web site at
http://www.crcpress.com

JOHN EDWARD KINSELLA
1938–1993

We think not a friend lost because he is gone into another room, nor because he is gone into another land, and into another world no man is gone, for that Heaven which God created and this world are all one.

William Penn

Dr. John E. Kinsella's contribution to the study of protein functionality is revealed in the reference list of nearly every chapter of this book. Among the contributors are several of John's former graduate students and postdoctoral scholars. All considered him a valued colleague. In recognition of this, we dedicate this book to his memory.

Preface

The Institute of Food Technologists (IFT) and the International Union of Food Science and Technology (IUFoST) sponsor an annual, 2-day Basic Symposium, held in conjunction with the IFT Annual Meeting. The Basic Symposium deals in depth with fundamental aspects of selected topics of interest to food scientists, with some applications of the fundamental concepts to the solution of problems facing the food scientist and the food industry.

This symposium, Protein Functionality in Food Systems, the 17th in the series, took place July 9 and 10, 1993, prior to IFT's 53rd Annual Meeting in Chicago. This topic was selected by the Basic Symposium Committee to help meet the demands of the food industry, to enhance the teaching of advanced courses in proteins, and to serve as reference material on protein functionality in food systems.

Aside from their biological activity and their obvious role in nutrition, proteins contribute significantly to the technological and organoleptic characteristics of foods. The *functional properties* of food proteins include solubility, viscosity, gelation, emulsification, and foam formation. Additionally, proteins exhibit the ability to form films and glasses, and contribute to color and flavor. Egg white, gelatin, soy protein, whey protein, and caseinates are all utilized as functional food ingredients.

The chapters in this book feature the latest information on fundamental structure–function relationships, protein-separation technologies, computer-aided techniques for predicting quality parameters of products directly from ingredient proteins, interactions of proteins with other food components, modification of proteins for improved functionality, special protein formulations, and novel applications. These chapters provide not only a sound basis for understanding basic principles involved in food protein functionality, but also valuable fundamental information to creatively develop unique food products utilizing proteins as novel ingredients.

The help provided by present and former Basic Symposium Committee members—Patricia Kendall, Colorado State University; John Rushing, North Carolina State University; Rakesh Singh, Purdue University; Frank Flora, USDA-CSRS; Richard McDonald, FDA; Henry Schwartzberg, University of Massachusetts; Elsa Murano, Iowa State University; Anna Resurrecction, University of Georgia; Ralph Waniska, Texas A&M University; Fred Wolfe, University of Alberta, and Barbara Klien, University of Illinois—in organizing the symposium is gratefully acknowledged.

The symposium organizers thank David Lineback, 1992–93 IFT President, Daniel E. Weber, IFT Executive Director, John B. Klis, Director of Publications, Anna May Schenck, Associate Scientific Editor, and all the other IFT staffers for their support. Most of all, the Basic Symposium Committee members and the co-chairs gratefully acknowledge the contribution of the speakers. Without their dedication, expertise, and hard work, publication of these proceedings would not have been possible.

Navam S. Hettiarachchy
Gregory R. Ziegler

Contributors

James C. Acton Department of Food Science, Clemson University, Clemson, South Carolina

Guillermo E. Arteaga Department of Food Science, The University of British Columbia, Vancouver, British Columbia, Canada

Shai Barbut Department of Food Science, University of Guelph, Guelph, Ontario, Canada

Dirk D. Beekman Department of Animal Science, Iowa State University, Ames, Iowa

Srinivasan Damodaran Department of Food Science, University of Wisconsin—Madison, Madison, Wisconsin

Paul L. Dawson Department of Food Science, Clemson University, Clemson, South Carolina

Harold M. Farrell, Jr.   Eastern Regional Research Center, Agricultural Research Service, United States Department of Agriculture, Philadelphia, Pennsylvania

J. Bruce German   Department of Food Science and Technology, University of California, Davis, California

Navam S. Hettiarachchy   Department of Food Science, University of Arkansas, Fayetteville, Arkansas

Elisabeth Jane Huff-Lonergan   Department of Animal Science, Iowa State University, Ames, Iowa

Arun Kilara   Department of Food Science, The Pennsylvania State University, University Park, Pennsylvania

Thomas F. Kumosinski   Eastern Regional Research Center, Agricultural Research Service, United States Department of Agriculture, Philadelphia, Pennsylvania

David A. Ledward   Department of Food Science and Technology, University of Reading, Whiteknights, Reading, England

Eunice C. Y. Li-Chan   Department of Food Science, The University of British Columbia, Vancouver, British Columbia, Canada

Michael E. Mangino   Department of Food Science and Technology, The Ohio State University, Columbus, Ohio

Mark S. Miller   Department of Ingredient Technology, Kraft General Foods, Inc., Glenview, Illinois

Shuryo Nakai   Department of Food Science, The University of British Columbia, Vancouver, British Columbia, Canada

Frederick C. Parrish, Jr.   Department of Animal Science, Iowa State University, Ames, Iowa

Lance Phillips   Department of Food Science, Cornell University, Ithaca, New York

Khee Choon Rhee   Department of Food Science and Technology, Food

Protein Research and Development Center, Texas A&M University, College Station, Texas

Denise M. Smith   Department of Food Science and Human Nutrition, Michigan State University, East Lansing, Michigan

J. Antonio Torres   Department of Food Science and Technology, Oregon State University, Corvallis, Oregon

Fakhrieh Vojdani   Department of Food Science and Technology, University of California, Davis, California

John R. Whitaker   Department of Food Science and Technology, University of California, Davis, California

Gregory R. Ziegler   Department of Food Science, Pennsylvania State University, University Park, Pennsylvania

Contents

1
Structure-Function Relationship of Food Proteins

Srinivasan Damodaran

University of Wisconsin—Madison
Madison, Wisconsin

INTRODUCTION

Food preferences by human beings are often based on sensory attributes such as appearance, color, flavor, and texture. Proteins play several functional roles in the expression of sensory attributes of various foods. The curd-forming properties of casein micelles and soy proteins, the foaming, whipping, and heat-setting properties of egg white, the water-binding, emulsifying, and texture-forming behavior of meat proteins are important in many food products such as cheese, diary products, meat products, bakery, ice cream, etc. Traditionally, proteins of animal origin, e.g., milk, egg, and meat proteins, have been used in conventional and fabricated foods. The use of plant proteins, although cheap and abundant, in food products is very limited mainly because of lack of desirable functional performance of these proteins in foods. The major impediment to increasing the utilization of plant proteins in formulated foods is the lack of proper understanding of the molecular bases for protein functionality in foods.

1

The functional role of a protein that contributes to the sensory quality of the food product does not arise from a single physicochemical property, rather it is a manifestation of a complex interaction of multiple properties. For example, egg white possesses multiple functionalities, such as foaming, emulsification, heat setting, and binding/adhesion, which make it the most desirable protein in many food applications. Therefore, for a protein to perform well in a food system, it should possess multiple functionalities. This requirement further complicates proper understanding of the structure-function relationship of food proteins.

The important functional properties of proteins that are relevant to food systems are given in Table 1.1. These are fundamentally related to their physicochemical and structural properties, such as size, shape, amino acid composition/sequence, net charge, charge distribution, hydrophobicity/hydrophilicity ratio, secondary, tertiary, and quaternary structural arrangements, number of microdomain structures, and adaptability of domain structures or the structure of the whole molecule to changes in environmental conditions.

Many physicochemical properties that directly affect functional behavior of proteins are ultimately related to amino acid sequence. The amino acid sequence dictates the three-dimensional structure of a protein and thereby its thermodynamic stability, charge distribution pattern on the protein surface, symmetric or asymmetric distribution of hydrophilic and hydrophobic patches on the surface, and the topography of the protein surface. The folding of a protein is dictated by the thermodynamic requirement that a majority of the hydrophobic residues be buried inside and a majority of the hydrophilic and charged residues be located on the surface so that the global free energy of the molecule is at the lowest possible level. In most proteins, while almost all the hydrophilic and charged residues are located on the surface, not all hydrophobic residues are completely buried in the interior because of steric constraints imposed by the polypeptide chain. In many globular proteins, including several food proteins, about 40–50% of the protein surface is occupied by hydrophobic patches (Lee and Richards, 1971). The distribution pattern of these hydrophobic patches (cavities) influences the shape of the molecule as well as the topography of the protein surface. In food proteins, the mode of distribution of nonpolar and polar patches on the protein surface significantly influences several functional properties such as solubility, the tendency to form oligomeric and micellar structures, and surface-active properties. For instance, in α_s- and β-caseins all the serinephosphate residues and a majority of carboxyl groups are segregated at the N-terminal segment, and the remaining

TABLE 1.1   Functional Roles of Food Proteins in Food Systems

Function	Mechanism	Food system	Protein source
1. Solubility	Hydrophilicity	Beverages	Whey proteins
2. Viscosity	Water binding, hydrodynamic size, shape	Soups, gravies, salad dressing	
3. Water binding	H-bonding, ion hydration	Meat sausages, cakes, breads	Muscle proteins, egg proteins
4. Gelation	Water entrapment and immobilization, network formation	Meats, gels, cakes, bakeries, cheese	Muscle proteins, egg and milk proteins
5. Cohesion/ Adhesion	Hydrophobic, ionic and H-bonding	Meats, sausages, pasta, baked goods	Muscle proteins, egg proteins, whey proteins
6. Elasticity	Hydrophobic bonding, disulfide cross-links	Meats, bakery	Muscle proteins
7. Emulsification	Adsorption at interfaces, film formation	Sausages, bologna, soup, cakes, dressing	Muscle proteins, egg proteins, milk proteins
8. Foaming	Interfacial adsorption, film formation	Whipped toppings, ice cream, cakes, desserts	Egg proteins, milk protein
9. Fat and flavor binding	Hydrophobic bonding, entrapment	Simulated meats, bakery, doughnuts	Milk proteins, egg proteins

Source: Kinsella et al., 1985.

two thirds of the molecules are highly hydrophobic (Swaisgood, 1982). This asymmetric distribution of charged and hydrophobic residues in the amino acid sequence provides these caseins with detergent-like characteristics.

The solubility of a protein under a given set of conditions is the thermodynamic manifestation of the equilibrium between protein-protein and protein-solvent interactions. It is related to the net free energy change arising from interaction of hydrophobic and hydrophilic residues on the protein surface with the surrounding solvent. In other words, solubility is directly related to the physicochemical nature of the protein surface, which in turn is influenced by the folding pattern of the polypeptide chain. The degree of exposure of hydrophobic residues on the surface of a protein also influences its thermodynamic stability. Proteins that have higher amount of exposed hydrophobic surfaces are more susceptible to thermal and interfacial denaturation than those that have most of the hydrophobic residues buried in the interior.

The amino acid composition of proteins also has a bearing on several functional properties. Proteins that have high proline content tend to exist in a disordered state. For instance, about 17% of the amino residues in β-casein and about 8.5% of the residues in α_{s1}-casein are proline residues (Swaisgood, 1982). Similarly about 30% of the residues in gelatin are either proline or hydroxyproline residues. The uniform distribution of these residues in the amino acid sequence of these proteins effectively precludes formation of ordered structures, such as α-helix and β-sheet, in these proteins. Because of their high flexibility these proteins exhibit multiple functional properties such as gelation, foaming, and emulsifying properties.

In addition to the intrinsic molecular factors, several extrinsic factors such as the method of isolation, pH, ionic strength, the redox potential of the food system, and interactions with other food components also affect the functional properties of proteins (Kinsella et al., 1985). However, the effects of these extrinsic factors are simply manifestations of alterations in the conformation and other physicochemical properties of proteins.

In a phenomenological sense, the various functional properties of food proteins are manifestations of two molecular aspects of proteins (Damodaran, 1989): (1) protein surface–related properties and (2) hydrodynamic properties. The functional properties that are affected by these molecular aspects of proteins are listed in Table 1.2. The surface-related properties are governed by the hydrophobic, hydrophilic, and steric properties of the protein surface, and the properties related to hydrodynamic properties of proteins are governed by size, shape, and flexibility of proteins.

TABLE 1.2 Functional Properties of Food Proteins

Surface-related properties	Hydrodynamic properties
Solubility	Viscosity
Wettability	Thickening
Dispersibility	Gelation
Foaming	Texturization
Emulsification	
Fat and flavor binding	

Although the interrelationship between the molecular properties and functional properties of food proteins has been qualitatively understood, quantitative prediction of the functional behavior of proteins in food systems from the knowledge of their molecular properties has not been achieved. The current knowledge of the functional properties of proteins is based on the behavior of individual proteins in simple model systems. The results obtained from such model systems often fail to predict the behavior in real foods prepared under industrial processing conditions. The extensive conformational changes that occur under industrial processing conditions and the multilateral interactions of proteins with other food constituents often render it impossible to translate the results of model systems to predict the behavior of proteins in real food systems (Harper, 1984; deWit, 1989). In addition, the lack of standardized methods to evaluate both the molecular properties of proteins and the functional properties also has confounded proper understanding of structure-function relationship of proteins (Kinsella, 1982). Proper evaluation of the large volume of available data on protein functionality seems to be a formidable task because of variations in methodologies and procedures used from laboratory to laboratory.

The majority of formulated foods are either foams, emulsions, or gels. Generally, proteins are preferred over small molecular weight surfactants to act as surface-active and network-forming agents in these types of foods. In this regard, a fundamental understanding of the molecular properties that affect the surface-active and gelling properties of proteins is crucial for their utilization in food products.

SURFACE-ACTIVE PROPERTIES

Adsorption and Film Formation

The ability of proteins to act as surfactants and stabilize foams and emulsions innately depends upon their ability to adsorb at interfaces, greatly reduce the interfacial tension, and form a cohesive film. Since all proteins are amphiphilic, i.e., they contain both hydrophobic and hydrophilic residues, they show a tendency to adsorb at interfaces. However, the extent of adsorption and the ability to reduce interfacial tension and form a cohesive film at the interface differ widely among proteins. These differences arise primarily from differences in conformation as well as differences in the physicochemical properties of the protein surfaces that interact with the dispersed and continuous phases in foams and emulsions. Specifically, the factors that affect adsorption and film formation of proteins at interfaces are conformational stability and adaptability at phase boundaries and symmetric or asymmetric distribution of hydrophilic and hydrophobic groups on the protein surface.

In a quiescent system, adsorption of proteins at the air-water or oil-water interfaces is thought to be a diffusion-controlled process. It is assumed that when a fresh interface is created, the molecules at the subsurface instantaneously absorb to the interface (Ward and Tordai, 1946; MacRitchie, 1978). The depletion of concentration at the subsurface creates a concentration gradient between the subsurface and the bulk phase, which acts as the driving force for diffusion of molecules from the bulk phase to the subsurface and then to the surface (Ward and Tordai, 1946). If adsorption is diffusion-controlled, then the rate of adsorption should be dependent only upon the size and shape of the molecule, viscosity of the solvent, and the temperature. This diffusion-controlled mechanism assumes that every collision of the protein molecules with the interface leads to adsorption irrespective of the physicochemical nature of the colliding protein surface. However, many experimental studies on the kinetics of adsorption of globular proteins have shown that the apparent diffusion coefficients of proteins calculated from adsorption studies were significantly lower than those obtained from solution diffusion studies, indicating that most globular proteins exhibit an energy barrier to adsorption at interfaces. The exact physical nature of this energy barrier is not understood, but may include surface pressure and electrostatic barriers (MacRitchie and Alexander, 1963a,b). However, in the simplest case, i.e., even in the absence of surface pressure and electrostatic energy barriers, the success of every collision leading to adsorption should be fundamentally related to the surface hydrophilicity/hydrophobicity ratio and their distribution on the protein surface (Damodaran, 1989). If the protein surface is extremely hydrophilic (and highly charged) and devoid of hydrophobic patches, adsorption at the interface might not take place. On the other hand, if the protein surface contains a few hydrophobic patches, adsorption might take place when this hydrophobic patch collides with the interface (Fig.

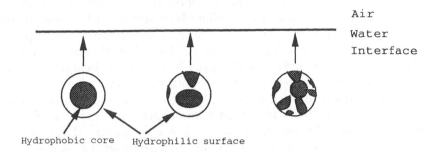

FIG. 1.1 Illustration of the role of surface hydrophobic patches on the probability of adsorption of proteins at interfaces. (*From Damodaran, 1989.*)

1.1). In other words, adsorption of a protein at the air-water and oil-water interfaces is related to the statistical probability of collision of a hydrophobic patch on the protein surface with the interface. The greater the number of hydrophobic patches, the greater would be the probability of each collision leading to adsorption. Thus, the rate of accumulation of a protein at an interface can be expressed as (Damodaran, 1989)

$$\Gamma = 2C_o \, (P_a) \, (D/\pi)^{1/2} \, t^{1/2}$$

where Γ is the protein concentration at the interface (mg/m^2), C_o is the protein concentration in the bulk phase, D is the diffusion coefficient, t is the time, and P_a is the probability factor. It should be pointed out, however, that in practical situations, foams and emulsions are formed by bubbling, whipping, or homogenization; under these turbulent conditions, the rate of adsorption of proteins at the freshly formed interfaces might not be dependent on diffusion coefficient, but the probability factor would still control the rate of adsorption.

The role of hydrophilicity and hydrophobicity of the protein surface in protein adsorption is evident from the kinetics of adsorption of phosvitin at the air-water interface (Fig. 1.2). Phosvitin is one of the major phosphorylated proteins in egg yolk. Phosvitin contains 216 amino acid residues and has a molecular weight of about 34,000 (Byrne et al., 1984). About 55% of the residues in phosvitin are phosphoserine residues. Because of the high phosphoserine residue content, the physicochemical behavior of phosvitin resembles that of a polyelectrolyte. Phosvitin is extremely negatively charged and exists in a random coil state at pH 7.0, whereas at pH 2.0 it assumes a folded structure (Renugopalakrishnan et al., 1985; Yasui et al., 1990). At pH 7.0 phosvitin exhibited no adsorption at the air-water interface from a dilute bulk solution when monitored over a period of 24 hr (Fig. 1.2). The rate and extent of adsorption increased as the pH was decreased from 7.0 to 2.0. Noticeable amount of adsorption was observed only below pH 4.0. The phosvitin film formed at pH 2.0 desorbed completely into the bulk phase when the pH of the bulk phase was readjusted to pH 7.0. These observations indicate that the extreme hydrophilicity and high net charge and the lack of well-defined hydrophobic patches in the random coil state at pH 7.0 prevent successful adsorption of phosvitin at the air-water interface. However, at pH 2.0 a reduction in the net charge and formation of ordered structure, which might involve creation of a well-defined hydrophobic patch, facilitate successful collision and adsorption of the protein at the interface. In spite of adsorption of about 1.2 mg/m^2 phosvitin at pH 2.0, the surface pressure (i.e., the net reduction in surface tension) of the phosvitin film was almost zero, indicating that the phosvitin molecules at the interface were in the gaseous state, i.e., present as individual molecules embedded at the interface, and there were no cohesive interactions between phosvitin molecules even at the monolayer coverage. These results clearly indicate that the hydrophilicity/hydrophobicity characteristics of a protein have enormous impact on adsorption and film-forming properties of proteins at interfaces.

Several studies have shown that under comparable solution conditions the rate and extent of adsorption of proteins at interfaces were greater in the denatured state than in the native state (Mitchell et al., 1970; Adams et al., 1971; Damodaran and Song, 1988; Xu and Damodaran, 1993b), implying that the initial structure and stability of the protein in the bulk phase influences the kinetics and thermodynamics of its adsorption at an interface. Recent studies on adsorption of several structural intermediates of bovine serum albumin at the air-water interface have shown that the rate and extent of adsorption of the intermediates increased as the degree of unfolding of the protein in the solution phase was increased (Damodaran and Song, 1988). However, contrary to the notion that a highly flexible random coil protein would occupy greater surface area at the interface than a compact folded protein, it was found that one of the folded intermediates of serum albumin apparently occupied greater area at the air-water interface than either the completely unfolded or the compact native state of serum albumin (Damodaran and Song, 1988). In other words, it seems that in order for a protein to occupy a greater area at the interface and exert maximum effect on the

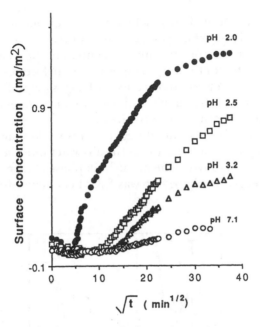

FIG 1.2 Kinetics of adsorption of phosvitin at the air-water interface at various pH and at 25°C. The bulk phase concentration of phosvitin was 1.5 μg/mL.

surface or interfacial force field (i.e., reduction of surface tension), it should possess an optimum degree of folded and unfolded structure.

To elucidate the influence of protein conformation on adsorption at interfaces, Graham and Phillips (1979a,b) studied the adsorption and film-forming characteristics of β-casein and lysozyme at the air-water interface. Recently, we reexamined the adsorption characteristics of these proteins at the air-water interface using a more refined surface radiotracer method (Xu and Damodaran, 1992, 1993a,b). It was shown that during the first 60 min after creation of a fresh air-water interface, native egg-white lysozyme exhibited a negative surface excess, implying that the protein molecules that were originally at or near the interface migrated to the subsurface (Fig. 1.3). After this initial desorption, the negative surface excess remained constant for a period of time, followed by a rapid positive adsorption at the interface at later stages. The surface concentration reached an apparent steady-state value of only about 0.6 mg/m^2 after about 1000 min. In addition, the development of surface pressure during adsorption exhibited a longer lag time than the surface concentration; the surface pressure remained zero for more than 400 min of adsorption, and then reached a value of about 3 mN/m after about 1500 min. The surface pressure did not reach a steady-state value even after the surface concentration reached an apparent steady-state value (Fig. 1.3). The continuing changes in the surface pressure indicate that the adsorbed lysozyme molecules undergo slow unfolding and re-arrangement of the hydrophobic and hydrophilic segments at the inter-face, which affects the surface force field.

We have shown that the initial negative adsorption and the lag time for the onset of positive adsorption at the air-water interface was affected by the conformation of lysozyme in the bulk phase (Xu and Damodaran, 1993b). When egg-white lysozyme was fully heat denatured in the pres-

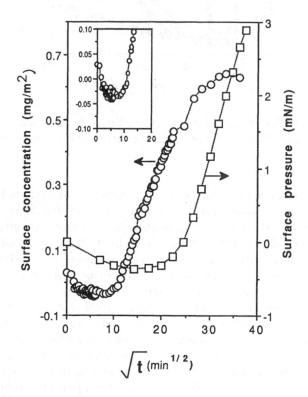

FIG 1.3   Kinetics of adsorption of native lysozyme at the air-water interface. The bulk concentration of lysozyme in both cases was 1.5 μg/mL in phosphate buffered saline solution, $I = 0.1$, pH 7.0. The inset shows changes in surface concentration during the initial periods of adsorption. (*From Xu and Damodaran, 1993 b.*)

ence of a disulfide-reducing agent followed by blocking of the sulfhydryl groups, no initial desorption from the interface was observed, and the lag time for positive adsorption was only about 5 min. Furthermore, the rate of adsorption was very fast and reached a steady-state surface concentration of about 1.2 mg/m² in about 2 hr (Fig. 1.4). The rate of increase of surface pressure paralleled that of the surface concentration and reached a steady-state value of 19 mN/m, indicating that, unlike the native lysozyme, the denatured lysozyme was able to unfold and re-arrange immediately upon adsorption at the interface. In contrast to lysozyme, β-casein showed no initial desorption from the interface; positive adsorption at the air-water interface commenced soon after a fresh air-water interface was created (Fig. 1.5). The rate of adsorption of β-casein was much greater than that of lysozyme and reached a steady-state value of about 1.9 mg/m² after about 900 min. The rate of surface

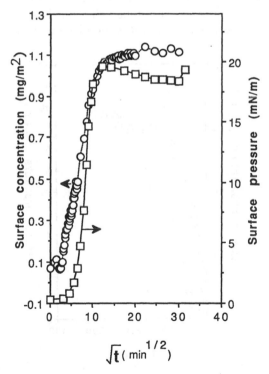

FIG 1.4   Kinetics of adsorption of reduced and urea denatured lysozyme at the air-water interface. The conditions were same as for Fig. 1.3. (*From Xu and Damodaran, 1993b.*)

pressure development showed only a lag time of about 30 min (compared to 400 min for lysozyme) and reached a steady-state value of about 16 mN/m. The fact that both the surface concentration and surface pressure attained steady-state values at the same time (Fig. 1.5) indicates that β-casein unfolds and rearranges its hydrophobic and hydrophilic segments almost immediately upon adsorption at the interface.

The fact that the native lysozyme was initially desorbed from the interface, whereas the denatured lysozyme did not desorb, indicates that there is an energy barrier for adsorption of native lysozyme at the interface. This energy barrier emanates from higher electrochemical potential of native lysozyme at the interface than at the subsurface (Xu and Damodaran, 1992, 1993b). The adverse effects of electrostatic free energy of proteins on adsorption at interfaces are depicted in Fig. 1.6. Let us assume that e is the net charge of lysozyme and ϵ_o and ϵ are the dielectric constants of the aqueous and the gas phases, respectively. The electrostatic theory (Perutz, 1978) stipulates that as lysozyme moves towards the air-water interface, an image charge, $e' = e(\epsilon_o - \epsilon)/(\epsilon_o + \epsilon)$, would appear in the gas phase. If d is the distance of the protein from

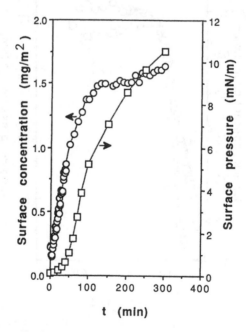

FIG 1.5   Kinetics of adsorption of β-casein at the air-water interface. The bulk concentration of β-casein was 1.5 μg/mL in phosphate buffer saline solution, $I = 0.1$, pH 7.0. (*From Xu and Damodaran, 1993.*)

the air-water interface, the electrostatic repulsive potential between the real and image charges would be (Xu and Damodaran, 1992)

$$\mu_{ele} = \frac{ee'}{2d\epsilon_o} = (e^2/2d\epsilon_o)\frac{(\epsilon_o - \epsilon)}{(\epsilon_o + \epsilon)}$$

If lysozyme molecules were present at the freshly formed air-protein solution interface (i.e., $d = 0$), in the absence of any hydrophobic interaction between the protein and the gas phase, the lysozyme molecules would desorb into the bulk phase because of the strong electrostatic repulsive potential at the interface. We have shown that the minimum distance at which the electrostatic repulsive potential between the real and image charges is equal to the thermal energy, kT (where k is Boltzman constant and T is the temperature), of the molecule at 25°C would be about 277 Å from the air-solution interface (Xu and Damodaran, 1992). In other words, because of the high electrostatic repulsive potential, native lysozyme would spontaneously desorb to a distance of about 277 Å from the interface. We have also shown that at the subsurface

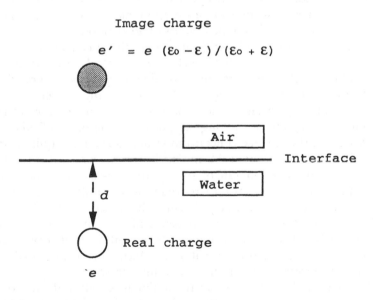

FIG 1.6   Schematic representation of the role of electrostatic energy barrier in adsorption of proteins at the air-water interface.

lysozyme undergoes partial denaturation due to reaction of the high-energy water molecules with hydrogen bonds in lysozyme (Xu and Da-modaran, 1992). This partial denaturation and consequent exposure of hydrophobic residues increases the hydrophobic chemical potential, which drives the molecule toward the interface. In other words, the desorption and adsorption phases are the result of the influence of various chemical potentials of the protein molecule, which are manipulated by the conformation and the sign and magnitude of interaction of these forces with the interfacial force field (Xu and Damodaran, 1993b). Based on our observations a general mechanism for protein adsorption and retention at interfaces was proposed, which invokes that it is the chemical potential gradient emanating from interaction of the interfacial force field with various molecular potentials, such as hydrophobic, electrostatic, hydration, and conformational (entropic) potentials, rather than concentration gradient alone that acts as the driving force for protein adsorption (Xu and Damodaran, 1992, 1993b).

Since food proteins contain a mixture of several proteins with a wide range of physicochemical properties, it is highly likely that the kinetics and extent of adsorption of proteins in a multi-component system might be different from that in a single-component system. To elucidate the influence of one protein on the kinetics of adsorption of another protein in a binary mixture, the kinetics of competitive adsorption of β-casein and lysozyme at the air-water interface was recently studied (Fig. 1.7). In the absence of β-casein in the bulk phase, lysozyme exhibited initial desorption and a long lag phase. However, in the presence of equal concentration of β-casein in the bulk phase, lysozyme exhibited neither the initial desorption nor a long lag phase for the onset of adsorption. Furthermore, the surface concentration of lysozyme at equilibrium decreased from 0.65 to about 0.2 mg/m^2 in the presence of β-casein. On the other hand, the initial rate of adsorption of β-casein was similar both in the presence and absence of lysozyme in the bulk phase. However, the extent of adsorption at equilibrium decreased slightly from about 1.85 to about 1.6 mg/m^2 in the presence of lysozyme (Fig. 1.7). These results clearly indicate that the kinetic and equilibrium adsorption behavior of one protein at the air-water interface can be modified by the presence of another protein in protein mixtures.

Many studies have shown that the initial area required for most proteins to penetrate and anchor themselves at an interface is about 60–100 Å^2 (Ter-Minassian Saraga, 1981; Damodaran and Song, 1988). This implies that only a peptide segment containing four to six nonpolar amino acid residues strategically located on the protein surface is involved in the initial binding step. Once adsorbed at an interface, proteins

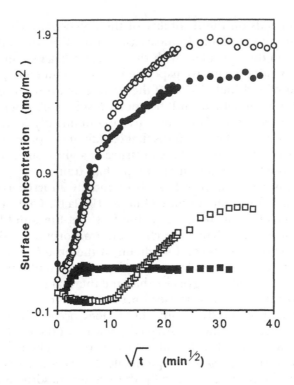

FIG 1.7   Kinetics of adsorption of β-casein (circles) and lysozyme (squares) in single-component (open symbols) and in 1:1 mixture (filled symbols) experiments. Concentration of each protein in both the experiments was 1.5 μg/mL in phosphate buffered saline solution, $I = 0.1$, pH 7.0.

invariably undergo conformational change because of the prevailing thermodynamic conditions at the interface. This unfolding and re-arrangement at the interface is an entropy-driven process in which transfer of the nonpolar residues from the aqueous phase to the nonaqueous phase and the consequent release of water molecules from the hydration shells of these groups to their higher entropic state act as the driving force. However, the rate and the extent to which proteins unfold and rearrange at the interface depends upon the intramolecular constraints present in proteins. Proteins that lack intramolecular disulfide bonds and extensive segment-segment interactions in the folded state often readily unfold at interfaces. Upon unfolding, the molecule spreads and the newly exposed hydrophobic segments attach themselves at the interface. The number of residues or segments that ultimately bind to the interface depends on the flexibility of the polypeptide chain as well as the amino acid distribution. In most cases, only a fraction of the polypeptide chain is in direct contact with the interface (trains), and a majority of the polypeptide chain is suspended into the aqueous phase in the form of "loops" and "tails." Most of the residues in the train configuration are hydrophobic in nature, and the apolar residues are oriented towards the nonaqueous phase. The loops are predominantly hydrophilic segments of the protein. Protein films that contain more train configuration than loops exhibit higher surface pressure at a given surface concentration than those that contain more loops than trains.

Protein adsorption at interfaces is essentially an irreversible process (MacRitchie, 1978; Graham and Phillips, 1979a,b). One of the fundamental reasons for this behavior is the fact that the unfolding and rearrangement of the polypeptide chain after adsorption at the interface facilitates anchoring of several segments of the protein at the interface. Since the sum total of interaction energies of these multiple segments at the interface is usually far greater than the thermal energy of the molecules, the activation energy barrier for desorption tends to be very high. Desorption of proteins from interfaces does occur when the protein film is compressed to higher surface pressures (MacRitchie and Ter-Minassian Saraga, 1983, 1984; MacRitchie, 1985). The energy input in the form of surface pressure sequentially displaces adsorbed segments from the interface, and eventually the protein desorbs into the bulk phase. However, the conformation of the desorbed protein might not be the same as the native protein's. MacRitchie (1985) showed that the critical surface pressure at which desorption of proteins from the air-water interface occurs was related to the molecular weight. The higher the molecular weight, the greater was the surface pressure needed to desorb the protein. When adsorption takes place from a very dilute solution, i.e., 0.0001–0.01% protein solution, it is likely that the extent of unfolding and denaturation of the protein as it adsorbs at the interface would be high because of the slow adsorption and slow development in the surface pressure. However, when adsorption takes place from a concentrated protein solution, it is highly likely that the extent of denaturation of the protein at the interface would be minimal because of rapid increase in surface pressure, which might inhibit the rate of unfolding. Therefore, in real foams and emulsions, where the concentration of protein solution is high, proteins may retain a greater amount of folded structure. This might influence the rheological behavior and stability of foams and emulsions.

The foamability and emulsifying capacity of proteins is innately related to their ability to adsorb and instantaneously reduce the interfacial tension as new interfacial area is being created. However, the stability of foams or emulsions is dependent not only upon the extent of reduction of the interfacial tension, but also on the rheological properties of the protein film (Graham and Phillips, 1976a; Mita et al., 1977; Izmailova, 1979). The rheological properties of protein films arise from intermolecular interactions between proteins at the interface that facilitate formation of a cohesive continuous protein network. The forces that contribute to gelation of proteins are obviously involved in the formation of viscoelastic protein films at interfaces. These include noncovalent interactions such as hydrophobic, van der Waals, electrostatic, and hydrogen-bonding interactions. The magnitude of these interactions in the film depends on the extent of conformational change in proteins. However, the optimum amount of protein unfolding at the interface that develops desirable viscoelastic properties in the film and the molecular properties that are critical to effect these interactions are not well understood.

FOAMING PROPERTIES

The basic requirements for a protein to be a good foaming agent are the ability to (1) rapidly adsorb at the air-water interface during whipping or bubbling, (2) undergo rapid conformational change and rearrangement at the interface, and (3) form a cohesive viscoelastic film via intermolecular interactions. In speaking of foaming properties of proteins, it is necessary to make a distinction between foamability and foam stability. While foamability relates to the amount of interfacial area that a protein is able to create per unit weight or concentration, foam stability relates to the ability of the protein to stabilize the foam against gravitational and mechanical stresses. The first two criteria are essential for better foamability, whereas the third is important for the stability of the foam. For instance, while β-casein exhibits good foamability because of its molecular flexibility, the β-casein foams are less stable because of its poor viscoelastic properties (Graham and Phillips, 1976a). On the other hand, globular proteins such as lysozyme exhibit poor foamability, and the foam stability is better than that of β-casein (Graham and Phillips, 1976). Thus, it appears that the molecular properties requirements for good foamability and good foam stability are different.

The foam stability is affected by several factors. These include rheological properties, such as viscosity and shear resistance, film elasticity, and the magnitude of disjoining pressure between the protein layers. These factors affect foam stability via affecting liquid drainage from the lamellar film. The rate of liquid drainage is given by Reynolds' equation,

$$V = -dh/dt \doteq (2h^3/3\mu R^2) \, \Delta P$$

where h is film thickness, t is time, μ is dynamic viscosity, R is the radius of the bubble, and $\Delta P = \pi_h - \pi_d$, where π_h and π_d are the capillary hydrostatic pressure and the disjoining pressure, respectively. According to the above equation, the greater the disjoining pressure and the dynamic viscosity of the film, the slower would be the rate of liquid drainage and collapse of the lamellar film. The magnitude of disjoining pressure between the two protein layers is related to the sum of steric (π_s), electrostatic (π_e), and dispersion (π_v) forces, including hydrophobic force, between the protein layers. In general, π_v contributes negatively, whereas π_s and π_e contribute positively to the disjoining pressure. The electrostatic repulsion between the protein films will increase the disjoining pressure. However, excessive lateral electrostatic repulsion between protein molecules within each protein film will impair film integrity and thus might cause collapse of the film. The steric effects arise mainly from the loops and tails and hydration repulsion forces between the adsorbed protein films. Films in which π_v is greater than the sum of π_s and π_e will thin rapidly. This is the case in β-casein films in which the hydrophobic attraction between the protein layers is high, the electrostatic repulsion is marginal, and the steric and hydration repulsion is minimum because of the predominance of the "train" configuration over the "loops."

The rheological properties of protein films, which contribute to foam stability, relate to intermolecular interactions and mechanical strength. The intermolecular interactions in the film are primarily dictated by protein conformation in the film and the magnitude of noncovalent interactions, such as electrostatic, hydrophobic, and hydrogen bonding, between protein molecules. Proteins that only partially unfold and retain a certain amount of tertiary structure usually form a thicker film at the air-water interface and a highly stable foam. In such films much of the folded structure is suspended into the subsurface in the form of loops; the greater degree of exposed functional groups in these partially unfolded structures promotes a greater extent of protein-protein interactions leading to formation of a cohesive protein network. The physical state of this network is similar to that found in gels and contributes to water binding and the high mechanical strength of the film. Although the requirements for rheological properties of films have been qualitatively understood, the optimum extent of unfolding and intermolecular interactions necessary for imparting superior rheological properties are not known.

One of the most important factors that affects foam stability is the elasticity of protein films. Elasticity relates to the change in surface pressure for a unit change in interfacial area of the film and is expressed as $\epsilon = -A(d\pi/dA)$, where A is the area of the film, and π is the surface pressure. The elasticity is affected by the flexibility of the protein molecules in the film. Highly flexible proteins exhibit poor elasticity, i.e., minimum change in surface pressure per unit change in the interfacial area during compression or expansion of the film. A good example is β-casein film (Phillips, 1981). The low elasticity of highly flexible proteins is attributed to rapid configurational changes from trains to loops and vice versa when the film is compressed or expanded, respectively. Poor film elasticity contributes to continuous expansion of the foam, resulting in liquid drainage and rapid thinning of the lamella. In contrast, films of rigid proteins, such as lysozyme, exhibit high elasticity (Phillips, 1981). The lack of molecular flexibility prevents configurational changes from trains to loops and vice versa during compression and expansion of the film. Because of this, local stresses on such films cause greater changes in local interfacial tension, which in turn causes rapid flow of the protein film toward higher interfacial tension regions. Such constant, rapid flow of the film exerts viscous drag on the liquid beneath it, which causes retardation of liquid drainage and film thinning.

The molecular properties that dictate the foamability of proteins are related to high molecular flexibility, charge density, and hydrophobicity. The hydrophobicity of a protein can be further classified into surface hydrophobicity and molecular hydrophobicity. Whereas surface hydrophobicity relates to the extent of hydrophobic patches on the protein surface, the molecular hydrophobicity refers to average hydrophobicity of the amino acid residues in the protein as defined by Bigelow (1967). Studies have shown that while the relationship between molecular hydrophobicity and foamability apparently follow a linear relationship (Townsend and Nakai, 1983), the foamability exhibits a curvilinear relationship with surface hydrophobicity (Bacon et al., 1988; Kato et al., 1983). The departure from linearity of surface hydrophobicity versus foamability occurs at a surface hydrophobicity value of 1000 as determined from the extent of binding of cis-parinaric acid (Kato et al., 1983). These findings imply that a surface hydrophobicity value of 1000 is sufficient to ensure adsorption of the protein at the air-water interface; but once adsorbed, the ability of the protein to unfold and expose all hydrophobic residues to the interface is critical for creating more interfacial area as the foam is being formed. In other words, the ability of the protein to instantaneously reduce the surface tension upon adsorption is critical for its foamability. This is possible only when the protein undergoes rapid conformational change and reorientation at the interface.

One of the major limitations to understanding microscopic as well as macroscopic events that lead to foam collapse and the contribution of various molecular factors to foam stability is the lack of a reliable method to study the kinetics of foam decay (Yu and Damodaran (1991a). The methods frequently used involve measurement of liquid drainage as a function of time, and the relative stability is expressed as the time required for 50% liquid drainage (Mita et al., 1977; Halling, 1981). Recently, we have developed a more reliable method to study stability of protein foams (Yu and Damodaran, 1991a). This method is based on monitoring of the pressure change that occurs as the foam inside a closed foam column breaks as a function of time. A schematic diagram of the apparatus used for measuring foam stability is shown in Fig. 1.8. The relationship between the pressure change inside the foam column and the interfacial area of the foam is given by

$$A_t = (3V/2\gamma)(\Delta P_n - \Delta P_t)$$

where A_t is the interfacial area of the foam at time t, V is the total volume of the foam column apparatus, γ is the surface tension, ΔP_t is the net pressure change at time t, and ΔP_n is the net pressure change at infinite time when the entire foam is collapsed. Since $\Delta P_t = 0$ at t = 0, the initial

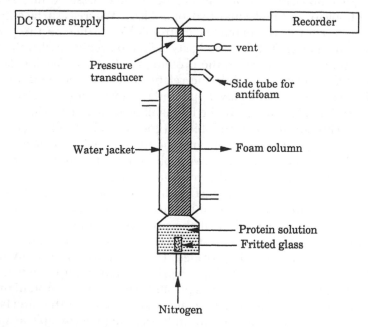

FIG. 1.8 Schematic diagram of the foam apparatus used for measuring foam stability. (*From Yu and Damodaran, 1991a.*)

interfacial area of the foam, which is related to its foamability, can be calculated from the expression

$$A_0 = 3V\Delta P_n/2\gamma$$

Using the above methodology, we have shown that the decay of protein foams essentially follows a biphasic first-order kinetics, indicating involvement of two macroscopic processes (Yu and Damodaran, 1991a). The two macroscopic processes that contribute to foam decay are the gravitational drainage of liquid from the lamella and gas diffusion across the bubble. That is,

$$A_t/A_0 = Q_g \exp(-k_g t) + Q_d \exp(-k_d t)$$

where k_g and k_d are first-order rate constants for the gravitational drainage and gas diffusion processes, respectively, and Q_g and Q_d are the amplitude parameters. The first kinetic phase of foam decay is essentially due to decay caused by liquid drainage, and the second kinetic phase is attributed to decay by interbubble gas diffusion and disproportionation.

Investigations on the foaming properties of bovine serum albumin using the above approach have revealed several interesting behaviors. The results of the effect of pH on the decay of serum albumin foam is shown in Fig. 1.9. It should be noted that the overall foam stability is greatest at pH 4.5–5.0. This is to be expected because it is known that for most protein-stabilized foams the stability is maximum at or near the isoelectric pH of the proteins (Buckingham, 1970; Mita et al., 1977; Kim and Kinsella, 1985). However, what is interesting in the data presented in Fig. 9 is that at pH 4.5 and 5.0 the nonlinear (biphasic) first-order plots were convex in shape, whereas those at pH 4, 6, and 7 were concave in shape. The concave shape at pH 4, 6, and 7 indicates that the decay due to gravitational liquid drainage is faster than that due to interbubble gas diffusion and disproportionation. The convex shape at pH 4.5 and 5, i.e., near the pI of the protein, indicates that the liquid drainage rate is slower than gas diffusion. In other words, decay due to liquid drainage is the rate-limiting step in the convex-type decay process, whereas decay due to gas diffusion and disproportionation is the rate-limiting step in the concave-type decay process. This transition from concave to convex shape of the decay curve indicates a fundamental change in the viscoelastic properties of the protein film as a function of pH. In the isoelectric pH region, the absence of electrostatic repulsion between protein molecules in the film seems to enhance cohesive interactions within the film, which increases the mechanical strength of the film. This viscous and elastic film dramatically retards liquid drainage by hydrostatic and gravitational forces. However, when the protein films on either side of the

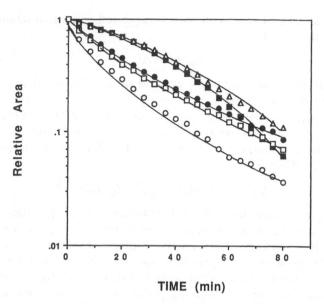

TIME (min)

FIG. 1.9 Surface area decay of BSA foam (1% protein, 25°C) at various pH: (O) pH 4.0; (■) pH 4.5; (Δ) pH 5.0; (●) pH 6.0; (□) pH 7.0. (*From Yu and Damodaran.*)

lamella approach each other below a critical distance owing to inescapable drainage by hydrostatic forces, the lack of interfilm electrostatic repulsion promotes attractive interaction between the films via hydrophobic and van der Waals forces. This might lead to coagulation/precipitation of the protein, which in turn might cause "holes" in the lamella film and thus enhance gas diffusion across bubbles (Yu and Damodaran, 1991a). On the other hand, at pH 4, 6, and 7, i.e., far from the isoelectric pH of the protein, electrostatic repulsion within the film might prevent formation of a cohesive film with high viscoelastic properties. Such a film may not have the ability to retard liquid drainage due to hydrostatic forces.

Cleavage of intramolecular disulfide bonds in proteins, which removes constraints for conformational changes at the interface, also affects the viscoelastic properties of protein films and thus the foam stability. For example, when disulfide bonds in serum albumin are progressively cleaved using an increasing concentration of dithiothritol (DTT), the decay of BSA foam changes from concave shape to convex shape at pH 7.0 (Fig. 1.10). The retardation of liquid drainage in the presence of higher concentrations of DTT is related to cleavage of increased number

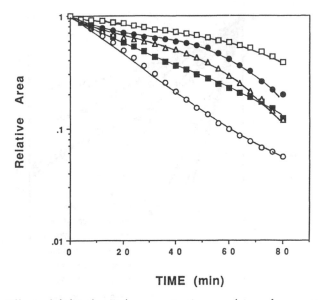

TIME (min)

FIG 1.10   Effect of dithiothreitol concentration on the surface area decay of BSA foam (1%, pH 7.0) at 25°C. (O) 0 mM; (■) 0.5 mM; (Δ) 1.0mM; (●) 5.0 mM; (□) 10 mM. (*From Yu and Damodaran, 1991b.*)

of disulfide bonds, and consequently, an increase in the hydrodynamic size of the protein, which might facilitate formation of a cohesive viscoelastic film. Studies on the foaming properties of 7S and 11S globulin fractions of soy proteins have shown that the foamability and foam stability of soy 11S was better than that of soy 7S (Yu and Damodaran, 1991b). While soy 11S foam exhibited a convex-type decay behavior, soy 7S foam exhibited a concave-type decay behavior. The poor stability of soy 7S foam was related to its inability to retard liquid drainage (Yu and Damodaran, 1991b).

Heat denaturation of proteins often results in improvement of foaming properties (de Wit, 1986; Haggett, 1976). This is primarily due to an increase in surface hydrophobicity and flexibility of denatured proteins. However, in some cases the extent of heat denaturation may have either a positive or a negative effect on foaming properties. For instance, when whey protein isolate (WPI) is progressively heat denatured at 70°C, the sample heated for 1 min exhibits better foam stability than either the unheated control or those that were heated for more than 5 min at 70°C (Fig. 1.11). On the other hand, foams of WPI samples heated for 5, 20, and 90 min at 90°C exhibit poorer stability than that of the control,

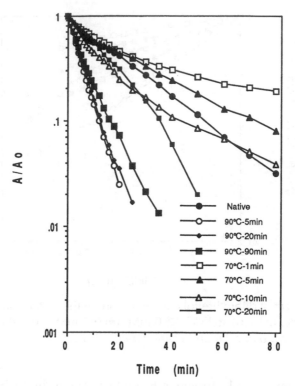

FIG. 1.11   Surface area decay of foams of native and various heat-treated whey protein isolate. Protein concentration was 5% in 20 mM phosphate buffer, pH 7.0.

indicating that extensive heat denaturation, and possibly polymerization via sulfhydryl-disulfide interchange reactions at high temperatures, adversely affects the foaming properties of WPI. When native WPI is mixed with WPI heated at 90°C for 20 min, the mixture exhibits maximum foam stability when the ratio of native to denatured WPI is about 40:60 and maximum foamability at about 60:40 ratio (Fig. 1.12). The data clearly indicate that to exert better foamability and foam stability, an optimum balance of denatured and folded state of the protein at the interface is essential.

In addition to the intrinsic properties of proteins, several extrinsic factors such as salts, lipids, and other food constituents also affect the foaming properties of food proteins. Addition of phospholipids increases the foamability, but decreases foam stability (Cooney, 1974). Since commercial protein preparations are mixtures of various proteins,

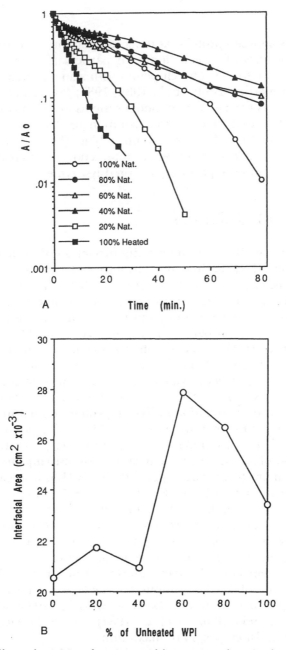

FIG. 1.12   Effect of mixing of native and heat-treated (90°C for 20 min) on (A) the stability and (B) foamability of whey protein isolate foams. The protein concentration was 5% in 20 mM phospahte buffer, pH 7.0. Foamability refers to initial interfacial area.

interactions between proteins at the gas-liquid interface affect the rheological properties of protein films. In particular, interactions of basic proteins (e.g., lysozyme and clupeine) with acidic proteins improve the foaming properties (Poole et al., 1984, 1987; Poole, 1989; Clark et al., 1988). This has been shown in model systems in which addition of lysozyme up to a level of 0.1% to a solution of 0.5% BSA dramatically improved both foamability and foam stability (Poole et al., 1984). This has been attributed to electrostatic interaction between the negatively charged BSA and the positively charged lysozyme.

EMULSIFYING PROPERTIES

The factors that affect the emulsifying properties of proteins are similar to those that affect the foaming properties of proteins. These include the rate of adsorption at the oil-water interface, the amount of protein adsorbed, conformational rearrangement at the interface, the extent of reduction in interfacial tension, and formation of a cohesive film.

The emulsifying properties of proteins are affected by the hydrophobicity of proteins. However, strong correlation is found only between emulsifying activity and surface hydrophobicity, but not with average molecular hydrophobicity (Kato and Nakai, 1980; Keshavarz and Nakai, 1979; Voutsinas et al., 1983a,b; Nakai, 1983). This tentatively suggests that, contrary to the situation in foams, proteins do not undergo extensive unfolding at the oil-water interface. Although a positive correlation between surface hydrophobicity and emulsifying properties exists, this is not an absolute one. For instance, the emulsifying properties of β-lactoglobulin are better above pH 7 than at pH 3, whereas the surface hydrophobicity is greater at pH 3 than at pH 7 (Shimizu et al., 1985). Such discrepancies indicate that molecular factors other than surface hydrophobicity are also important in the expression of emulsifying properties. These include the disjoining forces arising from electrostatic, hydration repulsion, and steric interaction between the loops of the adsorbed protein molecules at the interface (Kitchener and Mussellwhite, 1968). Protein-stabilized emulsions are stable at pH far away from the isoelectric pH of the protein (Das and Kinsella, 1989; Kitchener and Mussellwhite, 1968; Shimizu et al., 1981), where the electrostatic and hydration repulsion forces are maximum.

The role of rheological properties of protein films in emulsion stability is quite ambiguous. There seems to be no correlation between the viscoelastic properties of protein films at the oil-water interface and the stability of oil-in-water emulsions (Graham and Phillips, 1976b). For instance, the surface viscosities of β-casein, BSA, and lysozyme films at the oil-water interface were 1, 10, and 5000 mNs/m, respectively. The relative stability of emulsions of these three proteins followed the order BSA > lysozyme > β-casein (Phillips, 1981), suggesting that cohesive forces within the protein film are not critical for emulsion stability; interparticle forces seem to influence emulsion stability.

Solubility, in addition to surface hydrophobicity, plays a role in emulsifying properties of proteins (Kinsella et al., 1985). Highly insoluble proteins display very poor emulsifying properties. However, no strong correlation between solubility and emulsifying properties has ever been established (Voutsinas et al., 1983a, b; McWaters and Holmes, 1979). Nonetheless, since the stability of protein films at the oil-water interface requires favorable interaction of the protein chain with both the oil and aqueous phases, an optimum balance of hydrophilic and hydrophobic groups that maintains the protein in solution is needed for better emulsifying properties. Partial denaturation of proteins that does not cause insolubilization usually improves emulsifying properties of proteins (Kato and Nakai, 1980; Kato et al., 1981, 1983; Voutsinas et al., 1983a,b). This is attributable to an increase in the surface hydrophobicity. However, excessive denaturation that causes a decrease in solubility often results in poor emulsifying properties (McWaters and Holmes, 1979).

The emulsifying properties of protein isolates is affected by the molecular properties of its individual components. Shimizu et al. (1981) showed that the protein components in whey protein isolate exhibited selective adsorption at the oil-water interface. This selective adsorption was affected by pH. The fraction of α-lactalbumin at the oil-water interface decreased from 48% at pH 3 to about 10% at pH 9, whereas the fraction of β-lactoglobulin increased from 13% at pH 3 to about 62% at pH 9 (Shimizu et al., 1981). On the other hand, in the case of sodium caseinate-stabilized emulsions, the molar ratio of β-casein to α_{s1}-casein at the freshly formed emulsion interface was similar to that of sodium caseinate (Robson and Dalgleish, 1987). However, during aging, the ratio of β-casein to α_{s1}-casein at the interface increased presumably because of displacement of α_{s1}-casein by adsorption of β-casein from the bulk phase. This has been attributed to the greater surface hydrophobicity of β-casein (Robson and Dalgleish, 1987).

Highly hydrophobic proteins usually displace those that are less hydrophobic from liquid interfaces (Dickinson et al., 1985, 1987; Castle et al., 1986, 1987). Studies with gelatin/caseinate mixtures have shown that when the weight ratio of gelatin to caseinate was below 2:1, the interfacial film of a freshly formed emulsion contained only caseinate (Castle et al., 1986). However, in the case of gelatin/β-lactoglobulin mixtures, a significant amount of gelatin was able to adsorb to the oil-water interface even at high ratios of β-lactoglobulin to gelatin in the bulk phase (Castle et al., 1986). These results indicate that emulsions prepared from protein mixtures might undergo constant displacement and exchange between the bulk and the interface, in which the more hydrophobic proteins in the bulk phase might displace and exchange with the less hydrophobic proteins at the interface. Such time-dependent exchange might cause time-dependent changes in the stability of the emulsion.

GELLING PROPERTIES

Several excellent reviews on protein gels have been published (Ledward, 1986; Clark and Lee-Tufnell, 1986; Ziegler and Foegeding, 1990). Therefore, the objective of this section will be to highlight some of the major molecular properties that influence network formation in thermally induced protein gels.

The sequence of events involved in the transformation upon heating of a protein solution to gel state is shown in Fig. 1.13 (Damodaran, 1988). The first step in heat-induced gelation of proteins is the change in protein conformation from the native state to a progel state by heat energy. This transition involves dissociation and denaturation of the protein. During this transition, the functional groups that were engaged in intramolecular hydrogen bonding, electrostatic, and hydrophobic interactions in the native state become available for intermolecular interactions under appropriate conditions, resulting in a gel network. Generally, proteins form two types of gel networks. Proteins that contain high levels of nonpolar residues undergo random aggregation via hydrophobic interactions, resulting in an opaque coagulum-type gel network with low elasticity and water-holding capacity. On the other hand, proteins that contain low levels of nonpolar residues usually form an ordered gel network that is translucent, elastic, and has high water-holding capacity, e.g., gelatin gels. In most cases, coagulum-type gels are irreversible, i.e., cannot be melted back to progel state, whereas translucent gels are reversible. In the case of translucent-type gels, when the protein in the progel state is cooled, there is a possibility that the denatured molecules might partially refold (Damodaran, 1988). The extent of refolding and formation of intramolecular interactions would decrease the number of functional groups available for intermolecular interactions. In such cases, the gel would be weaker, and the minimum protein concentration needed to form a self-standing gel network would be higher than in the absence of protein refolding during cooling.

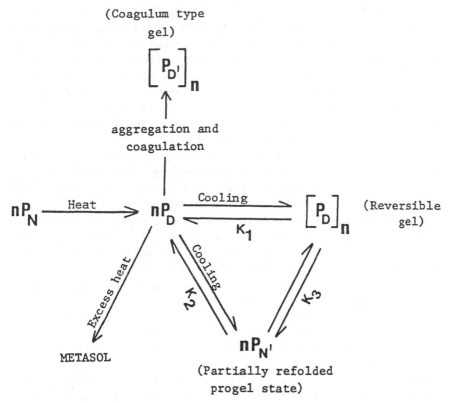

FIG. 1.13 A proposed mechanism for heat-induced gelation of globular proteins. n is the number of protein molecules, P_N, P_D, and $[P_{D'}]_n$ are protein in native, denatured (progel), and gel states, respectively; $P_{N'}$ is protein in partially refolded state, $[P_D]_n$ is coagulum-type gel state; K_1, K_2, and K_3 are equilibrium constants. (*From Damodaran, 1988.*)

    Shimada and Matsushita (1980) showed that proteins containing >31.5% of Val, Pro, Leu, Ile, Phe, and Trp form coagulum-type gel, whereas those proteins containing <31.5% of nonpolar amino acid residues form translucent gels. This empirical rule, although quite useful, is an oversimplification and does not take into account the influence of pH and ionic strength. For example, β-lactoglobulin, which has 32% of the above nonpolar amino acid residues, forms a translucent gel when the medium is water, whereas in 0.05 M NaCl it forms a coagulum-type gel. This suggests that both the attractive hydrophobic interactions and the repulsive electrostatic interactions affect the physical characteristics of protein gels. In this respect, both hydrophobicity and charge frequency of proteins need to be considered in predicting the physical characteristics of protein gels. Shimada and Matsushita's (1980) empirical rule also assumes that proteins completely unfold and expose all hydrophobic residues when heated above their denaturation temperature. However, evidence suggests that most proteins retain significant amount of folded structure in the gel network (Damodaran, 1988; Wang and Damodaran, 1991; Clark et al., 1981), and the extent of unfolding affects the physical and visual characteristics of gels.

A majority of cross-links formed in heat-induced protein gels are noncovalent in nature, e.g., hydrogen bonding, hydrophobic bonding, and electrostatic interactions. However, it has been suggested that in addition to noncovalent interactions, the ability of a protein to form intermolecular covalent disulfide cross-links during the heating regime is a prerequisite for gelation (Nakamura et al., 1984; Mori et al., 1986; Shimada and Cheftel, 1989). However, it has been argued that since phaseolin and gelatin do not have disulfide bonds, but form very good gels, formation of disulfide bonds is not quintessential for gelation (Wang and Damodaran, 1990). It has been pointed out that the positive role of disulfide bonds in gelation might not be related to its involvement in structure formation directly, but to its role in increasing the polypeptide chain length (Wang and Damodaran, 1990). A linear correlation was found between average molecular weight and square root of gel hardness (Wang and Damodaran, 1990). Based on experimental evidence, it has been postulated that the critical chain length for a globular protein to form a self-standing gel network is about 200 amino acid residues (i.e., about 23,000 molecular weight) (Wang and Damodaran, 1990). Proteins with less than 23,000 molecular weight can form gels, provided they contain at least one free sulfhydryl group so that when oxidized the molecular weight of the disulfide-linked dimer is above 23,000 (Wang and Damodaran, 1990). The longer the polypeptide chain formed via polymerization during heating, the greater would be the gel strength and the lower would be the minimum protein concentration needed to form a gel. In this regard, cross-links other than disulfide bonds, e.g., intermolecular ϵ(γ-glutamyl)lysine cross-links catalyzed by transglutaminase, may also promote gelation of proteins.

Several investigations have revealed that proteins retain a significant amount of folded structure in the gel state. Bovine serum albumin undergoes transconformational changes from α-helix to β-sheet structure during gelation (Wang and Damodaran, 1991). In the gel state BSA retained 42% α-helix, 26.5% β-sheet, and 31.5% aperiodic structure compared to 57.5% α-helix and 42.5% aperiodic structure in the native state (Wang and Damodaran, 1991). Investigations of soy 11S gels also have shown that about 26% β-sheet was retained by 11S in the gel state compared to 66% β-sheet and 28.5% aperiodic structure and very little of α-helix in the native state (Wang and Damodaran, 1991). Based on these findings it has been proposed that retention of a critical amount of β-sheet structure may be necessary for gel-structure formation in globular proteins. These β-sheets may be engaged in intermolecular hydrogen-bonded β-pleated sheets and may act as junction zones in the gel network (Wang and Damodaran, 1991). This is in contrast to gelatin gels in which partial reformation of collagen triple helices act as junction zones (Ledward, 1986).

Protein gelation is concentration dependent. Most globular proteins exhibit a critical concentration, known as least concentration endpoint (LCE), below which they cannot form a self-supporting gel network. The LCE for gelation differs between proteins. For soy proteins the LCE is about 8% (Catsimpoolas and Meyer, 1970), whereas for ovalbumin and gelatin it is about 3% and 0.6%, respectively (Shimada and Matsushita, 1980; Bello et al., 1962). These differences in LCE reflect differences in molecular properties such as net charge, molecular size, conformation, and the number density of functional groups available for cross-linking.

Several extrinsic factors, such as pH, ionic strength, heating temperature, etc., affect protein gelation. The gel strength is affected by heating temperature. Usually an increase in gel strength is observed until an optimum temperature range is reached. This temperature dependency is related to the extent of protein unfolding. For most proteins the optimum temperature range for gelation is found just above the thermal denaturation temperature of the protein. Excessive heating at high temperatures (>100°C) often causes scission of primary peptide linkages in proteins, which prevents gel-network formation.

SUMMARY

The manifestation of functional properties of food proteins is the result of a complex interplay of various intrinsic properties of proteins, such as size, hydrophilicity, hydrophobicity, charge, molecular flexibility, and the topography of the protein surface, and extrinsic factors, such as pH, ionic strength, temperature, and interactions with other food constituents. Quantitative understanding of the relative importance of the various intrinsic properties of proteins in the expression of a specific functional property is still elusive. The basic problem is that it is difficult to change each of these molecular properties individually and study its effect on a given functional property. Because of these difficulties much of the information in the literature on the structure-function relationship of food proteins is qualitative and descriptive in nature. A realistic approach to achieving a basic understanding of the structure-function relationship would be a multivariate approach in which the combined effects of various molecular factors on a functional property are studied simultaneously. For such an approach to be successful, precise methods need to be developed to quantify molecular descriptors such as flexibility, surface hydrophobicity, and protein conformation changes. Future research should focus on developing methods to quantify molecular descriptors that affect functional properties of proteins as well as standardized methods to measure functional properties that can be used by all researchers.

REFERENCES

Adams, D. J., Evans, M. T. A., Mitchell, J. R., Phillips, M. C., and Rees, D. A. 1971. Adsorption of lysozyme and some acetyl derivatives at the air-water interface. *J. Polymer Sci., Part C*. 34: 167.

Bacon, J. R., Hemmant, J., Lambert, N., Moore, R., and Wright, D. 1988. Characterization of the foaming properties of lysozymes and α-lactalbumins: a structural evaluation. *Food Hydrocolloids* 2: 225.

Bello, J., Bello, H. R., and Vinograd, J. R. 1962. The mechanism of gelation of gelatin: The influence of pH, concentration, time and dilute electrolyte on the gelation of gelatin and modified gelatins. *Biochim. Biophys. Acta* 57: 214.

Bigelow, C. C. 1967. On the average hydrophobicity of proteins and the relation between it and protein structure. *J. Theor. Biol.* 16: 187.

Buckingham, J. H. 1970. Effect of pH, concentration, and temperature on the strength of cytoplasmic protein foams. *J. Sci. Food Agric.* 21: 441.

Byrne, B. M., van het Schip, A. D., can de Klundert, J. A. M., Arnberg, A. C., Gruber, M., and Greet, A. B. 1984. Amino acid sequence of phosvitin derived from the nucleotide sequence of part of the chicken vitellogenin gene. *Biochemistry* 23: 4275.

Castle, J., Dickinson, E., Murray, A., Murray, B. S., and Stainsby, G. 1986. Surface behavior of adsorbed films of food proteins. In *Gums and Stabilizers for the Food Industry*, G. O. Phillips, D. J. Wedlock, and P. A. Williams (Ed.), p. 409. Elsevier Applied Sci., London and New York.

Castle, J., Dickinson, E., Murray, B. S., and Stainsby, G. 1987. Mixed-protein films adsorbed at the oil-water interface. In *Proteins at Interfaces: Physicochemical and Biochemical Studies*, J. L. Brash and T. A. Horbett (Ed.), p. 118. American Chemical Society, Washington, DC.

Catsimpoolas, N., and Meyer, E. W. 1970. Gelation phenomena of soybean globulins: I. Protein-protein interactions. *Cereal Chem.* 47: 559.

Clark, A. H. and Lee-Tufnell, C. D. 1986. Gelation of globular proteins. In *Functional Properties of Food Macromolecules*, J. R. Mitchell and D. A. Ledward, (Ed.), p. 203. Elsevier Applied Sci., New York.

Clark, D. C., Mackie, A. R., Smith, L. H., and Wilson, D. R. 1988. The interaction of bovine serum albumin and lysozyme and its effect on foam composition. *Food Hydrocolloids* 2: 209.

Clark, A. H., Saunderson, D. H. P., and Suggett, A. 1981. Infra-red and laser-Raman spectroscopic studies of thermally-induced globular protein gels. *Int. J. Peptide Protein Res.* 17: 353.

Cooney, C. M. 1974. A study of foam formation by whey proteins. *Diss. Abstr. Int.* 36/03, 1123-B.

Damodaran, S. 1988. Refolding of thermally unfolded soy proteins during the cooling regime of the gelatin process: Effect on gelation. *J. Agric. Food Chem.* 36: 262.

Damodaran, S. 1989. Interrelationship of molecular and functional properties of food proteins. In *Food Proteins*, J. E. Kinsella and W. G. Soucie (Ed.), p. 21. The American Oil Chemists' Society, Champaign, IL.

Damodaran, S. and Song, K. B. 1988. Kinetics of adsorption of proteins at interfaces: Role of protein conformation in diffusional adsorption. *Biochim. Biophys. Acta* 954:253.

Das, K. P. and Kinsella, J. E. 1989. pH dependent emulsifying properties of β-lactoglobulin. J. Dispersion Sci. Technol. 10: 77.

deWit, J. N. 1989. Functional properties of whey proteins. In *Developments in Dairy Chemistry-4*, P. F. Fox (Ed.), p. 285. Elsevier Applied Science, London and New York.

deWit, J. N., Klarenbeek G., and Adamse, M. 1986. Evaluation of functional properties of whey protein concentrates and whey protein isolates. 2. Effects of processing history and composition. *Neth. Milk Dairy J.* 40: 41.

Dickinson, D., Murray, A., Murray, B. S., and Stainsby, G. 1987. Properties of adsorbed layers in emulsions containing a mixture of caseinate and gelatin. In *Food Emulsions and Foams*, E. Dickinson (Ed.), p. 86. The Royal Society of Chemistry, London.

Dickinson, E., Murray, A., and Stainsby, G. 1985. Time-dependent surface viscosity of adsorbed films of casein + gelatin at the oil-water interface. *J. Colloid Interface Sci.* 106: 259.

Graham, D. E. and Phillips, M. C. 1976a. The fonformation of proteins at the air-water interface and their role in stabilizing foams. In *Foams*, R. J. Akers (Ed.), p. 237. Academic Press, New York.

Graham, D. E. and Phillips, M. C. 1976b. The conformation of proteins at interfaces and their role in stabilizing emulsions. In *Theory and Practice of Emulsion Technology*, A. L. Smith (Ed.), p. 75. Academic Press, New York.

Graham, D. E. and Phillips, M. C. 1979a. Proteins at liquid interfaces: 1. Kinetics of adsorption and surface denaturation. *J. Colloid Interface Sci.* 70: 403.

Graham, D. E. and Phillips, M. C. 1979b. Proteins at liquid interfaces: II. Adsorption isotherms. *J. Colloid Interface Sci.* 70: 415.

Haggett, T. O. R. 1976. The whipping, foaming and gelling properties of whey protein concentrates. *N. Z. J. Diary Sci. Technol.* 11: 244.

Halling, P. J. 1981. Protein-stabilized foams and emulsions. *CRC Crit. Rev. Food Sci. Nutr.* 15: 155.

Harper, W. J. 1984. Model food systems approach for evaluating whey protein functionality. *J. Diary Sci.* 67: 2745.

Izmailova, V. N. 1979. Structure formation and rheological properties of proteins and surface-active polymers of interfacial adsorption layers. *Prog. Surf. Membr. Sci.* 13: 141.

Kato, A. and Nakai, S. 1980. Hydrophobicity determined by a fluorescence probe method and its correlation with surface properties of proteins. *Biochim. Biophys. Acta* 624: 13.

Kato, A., Osako, Y., Matsudomi, N., and Kobayashi, K. 1983. Changes in emulsifying and foaming properties of proteins during heat denaturation. *Agric. Biol. Chem.* 47: 33.

Kato, A., Tsutsui, N., Matsudomi, N., Kobayashi, K., and Nakai, S. 1981. Effects of partial denaturation on surface properties of valbumin and lysozyme. *Agric. Biol. Chem.* 45: 2755.

Keshavarz, E. and Nakai, S. 1979. The relationship between hydrophobicity and interfacial tension of proteins. *Biochim. Biophys. Acta* 576: 269.

Kim, S. H. and Kinsella, J. E. 1985. Surface activity of food proteins: Relationship between surface pressure development, viscoelasticity of interfacial films and foam stability of bovine serum albumin. *J. Food Sci.* 50: 1526.

Kinsella, J. E. 1982. Relationship between structure and functional properties of food proteins. In *Food Proteins*, P. F. Fox and J. J. Condon (Ed.), p. 51. Applied Science, New York.

Kinsella, J. E., Damodaran, S., and German, J. B. 1985. Physicochemical and functional properties of oilseed proteins with emphasis on soy proteins. In *New Protein Foods: Seed Storage Proteins*, A. M. Altschul, and H. L. Wilcke (Ed.), p. 108. Academic Press, New York.

Kitchener, J. A. and Mussellwhite, P. R. 1968. The theory of stability of emulsions. In *Emulsion Science*, P. Sherman (Ed.), p. 77. Academic Press, London.

Ledward, D. A. 1986. Gelation of gelatin. In *Functional Properties of Food Macromolecules*, J. R. Mitchell and D. A. Ledward, (Ed.), p. 171. Elsevier Applied Sci., New York.

Lee, B. and Richards, F. M. 1971. The interpretation of protein structures: Estimation of static accessibility. *J. Molec. Biol.* 55: 379.

MacRitchie, F. 1978. Proteins at interfaces. *Adv. Protein Chem.* 32: 283.

MacRitchie, F. 1985. Desorption of proteins from the air/water interface. *J. Colloid Interface Sci.* 105: 119.

MacRitchie, F. and Alexander, A. E. 1963a. Kinetics of adsorption of proteins at interfaces: Part II. The role of pressure barriers in adsorption. *J. Colloid Sci.* 18: 458.

MacRitchie, F. and Alexander, A. E. 1963b. Kinetics of adsorption of proteins at interfaces. Part III. The role of electrical barriers in adsorption. *J. Colloid Sci.* 18: 464.

MacRitchie, F. and Ter-Minassian Saraga, L. 1983. Stability of highly compressed monolayers of I-labelled and cold BSA. *Prog. Colloid Polymer Sci.* 68: 14.

MacRitchie, F. and Ter-Minassian Saraga, L. 1984. Concentrated protein monolayers: Desorption studies with radiolabeled bovine serum albumin. *Colloids Surf.* 10: 53.

McWaters, K. H. and Holmes, M. R. 1979. Influence of moist heat on solubility and emulsification properties of heat denatured proteins. *J. Food Sci.* 44: 774.

Mita, T., Ishido, E., and Masumoto, H. 1977. Physical studies on wheat protein foams. *J. Colloid Interface Sci.* 59: 172.

Mitchell, J. R., Irons, L., and Palmer, J. 1970. A study of the spread and adsorbed films of milk proteins. *Biochim. Biophys. Acta* 200: 138.

Mori, T., Nakamura, T., and Utsumi, S. 1986. Behavior of intermolecular bond formation in the late stage of heat-induced gelation of glycinin. *J. Agric. Food Chem.* 34: 33.

Nakai, S. 1983. Structure-function relationships of food proteins with an emphasis on the importance of protein hydrophobicity. *J. Agric. Food Chem.* 31: 676.

Nakamura, T., Utsumi, S., and Mori, T. 1984. Network structure formation in thermally induced gelatin of glycinin. *J. Agric. Food Chem.* 32: 349.

Perutz, M. F. 1978. Electrostatic effects in proteins. *Science* 201: 1187.

Phillips, M. C. 1981. Protein conformation at liquid interfaces and its role in stabilizing emulsions and foams. *Food Technol.* (Chicago) 35: 50.

Poole, S. 1989. Review: The foam-enhancing properties of basic biopolymers. *Int. J. Food Sci. Technol.* 24: 121.

Poole, S., West, S. I., and Fry, J. C. 1987. Charge and structural requirements of basic proteins for foam enhancement. *Food Hydrocolloids* 1: 227.

Poole, S., West, S. I., and Walters, C. L. 1984. Protein-protein interactions: Their importance in the foaming of heterogeneous protein systems. *J. Sci. Food Agric.* 35: 701.

Renugopalakrishnan, V., Horowitz, P. M., and Glimcher, M. J. 1985. Structural studies of phosvitin in solution and in the solid-state. *J. Biol. Chem.* 260: 11406.

Robson, E. W. and Dalgleish, D. G. 1987. Interfacial composition of sodium caseinate emulsions. *J. Food Sci.* 52: 1694.

Shimada, K. and Cheftel, J. C. 1989. Sulfhydryl group/disulfide bond interchange reactions during heat-induced gelatin of whey protein isolate. *J. Agric. Food Chem.* 37: 161.

Shimada, K. and Matsushita, S. 1980. Relationship between thermocoagulation of proteins and amino acid compositions. *J. Agric. Food Chem.* 28: 413.

Shimizu, M., Kamiya, T., and Ymauchi, K. 1981. The adsorption of whey proteins on the surface of emulsified fat. *Agric. Biol. Chem.* 45: 2491.

Shimizu, M, Saito, M., and Yamauchi, K. 1985. Hydrophobicity and emulsifying activity of milk proteins. *Agric. Biol. Chem.* 49: 189.

Swaisgood, H. E., 1982. Chemistry of milk proteins. In *Chemistry of Milk Proteins*, P. F. Fox (Ed.), p. 1. Elsevier Applied Science Publishers, London and New York.

Ter-Minassian Saraga, L. 1981. Protein denaturation on adsorption and water activity at interfaces: An analysis and suggestion. *J. Colloid Interface Sci.* 80: 393.

Townsend, A. and Nakai, S. 1983. Correlations between hydrophobicity and foaming capacity of proteins. *J. Food Sci.* 48: 588.

Voutsinas, L. P., Cheung, E., and Nakai, S. 1983. Relationship of hydrophobicity to emulsifying properties of heat denatured proteins. *J. Food Sci.* 48: 26.

Voutsinas, L. P., Nakai, S., and Harwalkar, V. R. 1983. Relationship between protein hydrophobicity and thermal functional properties of food proteins. *Can. Inst. Food Sci. Technol. J.* 16: 185.

Wang, C.-H. and Damodaran, S. 1990. Thermal gelation of globular proteins: Weight-average molecular weight dependence of gel strength. *J. Agric. Food Chem.* 38: 1154.

Wang, C. -H. and Damodaran, S. 1991. Thermal gelation of globular proteins: Influence of protein conformation on gel strength. *J. Agric. Food Chem.* 39: 433.

Ward, A. F. H. and Tordai, L. 1946. Time-dependence of boundary tensions of solutions: 1. The role of diffusion in time effects. *J. Chem. Phys.* 14: 453.

Xu, S. and Damodaran, S. 1992. The role of chemical potential in the adsorption of lysozyme at the air-water interface. *Langmuir* 8:2021.

Xu, S. and Damodaran, S. 1993a. Calibration of radiotracer method to study protein adsorption at interfaces. *J. Colloid Interface Sci.* 157: 485.

Xu, S. and Damodaran, S. 1993b. Comparative adsorption of native and denatured egg-white, human, and T_4 phage lysozymes at the air-water interface. *J. Colloid Interface Sci.* 159: 124.

Yasui, S. C., Pancoska, P., Dukor, R. K., Keiderling, T. A., Renugopalakrishnan V., and Glimcher, M. J. 1990. Conformational transitions in phosvitin with pH variation: Vibrational circular dichroism study. *J. Biol. Chem.* 265: 3780.

Yu, M.-A. and Damodaran, S. 1991a. Kinetics of protein foam destabilization: Evaluation of a method using bovine serum albumin. *J. Agric. Food Chem.* 39: 1555.

Yu, M.-A. and Damodaran, S. 1991b. Kinetics of destabilization of soy protein foams. *J. Agric. Food Chem.* 39: 1563.

Ziegler, G. R. and Foegeding, E. A. 1990. Gelation of proteins. *Advan. Food Nutr. Res.* 34: 203.

2

Solubility of Proteins: Protein–Salt-Water Interactions

Thomas F. Kumosinski and Harold M. Farrell, Jr.

Agricultural Research Service
United States Department of Agriculture
Philadelphia, Pennsylvania

INTRODUCTION

Biotechnology holds the promise of developing new designer-type products for food science with tailor-made functionalities via genetic engineering of proteins and creation of new co-solutes, which may control functionality through chemical, biochemical, or genetic techniques (Richardson et al., 1992). However, the historic problem of developing quantitative measures for structure-function relationships still plagues the researcher. Without knowledge of these relationships, the new techniques are limited to costly hit-or-miss experiments, which have a low probability of success.

The caseins of bovine milk and their naturally occurring genetic variants provide an illustration of qualitative correlations between primary structure and protein functionality in food systems. For example, milks

---

Reference to a brand or firm name does not constitute an endorsement by the U.S. Department of Agriculture over others of a similar nature not mentioned.

containing α_{s1}-casein A rather than the more frequently occurring α_{s1}-casein B variant yield cheeses with a softer texture and body; at the same time these milks are more resistant to calcium-induced coagulation, i.e., they are more stable at elevated calcium concentrations (Thompson et al., 1969). Here the A variant is the result of the sequential deletion of 13 amino acids (residues 14 to 26) from the B variant. However, changes in protein secondary, tertiary, and quaternary structure as well as in the thermodynamic parameters resulting from this mutation have not been obvious. Thus, a complete thermodynamic and structural mechanism for the above functionality changes has not been elucidated, and the success rate of future technologically induced mutations cannot be predicted.

Finally, in recent years, the emergence of molecular modeling as a technique for refining existing three-dimensional molecular structures or building new predicted models has yielded a methodology with the capability of developing a molecular basis for structure-function relationships (Kumosinski et al., 1991a,b). Now, not only food proteins but preservatives, salts, stabilizers, etc., may be modeled for their potential effectiveness in structure-function relationships.

We have attempted to define and model one simple functionality test for the caseins: solubility as a function of calcium ion concentration. This system was selected for three reasons. From the point of view of the food industry, caseinate is an important commodity and milk and dairy products are widely consumed for their calcium content. Second, the interactions occurring in this important colloidal-transport system are still not well defined. Third, a wealth of information of a qualitative nature is available in the literature on calcium-induced casein solubility curves (Arakawa and Timasheff, 1984; Farrell and Kumosinski, 1988; Farrell et al., 1988; Kumosinski and Farrell, 1991). In order to better understand these calcium-protein interactions, the precipitation and re-solubilization of selected caseins were reinvestigated. The functionality data were analyzed with respect to computer-generated models; analysis of the data indicates that a thermodynamic linkage occurs between calcium binding and salting-out and salting-in reactions.

Thus, a quantitative thermodynamic mechanism could be established for the salting-in and salting-out of casein, and binding free energies for these ligand-induced protein solubility profiles may easily be calculated. Finally, molecular modeling techniques such as energy minimization and molecular dynamics were utilized to mimic protein–salt-water interactions, which explain the salt-induced solubility profiles of α_{s1}-casein. Here, a recently developed predicted energy-minimized three-dimensional structure of α_{s1}-casein was employed to discover the hydrophobic sites responsible for precipitation of the protein and potential salt-binding sites responsible for the salting-in process with added divalent salts.

THEORY

Thermodynamic Linkage

Wyman's theory of thermodynamic linkage (Wyman, 1964) is based on the concept that changes in an observable physical quantity (in this case solubility) can be linked to ligand binding. In previous studies on isolated caseins (Farrell et al., 1988; Kumosinski and Farrell, 1991) it was shown that the precipitation of the caseins in the presence of calcium is indeed linked to calcium binding, and that calcium binding is the driving force in both salting-out and salting-in.

Here, we assume that there are essentially two classes of binding sites for ligands responsible for the sequential salting-out and salting-in processes and, therefore, Wyman's linked functions equations (1964) can be used to treat these processes with the assumption that the following equilibria occur:

$$p + nX \overset{k_1^n}{\leftrightarrows} PX_n + mX \overset{k_2^m}{\leftrightarrows} PX_nX_m \qquad (1)$$
$$\quad (S_0) \qquad\quad (S_1) \qquad\quad (S_2)$$

where p is the unbound protein, X is the free salt, n and m are the number of X moles bound to species PX_n and PX_nX_m, and S_0, S_1 and S_2 are the solubilities of the species indicated. For this study S_1 and S_2 will be relative to S_0. The mathematical relationship representing the above stoichiometry can be represented according to the following:

$$S_{app} = S_0 f(p) + S_1 f(PX_n) + S_2 f(PX_nX_m) \qquad (2)$$

where S_{app} is the apparent protein solubility at a given salt concentration (X_T), $f(i)$ are the protein fractional component of species i and the S's are species previously defined. Incorporation of the salt-binding equilibrium constants (k_1 and k_2) as defined by Eq. (1) into Eq. (2) yields the following:

$$S_{app} = \frac{S_0 p}{p + k_1^n px^n} + \frac{S_1 k_1^n px^n}{p + k_1^n px^n} + \frac{(S_2 - S_1)k_2^m px^m}{p + k_2^m px^m} \qquad (3)$$

where p is the concentration in percent of the unbound protein and x is the concentration of unbound salt. Cancelation of common terms yields:

$$S_{app} = \frac{S_0}{1 + k_1^n x^n} + \frac{S_1 k_1^n x^n}{1 + k_1^n x^n} + \frac{(S_2 - S_1)k_2^m x^m}{1 + k_2^m x^m} \qquad (4)$$

It should be stressed here that the above expression is valid for sequential binding, i.e., $k_1 > k_2$, and n sites saturate prior to the binding of m sites on the protein and, for simplicity, that n and m do not interact. Also, for n or $m > 1$, k_1 and k_2 represent an average value for each class of the n or m binding sites. In reality n or m moles of salt will bind with only one equilibrium constant (K_1), i.e., $K_1 = k^n_1$ and $K_2 = k^m_2$.

Now, since the total salt concentration, X_T, is the sum of the free salt concentration, x, and the concentration of the bound salt of both species PX_n and PX_nX_m, it can be shown that

$$X_T = x \left(1 + \frac{nk_1{}^nP_Tx^{(n-1)}}{1 + k_1{}^nx^n} + \frac{mk_2{}^mP_Tx^{(m-1)}}{1 + k_2{}^mx^m} \right) \tag{5}$$

where P_T is the total concentration of protein. From Eq. (5) it can be seen that X_T approaches x when P_T is small relative to x. In some of our experiments this assumption is reasonable. For α_{s1}-casein, very low (1–2 mM) concentrations of calcium induce relatively strong aggregations (Waugh et al., 1971). This suggests that the kinetically active unit in precipitation is an aggregate, so that the number of moles of aggregate is smaller than the number of moles of total protein. In direct protein solubility studies this may not be the case. Using the values of K_a published by Dickson and Perkins (1971) for α_{s1}-casein along with P_T, free x can be calculated and used to generate adjusted k_i's which may be more properly related to an "apparent" binding constant (Kumosinski and Farrell, 1991). However, the latter is still an approximation, and from the point of view of the kinetically active species, X_T is also appropriate.

It could be argued that the calcium-induced protein self-associations are a complicating factor in these studies even though the monomer and aggregate are both soluble. Cann and Hinman (1976) developed equations for dealing with the effects of association on ligand binding. Using their concepts, it can be seen that:

$$\begin{array}{ccccc} S_0 & & k_0^j \; S_0 & & k^n_1 \; S_1 \\ kp + jx & \rightleftharpoons & P_kX_j + & nx \rightleftharpoons & P_kX_jX_n \end{array} \tag{6}$$

and the apparent solubility S_{app} again is equal to the sum of the solubilities of each species times the fraction of the protein in that species.

$$S_{app} = S_0 fp + S_0 \, (fP_kX_j - f \, P_kX_jX_n) + S_1 fP_kX_jX_n \tag{7}$$

Thus:

$$S_{app} = S_0 fp + S_0 fP_kX_j + (S_1 - S_0) \, fP_kX_jX_n$$

$$S_{app} = S_0 \left[\frac{p}{p + k_0^j P^k x^j} + \frac{k_0^j P^k x^j}{p + k_0^j P^k x^j} \right] + (S_1 - S_0) \frac{P_kX_jX_n}{P_kX_j + P_kX_jX_n} \tag{8}$$

and

$$S_{app} = S_0 \frac{k^j_0 P^k x^j + p}{p + k^j_0 P^k x^j} + (S_1 - S_0) \frac{k_1{}^n (P_k X_j) x^n}{(P_k X_j) + k_1{}^n (P_k X_j) x^n} \qquad (9)$$

Collection of terms yields:

$$S_{app} = \frac{S_0}{1 + k_1{}^n x^n} + \frac{S_1 k_1{}^n x^n}{1 + k_1{}^n x^n} \qquad (10)$$

It can be seen that Eq. (10) is now in the same form as Eq. (4) and that the association parameter k^j_0 has canceled out. Thus, only binding sites linked to changes in solubility will be seen in this analysis. This result is in line with the experimental data collected by Waugh et al. (1971), which also showed through sedimentation analysis that an associated aggregate and not the monomer participates in the precipitation reaction.

Salt-induced solubility profiles were directly analyzed using a Gauss-Newton nonlinear regression analysis program developed at this laboratory by Dr. William Damert. All profiles were analyzed by fixing the values of n and m and calculating the best least squares fit for the optimum evaluated k_1 and k_2 values. The n and m values were then fixed to new integer values and the entire procedure was repeated. The n and m values that yielded the minimum root-mean-square value for the analysis with the minimum error in k_1 and k_2 were then reported.

Molecular Modeling

All aggregate structures employed the α_s-casein B casein monomer structure previously refined via energy minimization (Kumosinski et al., 1994). Aggregates were constructed using a docking procedure on an Evans and Sutherland PS390 interactive computer graphics display driven by the Tripos Sybyl (St. Louis, MO) molecular modeling software on a Silicon Graphics 4200 Unix-based computer. The docking procedure allowed for individual manipulation of the orientation of up to four molecular entities relative to one another. The desired orientations could then be frozen in space and merged into one entity for further energy minimization calculations utilizing a molecular force field. The criterion for acceptance of reasonable structures was determined by a combination of experimentally determined information and the calculation of the lowest energy for that structure.

Force Field Calculation.   Studies concerned with the structures and/or energetics of molecules at the atomic level require a detailed knowledge of the potential energy surface (i.e., the potential energy as a function of the atomic coordinates). For systems with a small number of atoms, quantum mechanical methods may be used, but these methods become computationally intractable for larger systems (e.g., most systems of biological interest) because of the large number of atoms that must be considered. For these larger systems, molecular mechanics methods are used. Molecular mechanics is based on the assumption that the true potential energy surface can be approximated with an empirical potential surface consisting of simple analytical functions of the atomic coordinates. The empirical potential energy model treats the atoms as a collection of point masses that are coupled to one another through covalent (bonded) and noncovalent (nonbonded) interactions. The potential energy function (Weiner et al., 1986; Kollman, 1987) generally has the form:

$$E_{\text{total}} = \sum_{bonds} K_r(r - r_{eq})^2 + \sum_{angles} K_\theta(\theta - \theta_{eq})^2$$
$$+ \sum_{dihedrals} \frac{1}{2}K\,[1 + \cos(n\phi - \gamma)] \qquad (11)$$
$$+ \sum_{i < j} \frac{B_{ij}}{R_{ij}^{12}} - \frac{A_{ij}}{R_{ij}^{6}} + \frac{q_i q_j}{\epsilon R_{ij}}$$

The first three terms are due to covalent interactions and represent the difference in energy between the geometry of the actual structure and a geometry in which the bond lengths, bond angles, and dihedral angles all have ideal values. The remaining terms represent nonbonded van der Waals and electrostatic interactions. In Eq. (11), r, θ, ϕ, and R_{ij} are variables, determined by the atomic coordinates. All other entities are constant parameters chosen to reproduce experimental observables as closely as possible. Although empirical potential energy functions such as Eq. (11) are relatively crude, they have been applied successfully to the study of hydrocarbons, oligonucleotides, peptides, and amino acids, as well as systems containing a large number of small molecules such as water. The Tripos force field in Tripos' Sybyl software package uses the above functional form, plus a bump factor, which allows atoms within a fraction of the van der Waals radius for H-bond function if so chosen by the user. The parameters used for electrostatic calculations include atomic partial charges (q_i) calculated by the Kollman group (Weiner et al., 1986; Kollman, 1987) using a united atom approach with only essential hydrogens. All molecular structures were refined with an energy-minimization procedure using a conjugate gradient algorithm, in which the positions of the atoms are adjusted iteratively so as to achieve a minimum potential energy value. Energy-minimization calculations were terminated when the energy difference between the current and previous iterations was less than 1 kcal/mol. A nonbonded cutoff of 5 Å was used initially to save computer time and is an appropriate value for use of a function that varies with distance. A stabilization energy of at least -10 kcal/mol/residue was achieved for all structures, which is consistent with values obtained for energy minimized structures determined by x-ray crystallography.

Molecular Dynamics.   In the previous paragraph we considered only static structures. However, the dynamic motion of molecules in solution contributes to their functionality. The molecular dynamics approach is a method of studying motion and molecular configuration as a function of time (Andersen, 1980). All atoms in the molecule are assigned a kinetic energy through a velocity term, which can be related to the local temperature as well as to the average temperature of the system. These calculations can be performed in vacuum or in the presence of a desired number of solvent molecules such as water and at a constant temperature and volume using a periodic boundary condition to confine the calculation within a prescribed volume (van Gunsteren and Berendsen, 1977). For these calculations a force field describing the potential energy is combined with Newton's second law of motion.

$$F_i = m_i a_i(t) = m_i \frac{dv_i(t)}{dt} = m_i \frac{d^2 x_i(t)}{dt^2} = -\nabla_i E \qquad (12)$$

where F_i is the force on atom i, which has mass (m_i), velocity (v_i), acceleration (a_i), and position (x_i). ∇_i is the gradient or the derivative with respect to position, t is the time displacement, and E is the potential energy of the molecule described by the chosen force field. Equation (12) is integrated at various time intervals for the desired molecule using the chosen force field via a prescribed numerical integration method. The time interval chosen must be small in comparison with the period associated with highest frequency of motion within the molecule. This is usually stretching of a bond associated with a hydrogen atom, i.e., one femtosecond. Numerical integration of Eq. (12) over 1-fsec intervals to 100 psec for a protein molecule of 2000 atoms or more necessitates a fast computer with a large memory capacity. The results of these calculations can mimic the motions of molecules in solution and also time-dependent geometric parameters. For example, the distance from the center of moment for a set of atoms may be related to correlation times derived from NMR, EPR, and fluorescence experiments.

QUANTITATION OF THE SALT-INDUCED SOLUBILITY PROFILES OF THE CASEINS

Solubility at 37°C

Solubility determinations of α_{s1}-caseins A and B (α_{s1}-A, α_{s1}-B) and β-casein C (β-C) were performed at 37°C in 10 mM imidazole-HCl pH 7.0, 0.07 M KCl, at initial protein concentrations of 10 mg/mL (Farrell et al., 1988). As in the experiments of Noble and Waugh (1965), the proteins precipitate when added $CaCl_2$ exceeds 5 mM (Fig. 2.1A and B). Creamer and Waugh (1966) had suggested that about 13 sites of similar calcium binding strength exist in the α_{s1}-B at pH 6.6 and that when calcium ion concentration exceeds this critical binding level, charge neutralization occurs and precipitation results. Comparison of the solubility profiles of α_{s1}-A and -B indicates that at 37°C α_{s1}-A is more soluble than α_{s1}-B, while β-C is the most soluble. In order to quantify the data, nonlinear regression analyses were performed. The data of Fig. 2.1A were fitted by Eq. (4). Values of k_1 were obtained at fixed integer values of n; the correct value of n was taken to be the fit with the minimum root mean square (RMS). Fig. 2.1A shows the fit to $n = 2, 4,$ and 8 for α_{s1}-A; values for $n = 8$ gave the minimum RMS with the lowest error in k_1. Analysis of the solubility profiles of α_{s1}-A, α_{s1}-B, and β-C at 37°C, where hydrophobic interactions are maximized, showed no salting-in behavior so that k_2 and m were essentially zero. Values obtained for k_1 (salting-out) and n are given in Table 2.1.

Solubility at 1°C

Figure 2.2 shows that β-casein C is not precipitated at 1°C by Ca^{2+} at concentrations of up to 400 mM. It is known that hydrophobic forces are dominant in the association reactions of β-caseins (Schmidt, 1982). The solubility of β-casein C clearly distinguishes it from α_{s1}-B; it is known that β-casein binds Ca^{2+} at 1°C, but in this case binding is not linked to changes in solubility and so no analysis by thermodynamic linkage is possible. Aliquot addition of calcium chloride solutions to α_{s1}-casein results in a rapid decrease in solubility from 8 to 50 mM, where the protein is almost totally precipitated. When the calcium chloride concentration exceeds 100 mM, a gradual salting-in of the protein ensues at 1°C. The data for α_{s1}-B were fitted by Eq. (4) and the salting-out parameters k_1 and n, as well as the salting-in parameters k_2 and m were determined (Table 2.2). The α_{s1}-A, genetic variant, in contrast to the α_{s1}-B exhibits extraordinary solubility behavior over a broad range of calcium chloride concentrations. At 1°C (Fig. 2.3) α_{s1}-A, like α_{s1}-B (Fig. 2.2), is precipitated

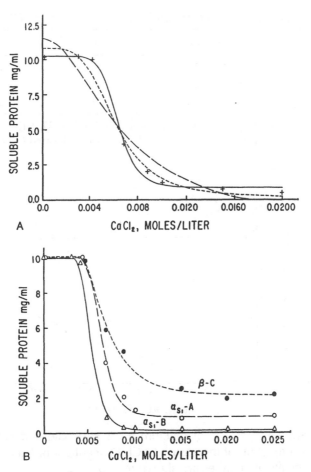

FIG. 2.1   Solubility at 37°C of the calcium salts of α_1-caseins A and B and β-casein C as a function of increasing $CaCl_2$ concentration. Solutions buffered at pH 7.0, 10 mM imidazole-HCl. (A) The experimental data for α_{s1}-A were fitted by Eq. (4) by nonlinear regression analysis with values of 2 (— —), 4 (---), and 8 (—) assigned to n. The best fit was obtained for $n = 8$. (B) Similar fits for α_{s1}-B and β-casein C; results of analyses are shown in Table 2.1.

with calcium at about 8 mM, whereupon the net electrical charge on the Ca-complexed protein may be close to zero. In the absence of electrolyte (KCl) or buffer, and after aliquot addition of $CaCl_2$, the protein is driven into solution at 90 mM. The Ca-complexed protein is now positively

TABLE 2.1   Calcium-Induced Insolubility of Casein at 37°C[a]

Casein	k_1(L/mole)	n	S_1[b](mg/mL)
α_{s1}-A	157 ± 3	8	0.9 ± 0.2
α_{s1}-B	186 ± 3	8	0.1 ± 0.1
β-C	156 ± 12	4	2.0 ± 0.3

[a]Solutions buffered at pH 7.0, 10 mM imidazole-HCL, 0.07M KCl.
[b]S_1 denotes the maximum value for soluble protein at elevated Ca^{2+} concentrations.

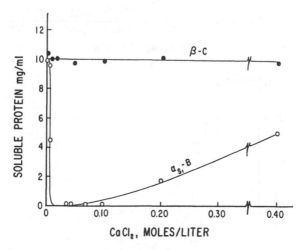

FIG. 2.2   Solubility at 1°C of calcium α_{s1}-B caseinate and calcium β-C caseinate as a function of increasing $CaCl_2$ concentration. Data were fitted by Eq. (4); results of analyses are shown in Table 2.2.

charged and is acting as a cation. This conclusion was verified by free-boundary electrophoresis at pH 7.0, 10 mM imidazole, 150 mM $CaCl_2$, where the protein is soluble at 1°C; it migrates (+ 1.36 cm² volt⁻¹ sec⁻¹ × 10⁻⁵) toward the cathode (Thompson et al., 1969). The evidence to this point favors direct salt-protein interactions as being responsible for the salting-in and salting-out of caseins by calcium rather than a salt-solvent interaction as previously proposed by Melander and Horvath (1977) following the cavity theory model of Sinanoglu (1968). The concept of salt binding is supported by the earlier work of Robinson and Jencks (1965), who studied salting-out of model compounds and concluded that binding was a factor.

TABLE 2.2   Calcium-Induced Insolubility and Solubility of Caseins at 1°C[a]

Casein	k_1[b]	n	k_2[b]	m
α_{s1}-B	123 ± 5	8	2.5 ± 0.2	4
α_{s1}-A	68 ± 1	8	10.6 ± 0.3	8
β-C	Totally soluble			

[a]Conditions as in Table 2.1.
[b]L/mole.

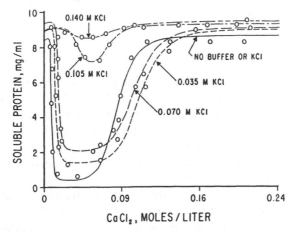

FIG. 2.3   Solubility at 1°C of calcium α_{s1}-A caseinate as a function of increasing $CaCl_2$ and KCl concentrations. Data were fitted by Eq. (4); results are shown in Table 2.3.

Influence of Electrolyte on Salting-Out and Salting-In Constants

Aliquot addition of $CaCl_2$ in the presence of various concentrations of KCl results in a shift in the solubility profile of α_{s1}-A at 1°C (Fig. 2.3). KCl was chosen as the electrolyte because it occurs at higher concentration than NaCl in milk serum (Farrell and Thompson, 1987). The data were analyzed and fitted by Eq. (4); parameters k_1, k_2, n, and m are given in Table 2.3. Note that as the KCl concentration increases, n drops to 4 for Ca^{2+}; the protein apparently requires less bound Ca^{2+} to precipitate. On the other hand, m for resolubilization remains at 8. Increasing ionic

TABLE 2.3 Ionic Strength Dependence of Calcium-Induced Solubility of α_{s1} – casein A at 1°C

KCl mM	$k_1{}^a$	$k_2{}^a$	$S_1{}^b$(mg/mL)
—	130 ± 3	13.3 ± 0.3	0.4 ± 0.2
35[c]	82 ± 1	10.1 ± 0.1	1.4 ± 0.1
70[c]	68 ± 1	10.6 ± 0.2	2.1 ± 0.1
105[d]	22 ± 1	15.9 ± 0.1	4.7 ± 0.4
140[d]	34 ± 1	15.7 ± 0.2	8.4 ± 0.1

[a]L/mole.
[b]S_1 denotes maximum value for soluble protein after precipitation but before total resolubilization.
[c]$n = 8$ and $m = 8$.
[d]$n = 4$ and $m = 8$.

strength by the addition of KCl for each calcium-induced solubility profile of α_{s1}-A at 1°C decreased k_1 and k_2 values. The variations of k_1 and k_2 with KCl are given in Fig. 2.4A and B. For k_1, a nearly monotonic decrease occurs; Fig. 2.4A can be analyzed as a potassium-binding isotherm itself and an association constant K_{a1} computed for KCl-protein interactions. The value given in Table 2.4 line 1 was found to be 20 ± 6 L/mole. For k_2, the variation appeared more complex with two transitions. Analyzing these data, K_{a2} and $K_{a2}{}^1$ were calculated (Table 2.4, lines 2 and 3). As compared with the primary k_1 values for Ca^{2+} precipitation, all derived constants related to KCl effects are substantially smaller. Changes are consistent with the electrostatic character of these binding isotherms and could be attributed to a competitive effect of elevated potassium ion for calcium-binding sites or to KCl-solvent interactions affecting solubility; again for this system direct salt binding to the caseins appears to be the driving force. The overall difference for the behavior of α_{s1}-A relative to α_{s1}-B, however, must reside in the structural differences due to the deletion mutation in the A variant. In all of the above analysis, the constants obtained approximate binding constants, but they may not be identical with average association constants, since only binding that leads to changes in solubility is disclosed by this analysis. Bear in mind that for β-casein (Fig. 2.2) binding occurs but no change in solubility accompanies this and so no analysis is possible.

Influence of Electrolyte on Soluble Protein (S_1)

Further insights into the source of the effects resulting in the α_{s1}-A solubility profiles can be obtained by analysis of the variation of the

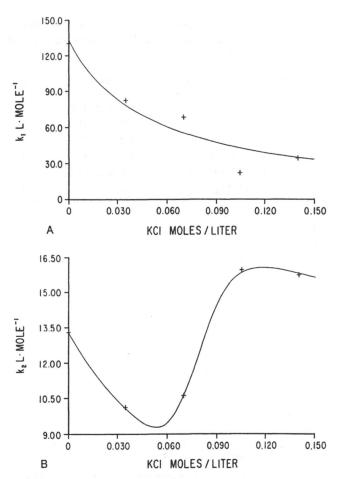

FIG. 2.4 Variations of k_1 (A) and k_2 (B) with KCl concentration. These parameters obtained from analysis of Fig. 2.3 (Table 2.3) were fitted by Eq. (4); results are shown in Table 2.4.

parameter S_1 of Table 2.3. This term is taken to represent the soluble form of α_{s1}-A in the valley between 2 and 8 mM $CaCl_2$ (Fig. 2.3).

Since the magnitude of salting-in for S_1 is larger than expected for weaker salt-solvent forces, it was decided to analyze the variations of S_1 with KCl as a binding isotherm (Fig. 2.5). Two cooperative transitions appear to occur. These were analyzed in terms of Eq. (4) and values of K_{as}, K_{as}^1 and n were calculated; these derived constants are compared with those derived from the variance of k_1 and k_2 with KCl in Table 2.4. All of the constants derived from k_2 or S_1 appear to be self-consistent in

TABLE 2.4   Comparison of Derived Constant from the Effect of KCl on k_1, k_2, and S_1 from the Calcium-Induced Solubility of α_{s1}-A at 1°C

Derived constant	Source of constant	Value (L/mole)	n	$\log K$
K_{a1}	k_1[a]	20.0 ± 6.0	1	1.30
K_{a2}	k_2-1st transition[b]	9.2 ± 0.6	1	0.96
K_{a2}^{1}	k_2-2nd transition[b]	12.5 ± 0.5	8	1.10
K_{as}	S_1-1st transition[c]	15.9 ± 0.8	1	1.20
K_{as}^{1}	S_1-2nd transition[c]	8.9 ± 0.3	8	0.94
K_a	$K^+ + HPO_4^{-2}$[d]	6.3	1	0.80

[a]See graph Fig. 2.4A.
[b]See graph Fig. 2.4B.
[c]See graph Fig. 2.5.
[d]Sillen and Martell (1971).

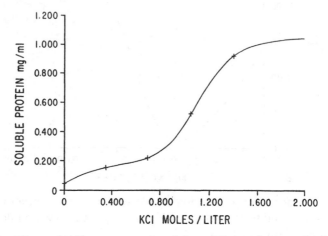

FIG. 2.5   Effect of KCl concentration (M × 10) on change of solubility of protein (α_{s1}-A) at 1°C relative to solubility in the absence of KCl. Data from Table 2.3 were fitted by Eq. (4); results are shown in Table 2.4.

order of magnitude and in effect (salting-in) and are in the order of magnitude of K_a for $K^+ + HPO_4^{-2}$. The most readily apparent conclusion is that salting-in, in this particular case, may be the result of direct salt (perhaps both anion and cation) interactions with charged groups (possibly phosphates) of the protein. This effect, as noted above, has somewhat wide-ranging implications. It had previously been postulated (Horvath and Mellander, 1977) that such calcium-protein-salt interactions occur primarily through salt-solvent interactions rather than direct salt binding to protein.

Effect of Other Cations at 1°C

Figure 2.6 illustrates the solubility of α_{s1}-A in the presence of various cations. Cu^{2+} and Zn^{2+} are the most effective precipitants. Coordinate complexes may be formed between α_{s1}-A molecules with Cu^{2+} and Zn^{2+} and Co^{2+}. Ca^{2+} is effective as a precipitant to a lesser extent than Cu^{2+} or Zn^{2+}, whereas Mg^{2+} is the least effective of the five cations studied. The salting-out and salting-in constants were estimated for each cation and are given in Table 2.5. Cation variation (i.e., use of magnesium, calcium, cobalt, copper, and zinc) of these profiles showed k_1 and k_2 behavior consistent with concepts of phosphate- and carboxylate-ligand coordination, respectively. Clearly, an inverse relationship exists between casein solubility (as quantified by changes in k_1) and the atomic number of the divalent cations studied. The salting-in constant k_2 appears to decrease and then increase with atomic number; no apparent correlation with ionic radius is evident, so that the best possible explanation is that as the atomic weight of the cation increases, the solubility decreases.

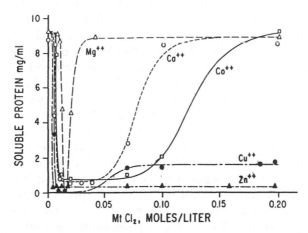

FIG. 2.6   Solubility at 1°C of various salts of α_{s1}-A caseinates as function of increasing concentration. Data were fitted by Eq. (4); results are given in Table 2.5.

TABLE 2.5   Cation-Induced Solubility of α_{s1}-Casein A at 1°C[a]

Cation	k_1[b]	k_2[b]	$(S_2 - S_1)$[c] (mg/mL)	Atomic no.	R[d](Å)
Mg^{2+}	76 ± 9	56 ± 7	8.4 ± 1.3	12	0.66
Ca^{2+}	150 ± 27	13 ± 2	8.2 ± 1.4	20	0.99
Co^{2+}	166 ± 4	8.1 ± 0.2	8.5 ± 0.2	27	0.72
Cu^{2+}	229 ± 2	18 ± 2	1.5 ± 0.1	29	0.72
Zn^{2+}	373 ± 27	202 ± 35	0.20 ± 0.03	30	0.74

[a]$n = m = 8$ for all calculations.
[b]L/mole.
[c]$(S_2 - S_1)$ = concentration of soluble α_{s1}-A in mg/mL.
[d]Cation atomic radius in Å.

Influence of Phosphate Groups on Salting-Out and Salting-In

As shown in Fig. 2.1, α_{s1}-A and -B and β-C readily precipitate at 37°C in 0.07 M KCl. Since under these conditions n was correlated with the number of phosphate residues in the native casein, the importance of these residues in the precipitation reaction could be tested. In previously conducted research, the phosphate groups of α_{s1}-B were removed enzymatically (Bingham et al., 1972) and the effects of KCl on the precipitation of native (N) and dephosphorylated (O-P) caseins had been compared but not quantitated (Fig. 2.7A shows a typical curve from Bingham et al., 1972). Analysis of these data by use of Eq. (4) (Farrell and Kumosinski, 1988) is summarized in Table 2.6. With no KCl present, dephosphorylation increases k_1 and some salting-in occurs for the O-P form; surprisingly, for both proteins (N and O-P) $n = 16$. Also, in the absence of KCl, salting-in occurs only for the dephosphorylated α_{s1}-casein B.

When α_{s1}-A is dephosphorylated, it becomes nearly completely soluble at 1°C and is salted-in even at 36°C (Fig. 2.7B). In contrast (Fig. 2.1), the native α_{s1}-A is not appreciably salted-in at 37°C. Results are compared in Table 2.6. For dephosphorylated α_{s1}-A without added KCl $n = 8$, mirroring the numbers found for the native protein, but for the O-P form at 1°C $n = 2$ and $m = 4$. The k_2 values observed for both native and O-P α_{s1}-A at both 1 and 36°C are similar to each other and to that of α_{s1}-B at 1°C. However, the small degree of salting-in that occurs for O-P of α_{s1}-B does so with an elevated k_2 (Table 2.6), showing another significant difference between the A and B variants.

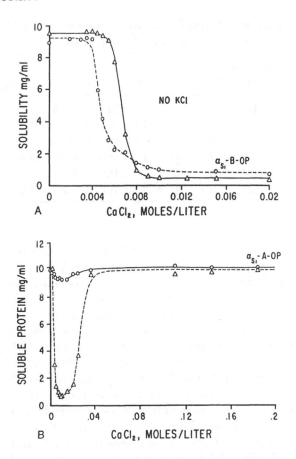

FIG. 2.7   (A) Solubility at 37°C of native and dephosphorylated α_{s1}-B (α_{s1}-B–OP) as a function of $CaCl_2$ concentrations at 10 mg/mL. (B) Solubility of α_{s1}-A O-P as a function of calcium ion concentrations at 1°C (—) and at 36°C (- - -). Data were fitted by Eq. (4). Results of analyses are given in Table 2.6.

Calcium-Induced Protein Self-Association Model

Finally, it would be advantageous to test the salt-induced protein self-association to determine whether or not the monomer of α_{s1}-casein is the molecular unit responsible for the salting-out process. Here, we used the sedimentation data of Waugh et al. (1971) at low calcium chloride concentrations at 37°C and at a protein concentration of 10 mg/mL.

TABLE 2.6　Calcium-Induced Solubility of Native (N) and Dephosphory-lated (O-P) α_{s1}-Casein B (B) and α_{s1}-Casein A (A)

Protein	Temp (°C)	k_1 (L/mole)	k_2 (L/mole)	S_1	S_2	n	m
NB	37	151 ± 1		0.49 ± 0.04		16	
NBO-P	37	219 ± 2	135 ± 12	2.6 ± 0.3	0.8 ± 0.06	16	8
NA	36	140 ± 3		0.9 ± 0.2		8	
NAO-P	36	326 ± 7	36 ± 1	0.6 ± 0.3	10.0 ± 0.6	8	8
NA	1	130 ± 4	13 ± 1	0.4 ± 0.2		8	
NAO-P	1	223 ± 59	46 ± 4	8.8 ± 1.8	10.0 ± 0.6	2	4

Two distinct peaks were observed, and Fig. 2.8 shows the variation of sedimentation coefficient $S_{20,W}$ for the slow peak (Fig. 2.8A), and fast peak (Fig. 2.8B) with added $CaCl_2$ up to a salt concentration of 0.007 M. At greater concentrations the protein precipitates. The appearance of a bimodal pattern in the sedimentation profile as the $CaCl_2$ is increased is indicative of a ligand-induced protein self-association system under equilibrium conditions (Cann, 1978). Thermodynamic linkage and non-linear regression analysis of the data using Eq. (4), but substituting the sedimentation constant(s) for S_{app}, was carried out. For the slow peak, Fig. 2.8A, a biphasic binding process was determined, whereas analysis of the fast peak, Fig. 2.8B, yields only one cooperative binding mechanism leading to protein self-association. The results of this analysis of Fig. 2.8 are given in Table 2.7. Here, for the slow peak the k_1 and k_2 values are 300 ± 50 and 166 ± 34 L/mole, respectively, with $n = 1$ and $m = 8$. The value of S_0 of 1.7 ± 0.3 for the slow peak is of the correct order of magnitude for α_{s1}-casein monomer with molecular weight of 24,000. The S_1 and S_2 values were 4.6 ± 0.5 and 12.6 ± 1.1 S, respectively. If we assume that S_0 is the sedimentation coefficient of monomeric α_{s1}-casein B, the degree of self-association can be approximated by dividing the S_1 and S_2 values by S_0 and raising that quotient to the 1.5 power, which yields values of 4.3 and 18.8 for the size of the calcium-induced aggregates of the slow peak (Cann, 1978). Using the same algorithm, an approximate degree of self-association of 8.18 and 46.5 can be calculated from S_0 and S_1 values of 7.4 ± 0.5 and 22.5 ± 1.5 S, respectively, obtained from thermodynamic linkage analysis of the fast component (Table 2.7). Thus, from the slow peak the apparent binding of one mole of $CaCl_2$ to α_{s1}-casein induces tetramer formation, which in turn leads to 18-mer when 8 moles of $CaCl_2$ bind cooperatively. (*Note*: Thermody-

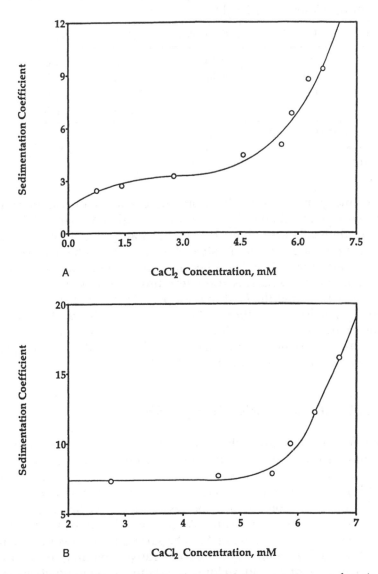

FIG. 2.8 Sedimentation velocity of α_{s1}-B at 10 mg/mL as a function of added $CaCl_2$. Two peaks were observed: a slow one (A) and a fast one (B). Analysis was by use of Eq. (4) and is given in Table 2.7.

TABLE 2.7 Thermodynamic Linkage Results of Sedimentation Coefficient of α_{s1}-Casein B vs $CaCl_2$[a]

Peaks	k_1(L/mole)	k_2(L/mole)	S_0	S_1	S_2
Slow[b]	300	166	1.7	4.6	12.6
	±50	±34	±.3	±.5	±1.1
Fast[c]	152		7.4	22.5	
	±32		±.5	±1.5	

[a]At 37°C with 10 mg/mL of protein.
[b]$n = 1$, $m = 8$.
[c]$n = 16$.

namic linkage reveals only those binding events that correlate with the observed physical changes.) The fast peak with a S_0 value of 7.4 or an octamer then appears to bind 16 moles of $CaCl_2$ cooperatively to yield a 46-mer aggregate with an S value of 22.5 ± 1.5. The analysis clearly demonstrates that α_{s1}-casein B becomes insoluble as an aggregate and not as a monomer unit. It is also interesting to observe that the k_2 value of the slow peak and the k_1 of the fast peak (Table 2.7) are 166 and 152 L/mol, respectively, and these values are comparable to the salting-out binding parameters ($k_1 = 186$) seen in Table 2.1. In contrast k_1 of the slow peak approaches the average K_A for binding of Ca^{2+} to α_{s1}-casein (380 L/mole) (Dickson and Perkins, 1971).

MOLECULAR MODELING OF α_{s1}-CASEIN

In the previous sections we have formed a quantitative thermodynamic mechanism for the salt-induced solubility profiles of α_{s1}-casein as a function of temperature, genetic variation, dephosphorylation, i.e., post-translational modification, and types of divalent cations employed. Here, the insolubility portion was described as a cooperative binding of n moles of salt with a constant of k_1 leading from a protein solubility of S_0 at zero salt concentration to a value of S_1. The salting-in process was quantitatively described by m moles of salt cooperatively binding to the protein with a constant of k_2 and yielding a limiting protein solubility of S_2. However, it must be stressed that the ligand-induced protein association model proposed by Cann and Hinmen (1976) cannot be used to describe the salt-induced precipitation profile used in this study (see theory section). Only molecular weight or sedimentation coefficient studies at lower salt concentrations such as those performed by Waugh et al. (1971) can totally describe the salt-induced self-association model as shown in the previous section. Hence, it would be prudent to assume that the oligomeric unit of α_{s1}-casein that leads to precipitation is a tetramer or octamer as concluded above.

In the molecular modeling section, we now attempt to construct tetramer and octamer models from an energy-minimized, predicted, three-dimensional structure of α_{s1}-casein (Kumosinski et al., 1994). Here, the antiparallel stranded sheets that occur in the hydrophobic portion of α_{s1}-B will be used as interaction sites for the construction of these tetrameric and octameric models. It is hoped that the presentation of such predicted structures will aid researchers in designing new chemical and biochemical modification experiments to test these working models.

To study the salt-induced solubility profiles of α_{s1}-casein by modeling, the A variant was chosen and its energy-minimized molecular model produced by deletion of the appropriate residues from the α_{s1}-casein B model. The α_{s1}-casein A structure was chosen to describe the salt-induced salting-in profiles since, as shown in the previous section, this genetic variant of α_{s1}-casein would be salted-in with added $CaCl_2$ at 1°C. Also, since molecular dynamic (MD) calculations in the presence of salt and water are very time consuming, it is more efficient to utilize only that portion of the molecule to which salt would most probably bind, i.e., the hydrophilic domain. Therefore, only residues 1 to 86 of α_{s1}-casein A were used in these simulations. In addition, MD calculations were performed on a dephosphorylated molecule to mimic the effect of phosphoserines on salt-induced solubility profiles.

Hydrophobic and Hydrophilic Dimers and Oligomers

The energy-minimized structure generated for α_{s1}-casein B as described above is shown in Fig. 2.9A, where it is displayed from carboxy- to amino-terminal (left to right). Analysis of this structure shows the molecule to be composed (right to left) of a short hydrophilic amino-terminal portion, a segment of rather hydrophobic β-sheet, the phosphopeptide region, and a short portion of α-helix, which connects this N-terminal portion to the very hydrophobic carboxyl-terminal domain containing extended antiparallel β-strands. For clarity the backbone without side chains is shown in Fig. 2.9A with prolines (P) indicated, and an accompanying chain trace stereo view (Fig. 2.9B) is given.

From the overall shape of the α_{s1}-B model (Fig. 2.9), it is apparent that it is impossible to approximate its structure with either a prolate or an oblate ellipsoid of revolution, as was done in the case of the β-casein refined structure (Kumosinski et al., 1993). Indeed, a rather large degree

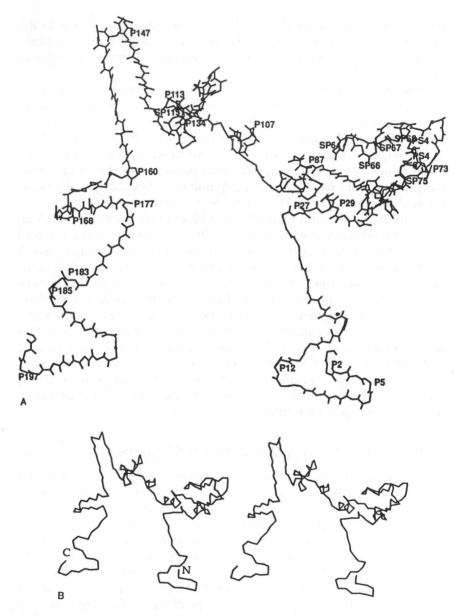

FIG. 2.9   (A) Backbone of the refined model of α_{s1}-casein, without side chains; prolines (P) and serine phosphates (SP) indicated. (B) Stereo view of the refined three-dimensional molecular model of α_{s1}-casein; the N- and C-terminal ends of the molecule are labeled.

of asymmetry is observed. As noted above, the hydrophilic and hydrophobic domains are joined by extended structures, whose central feature is an α-helix with its pitch perpendicular to the two domains. It is speculated that this α-helix would be important for preserving the integrity of the two domains when a dynamic calculation is finally performed.

Schmidt (1982) has summarized the light-scattering studies on variants of α_{s1}-casein under a variety of environmental conditions from which a stoichiometry of the α_{s1}-casein self-association is obtained at selected temperatures and ionic strengths. From the results, it can be concluded that α_{s1}-casein undergoes a concentration-dependent, reversible, hydrophobically controlled association from monomer to dimer then tetramer, hexamer, octamer, and even higher if the ionic strength is increased, especially as shown (Fig. 2.8) in the case of added $CaCl_2$. To mimic hydrophobic salt-induced self-association mechanisms, we now attempt to construct an energy-minimized dimer, tetramer, and octamer structure using hydrophobic and hydrophilic sites.

The first step is to create a dimer from the large-stranded β-sheets, which occur at residues 136–158. The side chains are predominately hydrophobic, and the hydrogen bonding of the sheet secondary structure yields rigidity to this site. After docking, a dimer can easily be formed if two of these stranded sheets are docked in an antiparallel fashion (Fig. 2.10A). Such an asymmetric arrangement minimizes the dipole-dipole interactions of the backbones while allowing the hydrophobic side chains to interact freely. In fact, this structure, following minimization (Fig. 2.10A), displays a stabilizing energy of -452 kcal/mole/residue, over monomer, i.e., $-520 = E_2 - 2 \cdot E_m$ where E_2 is the energy of the dimer and E_m is the monomer energy.

Another possible interaction site for hydrophobic dimerization resides is the deletion peptide of α_{s1}-casein A (i.e., the peptide that is deleted from α_{s1}-casein B to form α_{s1}-casein A). Closer inspection of the residues of this peptide show a β-sheet secondary structure with hydrophobic as well as acidic and basic side chains. Thus, by docking two molecules in an antiparallel fashion, a hydrophobically stabilized intermolecular ion pair (Tanford, 1967) can be formed upon the construction of a dimer. The dimer was then energy minimized and is shown in Fig. 2.10B. The formation of this hydrophobic ion pair has been used to explain the differences in the calcium-induced solubility and colloidal stability between α_{s1}-caseins B and A (Farrell et al., 1988; Kumosinski and Farrell, 1991). Small-angle x-ray scattering of micelles reconstituted from whole caseins containing α_{s1}-B or -A shows large differences with respect to

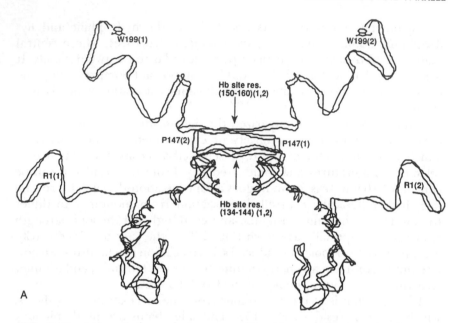

A

FIG. 2.10   (A) α-Carbon chain trace of backbone without side chains of hydrophobic stabilized (Hb) antiparallel sheet dimer; the large sheets centering on proline 147 are docked. (B) Backbone structure of hydrophobic ion-pair dimer (IPr) from α_{s1}-casein B; this area contains the α_{s1}-A deletion peptide. Key for labels: $R_1(1)$, W199 (1) represent N and C terminals of molecule 1 (arginine 1 and tryptophan 199 of molecule 1; the 2 in parentheses refers to molecule 2. Both dimeric structures energy minimized to -10 kcal/mole/residue.

submicellar packing density within the micelle structure, i.e., 3 to 1 versus 6 to 1 for B and A, respectively. This difference in packing density may be a result of destructive interference in the scattered intensity due to an asymmetrical structure with a center of inversion (Pessen et al., 1991). It was speculated that this asymmetrical structure was due to the formation of a hydrophobically stabilized intermolecular ion pair in α_{s1}-B at this deletion peptide site.

   With these two possible dimer structures, a tetramer can easily be modeled starting with the dimer formed by the intermolecular hydrophobic ion pair (Fig. 2.10B). To the two ends of this structure, two molecules of α_{s1}-B are added via the hydrophobic sheet-sheet interaction shown in Fig. 2.10A. Such a structure is presented in Fig. 2.11A. This tetramer structure is highly asymmetric and also contains two possible

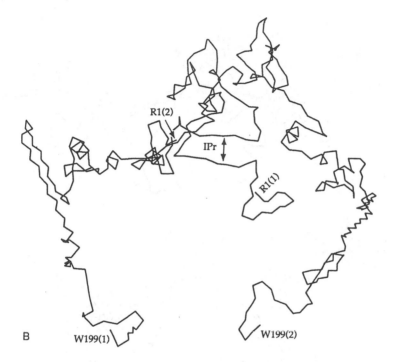

B   W199(1)   W199(2)

hydrophobic ion pair sites at each end of the molecule; these sites at either end of the tetramer could lead to further aggregation resulting in a large rod with a large axial ratio and dipole moment. It is noteworthy that the hydrophobic antiparallel stranded sheets of residues 163–174 could not be docked with the same site on another tetramer structure due to large steric factors. Such a structure could not possibly be made without major changes in the α_{s1}-casein B model. However, a more plausible site for octamer formation via simple hydrophobic interactions can be constructed from use of the hydrophobic side chains centered on prolines 177 and 185, which are located on the lower side of the asymmetric tetramer structure and are solvent accessible. With this in mind, two tetramers were docked in an antiparallel fashion with a center of inversion using these hydrophobic side chains as interaction sites. The octamer structure is shown in Fig. 2.11B. This octamer structure would still allow water to flow through part of the polypeptide chain yielding a high hydrodynamic hydration value. It would also be stable in solution since the hydrophobic side chains are predominately in the center of the model and all eight hydrophilic domains are solvent accessible—two on the

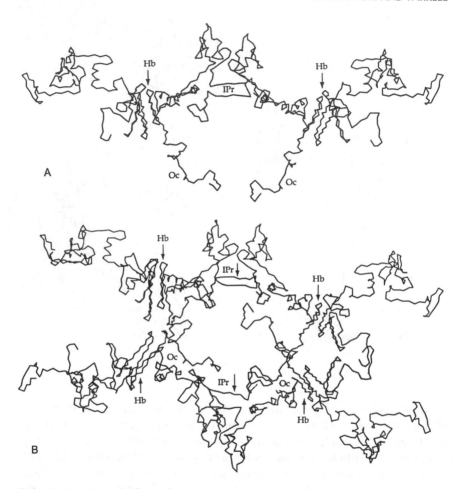

FIG. 2.11   (A) α-Carbon chain trace of α_{s1}-casein B tetramer resulting from the docking of two molecules (left and right) of α_{s1}-B through the hydrophobic (Hb) scheme shown in 2.10A with the (IPr) ion-pair dimer given in Fig. 2.10B. Sites for octamer formation (Oc) are noted. (B) Carbon chain trace of the octamer of α_{s1}-casein B resulting from dimerization of two tetramer structures, shown in 2.11A, via the two hydrophobic (Oc) antiparallel sheet-sheet interaction sites areas related to Hb and IPr sites are also noted.

upper and lower center part and two at either end of the structure (see Fig. 2.11B).

A very interesting feature of all the α_{s1}-caseins is the preservation of the C-terminal tryptophan. Ribabeau-Dumas and Garnier (1970) showed that carboxy-peptidase A could quantitatively remove the C-terminal tryptophan of α_{s1}-casein alone, and in native and reconstituted micelles. This was interpreted as a demonstration of the open network of the casein micelles that allowed penetration of the protease into the micelle. This is in accord with the model shown in Fig. 2.10A and B, where the C-terminal tryptophan is extended in space at the left side of the model. Thus, although residues 134–185 participate in hydrophobic interactions, a hydrophilic turn then intervenes and the C-terminal tryptophan can still be exposed in monomeric and polymeric structures, making it available to digestion with carboxypeptidase A.

Finally, it can be seen that deletion of residues 14–26 of the α_{s1}-B octamer structure to form α_{s1}-casein A would result in the elimination of the two hydrophobic ion pair–interaction sites. This mutation should result in a tetramer structure arising from only the left- or right-hand portions of the octamer α_{s1}-B structure of Fig. 2.11B.

Thus, we have now established a possible mechanism as well as structures for the calcium-induced salting-out profiles of α_{s1}-casein. Here, divalent cations can bind to serine phosphate or other negatively charged side chains, which are mostly in the hydrophilic domain of α_{s1}-casein (see Fig. 2.9A) causing the hydrophobic domains to interact and self-associate in a cooperative fashion to a limiting octamer structure in the case of α_{s1}-B or a tetramer for α_{s1}-A, respectively. These limiting oligomers are thermodynamically unstable with respect to their interaction with water because of their low charge and solvent exposure to water causing a further noncooperative aggregation leading to precipitation.

Molecular Dynamics of CaCl$_2$ and MgCl$_2$ in Water with the Hydrophilic Domain of α_{s1}-Casein A

In this section we shall attempt to form a structural basis for the thermodynamics of the salting-in process, which was quantitated using thermodynamic linkage and nonlinear regression analysis of the calcium-induced solubility profiles of α_{s1}-casein. Here we have chosen the hydrophilic N-terminal domain of the monomeric α_{s1}-casein A structure. This model was built from the α_{s1}-casein B model of Fig. 2.9A and B by excising residues 14–26 (Eigel et al., 1984) and then deleting residues 100–199 of α_{s1}-B. The resulting hydrophilic domain, i.e., residues 1–99 of the α_{s1}-casein B minus 14–26 (a total of 86 residues), was then energy minimized. Eleven hundred water molecules were added, and the resulting structure in solution was again energy minimized with a cutoff for nonbonded interaction of 5 Å utilizing a periodic boundary condition. The resulting structure with water was subjected to molecular dynamics (MD) calculation for 20 psec above equilibrium conditions, which

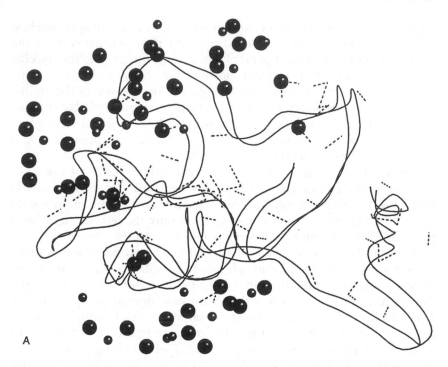

A

FIG. 2.12 (A) Backbone ribbon structure of the hydrophilic half of α_{s1}-casein A (residue 1 through 86) after molecular dynamics at 40 psec with 22 molecules of $CaCl_2$ in the presence of 1100 water molecules. Ca and Cl atoms are shown as ball models of radii equal to 0.15 times their known van der Waals radius. Dashed lines represent hydrogen bond formation. (B) Stereo view (relaxed) of (A).

were determined by the stabilization of potential energy, radius of gyration of the protein backbone, root-mean-square fluctuations of the backbone atoms, and change in second moment. Such an equilibrated dynamic structure should approximate the structure, energetics, and dynamic motion of this protein domain in solution.

To mimic the salt-binding mechanism, 22 molecules of calcium or magnesium and 44 molecules of chloride with appropriate ionic charges were added to the above system in a pseudo random fashion, energy minimized, and subjected to MD for a full 40 psec. Equilibrium was easily established once again at 15–20 psec. The resulting structure for the native half of α_{s1}-A in $CaCl_2$ is shown in Fig. 2.12A and B. The amount of salt added was chosen to comply with a condition that would result in

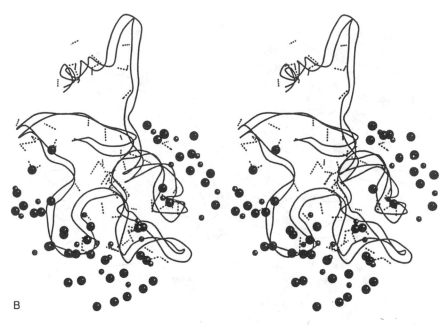

B

saturation of the Ca^{2+}-binding sites. Using a value of 380 M^{-1} for K$_A$ and 8 sites (Dickson and Perkins, 1971), 22 molecules of CaCl$_2$ per 1100 water molecules per molecule of protein is equivalent to greater than 99% occupancy of these calcium-binding sites and greater than 80% occupancy for 8 additional putative salting-in sites derived from k_2 of Table 2.2. A similar structure for the dephosphorylated half of α_{s1}-A was also studied and is shown in Fig. 2.13. In total, seven MD calculations were performed on the hydrophilic domain (H) of α_{s1}-A (residues 1–99) in the presence of 1100 water molecules to 40 psec: two in the absence of salt for native (H) and dephosphorylated (HO-P); two in the presence of CaCl$_2$ for H and HO-P; one with added MgCl$_2$ for H; and one each for MgCl$_2$ and CaCl$_2$ with no protein. Each calculation utilized a cutoff of 5 Å for nonbonded interactions, a Tripos force field, and a "bump" factor of 0.7 for simulating hydrogen bond formation (i.e., a distance of >0.7 Å means no bonding). MD calculations require a running time on the Silicon Graphics Unix computer system of at least 3 days. It should be noted at this time that the energetics and geometric parameters estimated by the MD calculation reflect the total salt binding to the protein, i.e., the sum of the free energies of salt binding for the protein salting-out, as well as the salting-in process. More detailed analysis of these MD calculations, which are beyond the scope of this chapter, must be

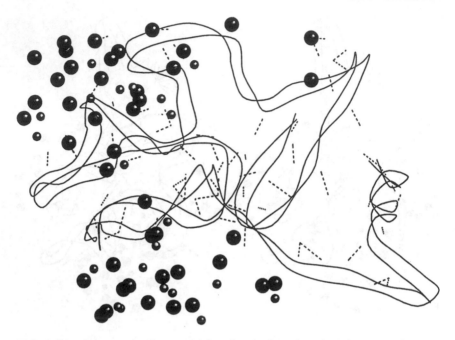

FIG. 2.13   Same as in Fig. 2.12A but for dephosphorylated α_{s1}-A. All serine phosphates mutated to serine with appropriate partial charges.

performed for separation of the protein precipitation and resolubilization processes.

The average results of several calculated geometric parameters with their corresponding errors are presented in Table 2.8. Here, the subscript 1 denotes the salt component, i.e., both Ca and Cl, while the subscript 2 describes the protein. R is the calculated dynamic radius of gyration, x is the average center of mass of the component atoms, and a is the root-mean-square fluctuation of component atoms from the center of mass. Here, a can be thought of as a dynamic Stokes radius of the chosen atoms, and x a spherical center of mass.

We now attempt to describe the distribution of the salt atoms at the end of MD calculations by inspection of the a and x values in Table 2.8 for both $CaCl_2$ and $MgCl_2$ alone and in the presence of the hydrophilic half of α_{s1}-A (H) and its dephosphorylated form HO-P. Little significant change is seen in the a_1 values, since the mass of the protein is far greater than that of the salt. The x_1 values are the best descriptors of the binding of salt atoms to the protein in these MD calculations. Here, in all cases the x_1 values have decreased in the presence of either protein component.

TABLE 2.8 Molecular Dynamics of Hydrophilic Half (H) of α_{s1}-casein A in Water with Salt, Geometric Parameters

Protein	Salt	a_1, Å²	x_1, Å	R_2, Å	a_2, Å²	x_2, Å
H	—	—		14.6 ± 0.2	8.2 ± 0.5	0.33 ± 0.04
HO-P	—	—	—	14.2 ± 0.04	7.0 ± 0.2	0.33 ± 0.03
H	CaCl$_2$	11.5 ± 0.7	0.8 ± 0.10	14.6 ± 0.2	8.7 ± 0.4	0.38 ± 0.04
HO-P	CaCl$_2$	9.38 ± 0.67	0.69 ± 0.06	15.5 ± 0.2	10.1 ± 0.4	0.51 ± 0.03
H	MgCl$_2$	7.8 ± 0.6	0.67 ± 0.07	14.6 ± 0.2	7.7 ± 0.5	0.37 ± 0.04
—	CaCl$_2$	9.1 ± 0.9	2.2 ± 0.1	—	—	—
—	MgCl$_2$	6.8 ± 1.1	1.5 ± 0.4	—	—	—

R, radius of gyration; a, RMS fluctuation of all atoms from center of mass (a dynamic Stokes radius); x, calculated spherical center of mass. Subscript 1 denotes salt, 2 denotes protein atoms.

Such a decrease in the average spherical center of mass of the salt atoms is a clear indication of protein-salt interactions, since the center of mass for salt atoms alone would be larger as they move randomly about. However, x_1 would of necessity be smaller for salt bound to protein where movement is restricted. A difference between x_1 for native H and for the O-P form appears, and it can easily be observed that the salt atoms associated with the H and HO-P forms (Fig. 2.12 and 2.13, respectively) have a different overall distribution. This needs to be verified but it is in agreement with the decreased x_1 in the presence of the protein fragments. It is assumed that the x_1 of the $MgCl_2$ was lower due to the lower value of the van der Waals radius for Mg^{+2} (0.66 Å) than for the Ca^{+2} ion (0.99 Å).

To observe the effect of salt binding on the dynamic structure of the hydrophilic half of the protein, we have calculated for the protein in the presence or absence of salt atoms its radius of gyration R_2, as well as the a_2 and x_2 values; these are presented as columns 5, 6, and 7 of Table 2.8. Virtually no changes within the calculated error are observed for the R_2, a_2, or x_2 values for the native H either in the absence or presence of $CaCl_2$ or $MgCl_2$; the HO-P form is not dramatically different either. However, large increases in these descriptors are observed in the dephosphorylated HO-P form when $CaCl_2$ is added in the MD calculations. This change in configuration may reflect a general swelling of the HO-P structure when $CaCl_2$ is added, which can also be observed by inspection of Fig. 2.14A and B. Here the two structures are compared by representations of the protein backbones only, with no side chains displayed. The H form is represented by a ribbon trace of the backbone, while the HO-P form is represented by a backbone–atomic stick model. The backbone model is much more swollen than the ribbon model. For easier observation of this conclusion in all dimensions, a stereo view is shown in Fig. 2.14B.

The reason for this phenomenon is most likely due to the hydrogen bonding of chloride ions to the serine side chains as well as to N-H atoms of the backbone. Such anion-protein hydrogen bonding can impose important dynamic structural changes on the protein component. In the H form these interactions may be "screened out" by the negatively charged phosphate groups. Whether this swelling phenomenon causes increased solubility of the protein, i.e., salting-in, cannot be established at this time, because the deleted residues could play a role in keeping the α_{s1}-B variant insoluble. More MD calculations in conjunction with solution structural physical chemical experiments must be performed in the future to test this hypothesis.

To further correlate the binding free energies calculated from ther-

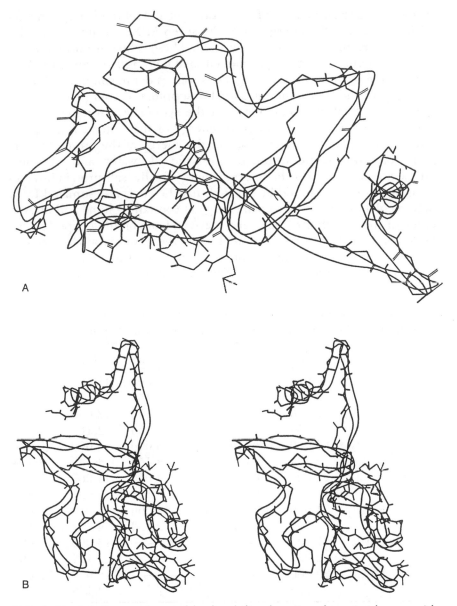

FIG. 2.14   (A) Comparison of hydrophilic domain of α_{s1}-A (shown with ribboned backbone) with dephosphorylated α_{s1}-A after molecular dynamics with 22 molecules of Ca and Cl_2 at 40 psec in the presence of 1100 water molecules. Dashes lines represent hydrogen bond formations. (B) Stereo view (relaxed) of (A).

modynamic linkage analysis of the salting-out and salting-in experiments (Tables 2.5, 2.6), the energetics of the seven MD experiments for the theoretical H and HO-P were calculated. These values are presented in Table 2.9 along with the experimental total binding free energies, ΔF_B, for native α_{s1}-A in $CaCl_2$ and $MgCl_2$ as well as dephosphorylated α_{s1}-A in $CaCl_2$. In the tables, the calculated descriptors of energy are E_T, for the average total potential energy of the system; E_W, for the internal energy of the water; E_{PN}, potential energy of the system; E_W, for the internal energy of the water; E_{PW}, for the energy of the protein, water, and protein-water interaction; E_{SW}, for the salt, water, and salt-water interaction; and E_{int}, for the internal energy for the salt-protein interaction. At this time it should be stressed that the energies estimated from MD calculations are the internal energy at constant volume and not the Gibbs or the Helmholtz free energy derived from binding experiments. However, a qualitative correlation between these two parameters can be utilized to describe variations in protein components or type of salt used. Inspection of column 4 and 5 of Table 2.9 shows that for native α_{s1}-A

TABLE 2.9 Molecular Dynamics of Hydrophilic Half (H) of α_{s1}-casein A in Water with Salt: Energetics and Comparison with Experimental Data

Protein	Salt	$-E_T$	$-E_{int}$	ΔF_B	$-E_{PW}$	$-E_W$	$-E_{SW}$
H	—	20,000 ± 152	—	—	20,000 ± 152	13,000 ± 64	—
HO-P	—	19,000 ± 54	—	—	19,000 ± 54	13,000 ± 101	—
H	$CaCl_2$	31,000 ± 134	8,746 ± 162	33.2	13,000 ± 154	11,000 ± 142	20,000 ± 128
HO-P	$CaCl_2$	28,000 ± 132	4,612 ± 198	14.3	14,000 ± 26	11,000 ± 91	22,000 ± 155
H	$MgCl_2$	34,000 ± 92	11,000 ± 175	36.6	12,000 ± 133	11,000 ± 142	22,000 ± 155
—	$CaCl_2$	27,000 ± 266	—	—	—	13,000 ± 122	27,000 ± 266
—	$MgCl_2$	29,000 ± 198	—	—	—	12,000 ± 198	29,000 ± 198

E, potential energy in kcal/mole, i.e., internal energy at constant volume subscripts; T, total atoms; P, protein atoms; W, water atoms; S, salt atoms; int, protein-salt interaction. ΔF_B, total cooperative salt-binding free energy in kcal/mole calculated from Tables 2.5 and 2.6 for the solubility data for α_{s1}-casein A (NA) and its dephosphorylated form (NAO-P).

and AO-P in $CaCl_2$ and native A in $MgCl_2$, the experimental changes of ΔF_B correlate well with changes in ΔE_{int} when these environmental factors are varied; the absolute values for the two parameters are, however, significantly different. These large differences in absolute values between ΔF_B and ΔE_{int} are due to the bump factor of 0.7 used in these calculations to quantitate hydrogen bonding in MD. If this factor were optimized then perhaps better correlations could be achieved. However, for this study, the value of 0.7 was used since it is the default value used by Sybyl for the Tripos force field. Such a small bump factor most likely allows for inordinately high electrostatic energy terms. Future work will tend to optimize this parameter. No conclusions can be made at this time concerning the meaning of the other energetic descriptors of Table 2.9, but they are presented along with their error for inspection by the reader.

SUMMARY AND CONCLUSIONS

It can be concluded that thermodynamic linkage in conjunction with molecular modeling techniques such as energy minimization and molecular dynamics can provide a powerful multifaceted approach for developing structure-function relationships such as the salt-induced solubility profiles of this study. Here, the stoichiometry and thermodynamics of the salt-induced solubility can be quantitated using thermodynamic linkage in conjunction with nonlinear regression analysis. The dynamic changes in the protein structure responsible for the salting-out and salting-in processes can be established using energy minimization and molecular dynamics calculations of the protein in water and in the presence and absence of salt atoms. In particular, predictions concerning the type and amount of protein modification that occurs can be utilized to increase the desired protein functionality in a rational way through chemical or genetic modification (Richardson et al., 1992).

For example, we have structurally defined the salting-out process in terms of a hydrophobically controlled octamerization for α_{s1}-B in compliance with the experimental results of Waugh and et al. (1971) and our quantitation of their results using thermodynamic linkage. This oligomerization is most likely caused by binding of the positive divalent cations to the phosphoserine and other negatively charged side chains on the hydrophilic domain of α_{s1}-casein (see Fig. 2.9A) resulting in the minimization of protein charge and the protein-protein electrostatic repulsion term. The first step in this self-association is most probably caused by the dimerization of both α_{s1}-B and α_{s1}-A, which proceeds by the docking of two large hydrophobic stranded antiparallel sheets in an asymmetric fashion (see Fig. 2.10A). Disruption of this interaction via modification of these large antiparallel sheets by chemical, genetic, or biochemical means may significantly change the solubility of α_{s1}-B or -A in the presence of $CaCl_2$. An example of this could be plasmin cleavage of α_{s1}-B to yield two large peptides (H. E. Swaisgood, personal communication). This potential enzymatic cleavage site (residues 101–104) is between the hydrophobic and hydrophilic domains of α_{s1}-casein (see Fig. 2.9A). The salting-in process appears to be more complicated; however, it can be naively described in a general fashion as further salt binding by the protein's hydrophilic domain allowing resolubilization by increased protein charge, resulting in a large protein-protein electrostatic repulsion term. The difference in the calcium-induced salting-in between the α_{s1}-B variant and the α_{s1}-A variant was structurally defined in terms of an intermolecular hydrophobically stabilized ion pair (see Fig. 2.10B) which is present in α_{s1}-B and which disrupts the resolubilization process. The deletion of residues 16–24 of α_{s1}-B results in the α_{s1}-A variant, which is more soluble with added calcium. The α_{s1}-A genetic variant has a gene frequency of 0.01% and is, therefore, not present in most cows. Hence, the possible enzyme cleavage of residues 1–24 of α_{s1}-B or chemical modification may aid in making this variant more soluble with added $CaCl_2$.

Finally, the molecular dynamics calculations could not at the present time distinguish between the salting-out and salting-in binding free energy. Only the total binding could be correlated in the MD results. In fact, significant differences in absolute values occur when the total salt-binding free energy calculated from thermodynamic linkage, ΔF_B, is compared with the interaction energy derived from MD, ΔE_{INT} (see Table 2.8). This discrepancy is most likely due to the bump factor value chosen for the MD calculations; a better value could be chosen for future works. However, the changes in ΔF_B with the change in salt from Ca to Mg for the native systems as well as the change with calcium between native and the dephosphorylated form show good correlations with ΔE_{INT}. More importantly, the geometric parameters for the protein from MD calculations, i.e., R_2, a_2, and x_2 of the HO-P, change dramatically over the H form in the presence of $CaCl_2$. This phenomenon was interpreted as a swelling of the dephosphorylated form with added $CaCl_2$ (Fig. 2.14A and B). It is apparent that this structural change can also be labeled as a conformational change since many internal hydrogen bonds are disrupted. This could be tested experimentally by FTIR.

However, this general, swelling of the HO-P form with added $CaCl_2$ can lead to large thermal fluctuation as seen by the a_2 parameter of Table

2.8. This increase in motion would lead to an increase in the entropy of mixing and ultimately would increase the solubility of the protein. To what extent this entropy of mixing is important to the resolubilization process cannot be assessed at the present time. These large fluctuations of HO-P with added $CaCl_2$ may in the future also aid in increasing other desired functionalities such as gelation, whippability, or foam formation.

REFERENCES

Andersen, H. C. 1980. Molecular dynamics simulations at constant pressure and/or temperature. *J. Chem. Phys.* 72: 2384–2394.

Arakawa, T. and Timasheff, S. N. 1984. Mechanisms of protein salting-in and salting-out by divalent cation salts: balance between hydration and salt binding. *Biochemistry* 23: 5912–5923.

Bingham, E. W., Farrell, H. M., Jr., and Carroll, R. J. 1972. Properties of dephosphorylated α_{s1}-casein. Precipitation by calcium ions and micelle formation. *Biochemistry* 11: 2450–2454.

Cann, J. R. 1978. Measurements of protein interactions mediated by small molecules using sedimentation velocity, In *Methods in Enzymology XLVIII*, C. H. W. Hirs and S. N. Timasheff (Ed.), pp. 242–248. Academic Press, NY.

Cann, J. R., and Hinman, N. D. 1976. Hummel-Dryer gel chromatographic procedures as applied to ligand-mediated associations. *Biochemistry* 15: 4614–4628.

Creamer, L. K., and Waugh, D. F. 1965. Calcium binding and precipitate solvation of Ca-α_s-caseinates. *J. Dairy Sci.* 49: 706.

Dickson, I. R., and Perkins, J. D. 1971. Studies on the interactions between purified bovine caseins and alkaline earth metal ions. *Biochem. J.* 124: 235–240.

Eigel, W. N., Butler, J. E., Ernstrom, C. A., Farrell, H. M., Jr., Harwalkar, V. R., Jenness, R., and Whitney, R. McL. 1984. Nomenclature of the proteins of cows' milk: 5th Revision. *J. Diary Sci.* 67: 1599–1631.

Farrell, H. M., Jr., and Kumosinski, T. F. 1988. Modeling of calcium-induced solubility profiles of casein for biotechnology. *J. Industrial Micro.* 3: 61–71.

Farrell, H. M., Jr., and Thompson, M. P. 1988. The caseins of milk as calcium binding proteins. In *Calcium Binding Proteins*, M. P. Thompson (Ed.), pp. 117–137. CRC Press, Boca Raton, FL.

Farrell, H. M., Jr., Kumosinski, T. F., Pulaski, P., and Thompson, M. P. 1988. Calcium-induced associations of the caseins: A thermodynamic linkage approach to precipitation and resolubilization. *Arch. Biochem. Biophys.* 265: 146–158.

Kollman, P. A. 1987. Application of force fields to molecular models. *Ann. Review Phys. Chem.* 38: 303–333.

Kumosinski, T. F., and Farrell, H. M., Jr. 1991. Calcium-induced associations of the caseins: Thermodynamic linkage of colloidal stability of casein micelles to calcium binding. *J. Protein. Chem.* 10: 3–16.

Kumosinski, T. F., Brown, E. M., and Farrell, H. M., Jr. 1991a. Molecular modeling in food research: Technology and techniques. *Trends Food Sci. Technol.* 2: 110–115.

Kumosinski, T. F., Brown, E. M., and Farrell, H. M., Jr. 1991b. Molecular modeling in food research: Applications. *Trends Food Sci. Technol.* 2: 190–195.

Kumosinski, T. F., Brown, E. M., and Farrell, H. M., Jr. 1993. Three dimensional molecular modeling of bovine caseins: An energy-minimized β-casein structure. *J. Dairy Sci.* 76: 931–945.

Kumosinski, T. F., Brown, E. M., and Farrell, H. M., Jr. 1994. Three dimensional molecular modeling of bovine caseins: an energy minimized α_{s1}-casein structure. In *Molecular Modeling*, T. F. Kumosinski and H. N. Liebman (Eds.), ACS Symposium Series, Denver, CO.

Melander, W. and Horvath, C. 1977. Salt effects on hydrophobic interactions in precipitation and chromatography of proteins: An interpretation of the lyotropic series. *Arch. Biochem. Biophys.* 183: 200–215.

Noble, R. W. and Waugh, D. F. 1965. Casein micelles, formation and structure I. *J. Am. Chem. Soc.* 87: 2236–2245.

Pessen H., Kumosinski, T. F., Farrell, H. M., Jr., and Brumberger, H. 1991. Tertiary and quaternary structural differences between two genetic variants of bovine casein by small-angle X-ray scattering. *Arch. Biochem. Biophys.* 284: 133–142.

Ribadeau-Dumas, B. and Garnier, J. 1970. Structure of casein micelle. The accessability of subunits to various reagents. *J. Dairy Res.* 37: 269–278.

Richardson, T., Jiminez-Flores, R., Kumosinski, T. F., Oh, S., Brown, E. M., and Farrell, H. M., Jr. 1992. Molecular modeling and genetic engineering of milk proteins. In *Advanced Diary Chemistry 1: Proteins*, P. F. Fox (Ed.), pp. 545–578. Elsevier, Essex, UK.

Robinson, D. R., and Jencks, W. P. 1965. Effects of concentrated salt solutions on the activity coefficient of acetyltetraglycine ethylester. *J. Am. Chem. Soc.* 87: 2470–2479.

Schmidt, D. G. 1982. Association of caseins and casein micelle structure. In *Developments in Diary Chemistry*, P. F. Fox (Ed.), pp. 61–86. Applied Science Publications Ltd., London, UK.

Sillen, L. G. and Martell, A. E. 1971. *Stability Constants of Metal Ion Complexes*. Special Publication No. 25 of The Chemical Society of London. Alden Press, Oxford, UK.

Sinanoglu, O. 1968. Solvent effects on molecular association. In *Molecular Associations in Biology*, B. Pullman (Ed.), pp. 429–445. Academic Press, New York.

Tanford, C. 1967. *Physical Chemistry of Macromolecules*. John Wiley & Sons, New York.

Thompson, M. P., Gordon, W. G., Boswell, R. T., and Farrell, H. M., Jr. 1969. Solubility, solvation and stabilization of α_{s1}- and β-caseins. *J. Dairy Sci.* 52: 1166–1173.

Waugh, D. F. and Noble, R. W. 1965. Casein micelles, formation and structure II. *J. Am. Chem. Soc.* 84: 2246–2257.

van Gunsteren, W. F. and Berendsen, H. J. C. 1977. Algorithms for molecular dynamics and constraint dynamics. *Mol. Phys.* 34: 1311–1327.

Waugh, D. F., Slattery, C. W., and Creamer, L. K. 1971. Binding of calcium to caseins. *Biochemistry* 10: 817–823.

Weiner, S. J., Kollman, P. A., Nguyen, D. T., and Case, D. A. 1986. Force field calculations in computational chemistry. *J. Comput. Chem.* 7: 200–230.

Wyman, J., Jr. 1964. Linked functions and reciprocal effects in hemoglobin: A second look. *Adv. Protein Chem.* 19: 223–286.

3

Protein Separation and Analysis of Certain Skeletal Muscle Proteins: Principles and Techniques

Elisabeth Jane Huff-Lonergan, Dirk D. Beekman, and Frederick C. Parrish, Jr.

Iowa State University
Ames, Iowa

PROTEIN SEPARATION AND ANALYSIS: PRINCIPLES AND TECHNIQUES

Skeletal muscle and its proteins are chemically, structurally, and functionally complex biological materials. Moreover, the conversion of muscle to a food, meat, involves many complex reactions and processes. Muscle is made up of fibers, the physiological unit of the cell, and fibers consist of myofibrils. Interestingly, myofibrils are where the action is in both living and postmortem muscle systems. In living muscle the myofibrillar/cytoskeletal proteins are the essential constituents of the structural and contractile elements, and in postmortem muscle, modifications of the myofibril and its associated meat proteins are essential for sensory product qualities and consumer acceptance. Furthermore, these proteins make up 50% of cellular proteins in skeletal muscle. Therefore, it is imperative that meat scientists have a clear understanding of the chemical, structural, and functional role of skeletal muscle proteins, especially the myofibrillar/cytoskeletal proteins, and their relationship to meat as a food.

Tenderization and tenderness of meat, especially beef, is a very important palatability and economic problem. From a consumer perspective, beef needs to be more consistently tender. Indeed, tenderness is complex as many ante- and postmortem factors have an impact on it during the production and processing of beef. We know postmortem aging significantly improves tenderness, and we know the degradation of the Z disks and certain myofibrillar/cytoskeletal proteins is of major consequence to tenderness improvement. There is a need, however, to better understand the mechanisms of tenderization with the objective of reducing variation in tenderness.

The understanding and regulation of the degradation of the structural elements during postmortem aging, termed myofibril fragmentation, seem to hold promise in producing more consistently tender beef. This fragmentation is accomplished by the natural proteases, calcium-activated factor (CAF) (Olson et al., 1977). These natural proteases, now called calpains, are inhibited by the natural inhibitor of calpains, calpastatin. Indeed, because calcium is an activator of these proteases, practical application of this basic information is currently being researched by injecting calcium into meat (Koohmaraie et al., 1988, 1989; Morgan et al., 1991; Wheeler et al., 1991) to improve tenderness.

Fresh lean meat contains about 75% water, and the retention of inherent and added water during storage and further processing is of interest to meat scientists and the meat industry. First, water retention is essential for meat product palatability attributes of tenderness and juiciness. Second, water retention is economically important because loss of water results in a decreased amount of marketable product. Retention of inherent and added water in meat is termed water-holding capacity (WHC). Evidence indicates that the state of myofibrillar/cytoskeletal proteins play a primary role in water-holding capacity. That is, the removal of transverse structural constraints (myosin cross bridges, M and Z lines) by salt and phosphate effectively improves WHC (Offer and Trinick, 1983). We think the solubilization of titin and nebulin filaments also improves WHC (Paterson et al., 1988). The addition of nonmeat ingredients like sodium chloride and alkaline phosphates in meat processing plays a very significant role in improving water-holding capacity. The mechanisms involving water-holding capacity of fresh and especially low-fat processed meats are challenging areas of research.

Other roles of the myofibrillar/cytoskeletal proteins include meat binding and emulsification. Meat binding and emulsification are critical to high-quality processed meat products. Although there was little knowledge about myofibrillar proteins, the sausage maker recognized long ago that salt-soluble proteins (myosin and actin) were essential in producing quality products, i.e., those products with fat-emulsifying and water-holding attributes. Scientific evidence is available now to indicate the importance of the various protein fractions to processed meat manufacture (Siegel and Schmidt, 1979). There are still many questions, however, about the role of myofibrillar/cytoskeletal proteins in processed meats.

To learn about how myofibrillar/cytoskeletal proteins function in meat products, a number of sophisticated physical and chemical techniques are required. The following will present some of the principles and techniques for the study of certain skeletal muscle proteins, in particular, the myofibrillar/cytoskeletal proteins titin and nebulin, and the calpain enzyme system.

Principles and Techniques of Differential Centrifugation

As a technique, centrifugation is widely used by muscle biologists and meat scientists in the isolation and purification of muscle/meat components and their proteins. For a cogent review on centrifugation, the review by Goll et al. (1974) is recommended. Indeed, differential centrifugation is essential in the isolation and preparation of myofibrils for gel electrophoresis, chromatography analysis, and microscopy.

Differential centrifugation is based on the principle of sedimentation of different components within a cell homogenate in a gravitational field. Initially, all the components in the isolating medium are uniformly distributed throughout the centrifuge tube. As centrifugation proceeds, all subcellular components of the cell homogenate sediment at their respective sedimentation rates, resulting first in complete sedimentation of the largest particles present, followed by sedimentation of progressively smaller particles. The degree of contamination of larger particles in the pellet by smaller particles at the exact moment when the largest particles have just been sedimented is approximately proportional to the ratio of the sedimentation coefficients of the two particles. Another way of looking at the technique of differential centrifugation is to centrifuge (spin) at low speeds permitting the largest component to sediment first with slight contamination by lighter components. Decanting the supernatant and centrifuging at greater gravitational forces and longer times (usually) allows the separation of lighter cell components of the homogenate. Repeated washing, homogenizing, and centrifuging steps result in purified components.

By increasing centrifugal speed within appropriate periods of time, greater sedimentation forces are developed and even the lightest subcellular components can be isolated. Based on theoretical calculations, sedimentation of particles due to centrifugal forces follows the same physical laws as sedimentation of particles due to gravitational forces, and that difference in rates of sedimentation of two particles in the same solvent or suspending medium depends only on differences in the size, shape, and density of the two particles. Because of the direct relationship between sedimentation of subcellular components in a centrifugal field and the differential settling of components in a gravitational field, separation of subcellular components by a series of centrifugation cycles, each done at different speeds, is termed differential centrifugation. Different forces developed during centrifugation are usually given in times greater than the force of gravity (g), e.g., $1000 \times g$.

Solvent densities commonly used in differential centrifugation are usually only slightly greater than 1.0, and particle densities range between 1.15 and 1.75. The radius of the same particles may vary from 20 to 500 nm, therefore, the rate of sedimentation of particles is more dependent on particle size and shape and less on particle density. Hence, particle size and shape are characteristics of the components isolated by differential centrifugation.

Using differential centrifugation to separate the different components of the muscle cell is indeed unique because of the complex nature of its subcellular components. That is, skeletal muscle cells (fibers) are composed of at least nine different kinds of identifiable subcellular components. These are the nuclei, sarcosomes, sarcolemma or outer cell membrane, myofibrils or contractile elements, transverse tubules (T tubules), the sarcoplasmic reticular (SR) membranes, glycogen granules, ribosomes, and sarcoplasm. The cell may also contain membranes of the Golgi complex, lysosomes, and lipid inclusions. These subcellular components can be identified by electron microscopy, but to understand the biochemical and functional nature of these subcellular components they must be separated into a purified form.

Before differential centrifugation is undertaken for isolating and purifying myofibrils, muscle/meat tissue must be trimmed of visible fat and connective tissue and then minced into finer particles by passing it through a household-type grinder causing rupture of connective tissue layers surrounding the fasiculi and fibers and cellular membranes (sarcolemma) to release subcellular components such as myofibrils. Ground tissue is then homogenized in an isolating medium usually consisting of phosphate-buffered (pH 6.8) potassium chloride by using a Waring Blendor. Buffered medium is a must in obtaining reproducible results. A biological buffer such as Tris [tris(hydroxymethy)aminomethane] can be a very acceptable substitute for potassium phosphate. The medium is near or slightly higher in cellular osmolarity to avoid swelling, rupture, or loss of structure and function of subcellular components.

A Ca^{2+} chelator (EGTA [ethylene glycol bis (B-aminoethyl ether)-N,N,N^1,N^1-tetra acetate]) and Mg^{2+} are included to limit the amount of myofibril shortening, especially of prerigor myofibrils, caused by grinding, cold temperature, and homogenization. Cell rupture and suspension and all subsequent steps in the isolation scheme of myofibrils by differential centrifugation must be done at cold temperatures (2–4°C) (rooms and ice buckets, precooled rotors, tubes, suspending medium), with a refrigerated centrifuge. Cold temperatures and materials are imperative in preserving the integrity of the myofibril.

Isolation of myofibrils by differential centrifugation is based on the principle of their very high sedimentation coefficient (100,000 to several million Svedberg units) relative to other subcellular components of the muscle cell. Hence, their isolation requires very low g forces for short times and numerous washes to separate them from many "lighter" components (Table 3.1). At 2000 × g, nuclei, unbroken cells, connective tissue, sarcolemmal sheaths, as well as myofibrils are sedimented. Obviously, many contaminants are present in the first step of isolating myofibrils by differential centrifugation. Consequently, special techniques are required to obtain "purified" myofibrils. Passing myofibrils through a

TABLE 3.1 Conditions for Sedimentation of Subcellular Components of the Muscle Cell by Differential Centrifugation

Conditions	Components sedimented
0–2000 × g for 10–15 min	Nuclei, myofibrils, unbroken cells, connective tissue, and sarcolemmal sheaths
2000–10,000 × g for 10–20 min	Sarcosomes
10,000–25,000 × g for 20–30 min	Lysosomes and "heavy" microsomes
25,000–100,000 × g for 120–180 min	Microsomes, polysomes, ribosomes and fragments of sarcolemma
100,000 × g supernatant	Sarcoplasmic proteins, t-RNA, and other soluble substances

Source: Modified from Goll et al. (1974) and reproduced by permission of the Am. Meat Sci. Assn. and the authors.

typical household nylon strainer facilitates removal of connective tissue and debri. That is, myofibrils will pass through the strainer, but connective tissue and debri will be retained. Another technique is the addition of Triton X-100, a nonionic detergent, to remove membranes normally adhering to the surface of myofibrils. Thorough and repetitive washing and changes in concentration of isolating medium are essential for securing a purified myofibril preparation. Furthermore, because nuclei sediment with myofibrils, it is essential to subject the homogenate to repeated blending in the Waring Blendor. This repeated homogenization ruptures the nuclear membranes, and the fragmented membrane and its contents are not sedimented at low centrifugal forces. A modified procedure we have successfully used for many years can be observed in Table 3.2 (Goll et al., 1974). Protein content of the purified myofibril prep is determined with the modified buiret procedure described by Robson et al. (1968).

TABLE 3.2   Isolation of Purified Myofibrils by Differential Centrifugation

I. Ground or minced muscle
  1) Suspend in 10 volumes (v/w) 100 mM KCl, 20 mM K phosphate, pH 6.8, 2 mM $MgCl_2$, 1 mM EGTA, 1 mM NaN_3 (standard salt solution) by homogenizing for 10 sec in Waring Blendor.
  2) Centrifuge at 1000 × g for 10 min.

II. Sediment or pellet retained
Supernatant
(decanted)
  1) Suspend in 6 volumes (v/w) of standard salt solution by homogenizing for 10 sec in Waring Blendor.
  2) Centrifuge at 1000 × g for 10 min.

III. Sediment retained
Supernatant
(decanted)
  1) Suspend in 8 volumes (v/w) of standard salt solution by homogenizing for 10 sec in Waring Blendor.
  2) Pass suspension through household nylon net strainer.
  3) Centrifuge at 1000 × g for 10 min.

IV. Sediment retained
Supernatant
(decanted)
  1) Suspend in 8 volumes (v/w) of standard salt solution by homogenizing for 10 sec in Waring Blendor.

TABLE 3.2 *(Continued)*

        2) Pass suspension through household nylon net strainer.

        3) Centrifuge at 1000 × g for 10 min.

V. Sediment retained

Supernatant
(decanted)   1) Suspend in 6 volumes (v/w) of standard salt solution plus 1% (v/w) Triton X-100 by homogenizing for 10 sec in Waring Blendor.

        2) Centrifuge at 1500 × g for 10 min.

VI. Sediment retained

Supernatant
(decanted)   1) Suspend in 6 volumes (v/w) of standard salt solution plus 1% (v/w) Triton X-100 by homogenizing for 10 sec in Waring Blendor.

        2) Centrifuge at 1500 × g for 10 min.

VII. Sediment retained

Supernatant
(decanted)   1) Suspend in 8 volumes (v/w) of 100 mM KCl by stirring vigorously with polyethylene stirring rod.

        2) Centrifuge at 1500 × g for 10 min.

VIII. Sediment retained

Supernatant   Repeat Step VII
(decanted)   1)

IX. Sediment retained

Supernatant
(decanted)   1) Suspend in 10 volumes (v/w) of 5 mM Tris HCl, pH 8.0 homogenizing for 3 sec in Waring Blendor.

        2) Centrifuge at 1500 × g for 10 min.

X. Sediment retained

Supernatant
(decanted)   1) Repeat step IX.

XI. Sediment retained

Supernatant
(decanted)   1) Suspend in 4 volumes (v/w) of 5 mM Tris, pH 8.0 homogenizing for 3 sec in Waring Blendor.

        2) Do protein analysis of suspension.

Purified myofibrils
free of membranes

Source: Modified from Goll et al. (1974) and reproduced by permission of the Am. Meat Sci. Assn. and the authors.

In addition, the use of the centrifuge is essential in the purification of the myofibrillar/cytoskeletal proteins and the calpain/calpastatin protease system.

Principles and Techniques of Titin and Nebulin Purification

Titin and nebulin are the subjects of much interest to many meat scientists and muscle biologists. These two megadalton ($M_r \cong 3000$ kDa) proteins discovered by Wang and Ramirez-Mitchell (1979) usually migrated in porous SDS-PAGE as a doublet and were named T_1 and T_2. Maruyama et al. (1981b) also has reported a protein with similar properties called α and β connectin, corresponding to T_1 and T_2, respectively. Nebulin was originally termed band 3, as it was the third (below T_1 and T_2 bands) band. Subsequently, it was named nebulin (Wang, 1981). Titin makes up approximately 10% of the total mass of myofibrillar protein (Kurzban and Wang, 1988). A single titin molecule is 900–1000 nm in length and has been shown to span from the Z-line to near the M-line (one half of the intact sarcomere) (Furst et al., 1988). Electron microscopy studies of rotary shadowed titin molecules have shown that titin monomers have a globular headlike region at the M-line end and a tail region with a diameter of approximately 4 nm (Nave et al., 1989; Soteriou et al., 1993). The exact function of this protein in skeletal muscle is not yet known. Indeed, little was known until the protein was purified by Trinick et al. (1984). One of the proposed functions is involvement in the regulation of the assembly of the thick filament during muscle development and growth (Whiting et al., 1989). Another proposed role of great interest to meat scientists is the loss of structural integrity of the sarcomere. Interestingly, titin has been identified in a third set of filaments, gap filaments, in the myofibril (LaSalle et al., 1983). The existence of gap filaments has been known for some time, but it was Locker (1982) who gave significance to these filaments because of their possible involvement in tenderization. Lusby et al. (1983) showed that titin degradation in bovine longissimus is both postmortem temperature and time dependent. It is this proposed function (postmortem degradation of titin) that has led to the hypothesis that degradation or loss of titin may be involved in beef tenderization and water-holding capacity (Paterson and Parrish, 1987; Paterson et al., 1988; Anderson and Parrish, 1989; Huff et al., 1993).

The second protein, nebulin, is also very large, with its size varying from 600 to 800 kDa depending on the muscle tissue it is extracted from (Wang and Wright, 1988). Nebulin makes up 3–4% of the total myofibrillar protein in skeletal muscle, and it has not been found in cardiac or smooth muscle. Nebulin is thought to be closely associated with the thin filaments and has been proposed to function as a template for the development of the thin filament. In postmortem muscle nebulin is degraded rapidly, often to the point of not being visible on gels by 3 days postmortem (Lusby et al., 1983; Paterson and Parrish, 1987; Paxhia and Parrish, 1988; Anderson and Parrish, 1989; Fritz and Greaser, 1991; Huff et al., 1993).

Purification of titin and nebulin is challenging as they are very prone to proteolysis and tend to be rather insoluble under aqueous conditions unless detergents such as SDS (sodium dodecyl sulfate) are present. The extremely large size of these proteins, however, does make gel filtration a natural choice for the final separation step.

Because these proteins are myofibrillar structural proteins, the first step in their purification is the preparation of highly washed myofibrils by differential centrifugation. Titin and nebulin are very prone to proteolysis by enzymes inherent in the muscle system; consequently, it is essential that muscle samples be removed as quickly as possible after exsanguination and be placed immediately in an isolating medium. The isolating medium used consists of 100 mM potassium chloride, 20 mM potassium phosphate (pH 6.8), 2 mM magnesium chloride, 2 mM EGTA, and 1 mM sodium azide. Just prior to use, PMSF (phenylmethylsulfonyl fluoride) is added to make the final concentration 0.1 mM with respect to PMSF. PMSF acts as an inhibitor of serine proteases and as a result reduces protein degradation. Fifty grams of muscle are homogenized in six volumes of the isolating medium by using 15-sec bursts in a Waring Blendor. Suspensions are then centrifuged for 5 min at 3840 $\times$ g. The supernatant is decanted, and the sediment is then resuspended in isolating medium and centrifuged twice more. Each time the supernatant is decanted. After the third homogenization step, the suspension is filtered through cheesecloth or a nylon strainer to remove any connective tissue that may be present. After the third centrifugation step, the pellet is resuspended in the isolating medium containing 1% Triton X-100 to solubilize biological membranes. The suspension is then centrifuged again at 3840 $\times$ g for 5 min. The pellet is resuspended in 10 volumes of the isolating medium (without Triton X-100) by vigorous stirring before centrifuging at 3840 $\times$ g for 5 min. This step is repeated, and following the last centrifugation, the sample is suspended in a minimum volume of the same isolating medium. Protein content of the suspension is determined using the modified biuret method of Robson et al. (1968), and 500 mg of washed myofibrils are removed and used for purification of titin and nebulin. The remaining myofibrils can be suspended in an equal volume of cold glycerol and stored at $-20°C$.

The 500 mg of myofibrils are washed twice with four volumes of 4°C 5 mM Tris-HCl, pH 8.0 and centrifuged at 3020 × g for 5 min. In all steps prior to and including this step, the temperature of the sample must be maintained at 4°C or below. The resulting sediment is suspended in 10 mL of sample buffer containing 200 mM Tris-HCl (pH 8.0), 20 mM EDTA, 80 mM dithiothreitol, 20% (w/v) SDS, and 4 mM PMSF that has been warmed to 50°C (Wang, 1982). The sample is then homogenized in a Teflon-glass tissue homogenizer with several quick strokes. After incubation at 50°C for 20 min, the sample is centrifuged at 78,900 × g for 1 hr at 20°C (this temperature is necessary to keep SDS from precipitating).

The next step involves the fractionation of titin from nebulin (Wang, 1982). This is accomplished by diluting the supernatant from the final centrifugation step with two volumes of 5 mM Tris-HCl, pH 8.0. Subsequently, 4 M NaCl in 5 mM Tris-HCl (pH 8.0) is added dropwise to the protein-containing solution while stirring with a glass rod. Enough should be added to bring the final concentration to 0.64 M NaCl. As the buffered NaCl is added, the presence of a white precipitate is observed with the addition of each drop, and the final solution is rather cloudy. The cloudy solution should be left to stand at room temperature for 30 min. After this stand time, the sample is centrifuged at 23,700 × g for 20 min at 20°C. The resulting sediment is highly enriched in titin, while the supernatant contains nebulin and should be retained. The sediment is resuspended in 0.64 M NaCl in 5 mM Tris-HCl and is centrifuged at 23,700 × g for 5 min. A minimum amount (less than 1 ml) of a 50°C solution of 200 mM Tris-HCl, 20 mM EDTA, 80 mM dithiothreitol, 20% (w/v) SDS, and 4 mM PMSF (pH 8.0) is added to the sediment. The sample is stirred with a glass rod and incubated at 50°C for 5 min. A minimum amount of 3% SDS in 5 mM Tris-HCl (pH 8.0) is added, and a further incubation at 50°C is done in order to solubilize the sediment completely. Subsequently, the sample is centrifuged at 183,000 × g for 30 min at 20°C. The supernatant is then collected and loaded onto a gel filtration column as described in a following section.

Nebulin is purified by adding enough of the 4 M NaCl solution to raise the salt concentration of the retained supernatant that is enriched in nebulin to 0.9 M. Again, as in the titin purification step, the presence of a white precipitate can be seen as the NaCl solution is added. It is important to stir the mixture well with a glass rod during this stage to avoid clumping of the precipitate. Nebulin will precipitate out over a range of 0.7–0.9 M NaCl (Wang, 1982). The mixture is allowed to stand at room temperature for 10 min and is then centrifuged at 183,000 × g for 30 min at 20°C. The sediment is dissolved in a minimum amount of a 50°C solution of 200 mM Tris-HCl, 20 mM EDTA, 80 mM dithiothreitol, 20% (w/v) SDS, and 4 mM PMSF (pH 8.0) and is then loaded onto a gel filtration column.

The final step in the purification of these two proteins involves the use of gel filtration (also known as gel permeation, gel exclusion, and molecular sieving). Gel filtration is a method of column chromatography whereby macromolecules in solution are separated by size. The gel or column matrix is made up of beadlike particles and consists of an open, cross-linked, three-dimensional molecular network. Within the beads of the column matrix are pores that are easily accessible to some molecules, but not to others that are above a certain size (Scopes, 1987). Those molecules that can enter the pores do so and thus are retarded. Those molecules that are too large pass around the pores and are less retarded, thereby eluting earlier.

There are several parameters that can be manipulated to optimize the resolution of protein peaks from the column. *Sample volume* should be no more than 1–3% of the total column bed volume. Smaller sample volumes (down to 1%) give better resolution, as the volume of elutant containing the protein will increase because of nonideal flow and diffusion (Scopes, 1987). For large molecular weight components, such as titin and nebulin, a lower *flow rate* will enhance the resolution of the peaks. *Column length* also influences the resolution between two peaks. In fact, the resolution between two bands increases as the square root of the column length (Pharmacia Biotech Inc., personal communication).

The procedure for gel filtration of titin and nebulin in our laboratory is as follows: 5 mL of sample are loaded onto a 2.5 × 90 cm Sephacryl S-500-HR column (Furst et al., 1988; Wang and Wright, 1988) that has been previously equilibrated with a buffer containing 40 mM Tris-HCl, 20 mM sodium acetate, 2 mM EDTA, 0.1 M dithiothreitol (DTT), 0.1% (w/v) SDS, pH 7.4. Equilibration is done by passing a minimum of two bed volumes through the column. This column is maintained at room temperature due to the presence of SDS. Sephacryl S-500-HR is a cross-linked co-polymer of allyl dextran and N,N′-methylene bisacrylamide. This cross-linked polymer produces rigid beads that can be operated under fairly high pressures and flow rates. This particular matrix is designed to separate very large molecules and has an effective separation range of molecular weights of 40,000 to 20,000,000.

Titin emerges as the first peak (measured by UV absorbance at 280 nm) and is eluted fairly well in advance of the retarded peak. When the nebulin-enriched fraction is loaded on an identical column and eluted, it also elutes in the first peak, only just ahead of a myosin peak. In fact, in most cases, the last of the nebulin peak will overlap with the beginning of the very broad myosin peak. Extreme care must be taken in order to collect only those fractions that contain only nebulin and not myosin.

Monitoring of the column fractions is done by UV absorbance at 280 nm and by sodium dodecyl sulfate polyacrylamide gel electrophoresis (SDS-PAGE) on 5 and 10% gels. On 5% gels (acrylamide/bisacrylamide ratio 100:1), titin from the column appears as a single band near the top of the gel, while nebulin appears about one third of the way into the gel. On 10% gels (acrylamide/bisacrylamide ratio 50:1), nebulin appears as a sharp band just below the interface of the 5% stack and the 10% resolving gel.

Once positive identification of the pure titin fractions has been made, the samples should be pooled and immediately concentrated and frozen ($-70°C$) to preserve them. Titin is unstable at ambient room temperatures. Evidence of this instability is that degradation of the sample is evidenced by an increasing amount of "smearing" of the bands as seen by SDS-PAGE analysis. This degradation of the sample can be retarded if the fractions are maintained at 4°C after they come off the column. Nebulin, on the other hand, tends to be rather stable at room temperature after it is eluted off the column, and little deterioration is seen by SDS-PAGE analysis even after several days at room temperature.

By following the above procedures, up to 10 mg of titin and 4 mg of nebulin can be easily purified from 500 mg of myofibrils. By extremely careful execution of the above procedures, these yields could conceivably be better than 10 and 4 mg for titin and nebulin, respectively.

Principles for Purification of μ- and m-Calpain and Calpastatin

The calpain system is a highly ubiquitous proteolytic enzyme system composed of several proteins known to date. The proteins in this system that have been studied include the enzymes μ-calpain, m-calpain, and the relatively newly discovered n-calpain. The system also includes the indigenous inhibitor of the calpains, calpastatin, and possibly a protein activator of the calpains (Goll et al., 1992). The m-calpain enzyme was the first component of the system to be discovered and characterized in skeletal muscle. Most of the early major work in this area was done at Iowa State University (Busch et al., 1972; Dayton et al., 1975; Dayton et al., 1976a, b; Reville et al., 1976).

The enzyme components of the system differ both in their structure and their [Ca^{2+}] requirements for half-maximal activity. μ- and m-Calpain are each composed of 80- and 28-kDa subunits. The 28-kDa subunits of both μ- and m-calpain are identical and are encoded on the same gene. The 80-kDa subunits are the catalytic subunits and are encoded on different genes and share a 50% sequence homology. The [Ca^{2+}] requirements for the two enzymes are different. μ-Calpain requires 3–50 μM [Ca^{2+}] for half-maximal activity, while m-calpain requires 200–1000 μM [Ca^{2+}] for half-maximal activity. Both enzymes seem to be very similar in their choice of substrates. n-Calpain (Sorimachi et al., 1989; Wolfe et al., 1989) is composed of a single polypeptide chain of 94 kDa and requires >3000 μM [Ca^{2+}] for half-maximal activity. All three of these enzymes are cysteine proteases and have a pH optimum near neutrality. These enzymes have been shown to be present in nearly every type of vertebrate cell examined. They are localized mostly with myofibrils, nuclei, and mitochondria in skeletal muscle cells and seem to be associated with subcellular organelles and the plasma membrane in all cells examined (Goll et al., 1992). Several roles have been proposed for the calpains. These include activation or alteration of the regulation of certain enzymes, cleaving of hormone receptors, and remodeling and/or disassembly of the cytoskeleton. This latter function is of most interest to meat scientists. Many scientists have hypothesized that calpain may be the major factor involved in the tenderization of fresh meat (Olson et al., 1977; Zeece et al., 1986; Koohmaraie, 1992).

The competitive inhibitor of the calpains, calpastatin is also present in cells (Takano and Murachi, 1982). Depending on the species from which it is isolated, calpastatin has been shown by cDNA analysis (Maki et al., 1990) to be composed of a single polypeptide of 73–77 kDa, but migrates with an apparent molecular mass of 107–117 kDa under SDS-PAGE conditions. One large calpastatin molecule from skeletal muscle can theoretically inhibit up to four calpain molecules, based on cDNA studies (Goll et al., 1992). Calpastatin appears to be co-localized with the calpains in the cell. This inhibitor is highly specific for the calpains and has no sequence homology to any other protease inhibitor that has been sequenced.

The major steps of the purification procedure for μ- and m-calpain and calpastatin involve adsorption chromatography procedures and gel filtration techniques. The principles of gel filtration chromatography were discussed previously in the section on the purification of titin and nebulin.

Adsorption chromatography techniques are based on the principle that proteins can and do adsorb or adhere to the surface of many types of solid matrices and are selective in the manner that they adsorb. In general, this family of methods gives a very high increase in purity and, in the case of enzymes, a high increase in the specific activity. Most adsorbents used include ion exchangers, chemically synthesized ligand adsorbents, and inorganic materials (Scopes, 1987). The choice of a specific technique or the adsorbent used is based on a certain parameter of the protein such as its relative charge or its degree of hydrophobicity (Roe, 1989). Some methods also use highly specific biological interactions to separate proteins (e.g., immunopurification or affinity chromatography techniques using monoclonal antibodies).

Chromatography methods themselves are based on the separation of components using a mobile and a stationary phase. In column chromatography the stationary phase is usually made up of a matrix of spherical particles that are packed into a column. In adsorption chromatography, the proteins to be separated are generally introduced in the mobile phase and separation is based on the affinity of the proteins in the solution to the column matrix (stationary phase). Those proteins that have a greater affinity for the column matrix migrate more slowly and are thus eluted later than those proteins that have lesser or no affinity for the column matrix (Roe, 1989). The column matrix is generally chosen based upon the affinity of the desired protein for the matrix, although in some cases one may choose to select a matrix based on the affinity of the contaminant proteins for the matrix.

Ion exchange chromatography is based upon the electrostatic attraction that occurs between charged groups on the proteins in the mobile phase and oppositely charged groups that are on the stationary phase. Proteins contain both positively and negatively charged groups that arise from the acidic and basic side chains of their constituent amino acids. The overall, or net, charge of a protein depends on the number of either positively or negatively charged groups and is affected by the pH of the solution the protein finds itself in. The isoelectric point (pI) of a protein is the pH at which the number of positively and negatively charged groups are equal. The pI for most proteins falls between pH 5 and 9. At pH values above their pI, proteins have a net negative charge, while below their pI, proteins have a net positive charge (Cheftel and Cuq, 1985).

Basically two types of ion exchangers are used. These are anion exchangers and cation exchangers. Anion exchangers are derivatized with positively charged functional groups to adsorb anionic (negatively charged) proteins. Cationic exchangers are derivatized with negatively charged functional groups to adsorb cationic (positively charged) proteins.

There are many types of matrices and functional groups available for use in ion exchange chromatography. Some of the more popular matrices include celluloses, agaroses, dextrans, Sephadex, TSK and Tris-acryl, to name a few. Functional groups that are available can be divided not only on the basis of charge, but also according to the strength or weakness of the groups. Those groups that are categorized as strong will remain ionized over the whole pH range used. Weak groups, on the other hand, are effective over a narrow pH range and for the most part are only partially ionized (Roe, 1989). Some of the functional groups commonly used include: *DEAE* (diethylaminoethyl), *QAE* (quaternary aminoethyl), *CM* (carboxymethyl), and *SP* (sulfoproxyl).

DEAE is the anion exchanger longest in use and tends to be the most popular. It is a weak anion exchanger with a pKa of 9.5. *QAE* is classified as a strong anion exchanger and has a pKa of 12.0. *CM* is one of the original cation exchangers and has a pKa of 4.0, thus it is classified as a weak cation exchanger. *SP* is a strong cation exchanger and has a pKa of 2.0.

The choice of a functional group depends upon the pH at which the protein is most stable. An anion exchanger would be the matrix of choice if a protein is most stable above its pI, while a cation exchanger would be used for those proteins that are most stable below their pI. Most proteins have pIs below 7, so anion exchangers tend to be the most common (Roe, 1989). One important point to note about the use of ion exchangers is that anion exchangers should not be used at pH values above the pKa of their functional group, while cation exchangers should not be used at pH values below the pKa of their functional group or these matrices will not bind proteins effectively.

Other important points to consider when using ion exchangers are sample volume and concentration. Unlike gel filtration columns, the volume of the sample loaded onto the column is of lesser importance, while concentration of the sample is of greater importance. Sample concentration is related to the capacity of the column matrix to bind proteins and thus plays a large role in purification by ion exchange chromatography.

Another important point to consider is the method used to adsorb the protein to the matrix. The two main methods used are batch adsorption and column adsorption. *Batch adsorption* is often used early in the purification protocol. It is a system in which the material to be separated is added to the matrix in solution and the proteins are adsorbed to the matrix. Elution can be done in the batch mode by washing and filtration, or the slurry can be packed into a column from which the protein is then eluted. *Column adsorption* utilizes a packed bed ion exchanger in which the protein solution is loaded onto the column and either the contaminants are adsorbed to the matrix and the desired protein is passed through the column without binding to it, or the desired protein is adsorbed to the matrix and the contaminants are washed off. In this latter method, the protein is eluted off by a change in the column environment (usually by an increase or decrease in the ionic strength or pH of the buffer), which changes the affinity of the matrix for the protein, allowing it to elute off. This latter method allows for greater concentration of the protein as well as being more conducive to collection of the protein in smaller fractions than does the former method.

There are three main methods used for elution of adsorbed proteins from the column matrix. These include isocratic elution, stepwise elution, and gradient elution. *Isocratic elution* is an elution method in which the composition of the elution buffer does not change during elution of the sample. *Stepwise elution* is a method in which the composition of the buffer is changed, in a discontinuous manner, during the course of elution of the protein to conditions that favor dissociation of the protein from the matrix. *Gradient elution* is a method whereby the composition of the buffer is changed continuously during the course of elution. In this method, the steepness or shallowness of the gradient must be carefully considered. A gradient that is too steep may result in loss of the protein, while a gradient that is too shallow can lead to greater dilution of the protein. In many cases, a gradient elution method is preferred over batch or stepwise elution procedures as greater resolution can be obtained. One may have to be willing to use longer elution times with this method in order to obtain greater resolution. In general, the recommended volume of the elutent for gradient elution methods is about five times that of the bed volume (Roe, 1989).

Other features of proteins exploited for purification purposes other than electrostatic properties are the hydrophobic regions that occur on the surface of proteins. In hydrophobic interaction chromatography the interaction between the hydrophobic regions on the surface of proteins can interact with aliphatic chains on the adsorbent and allow the protein to bind to the column matrix (Scopes, 1987). The number and the relative size of hydrophobic regions on the surface of proteins vary among proteins, thus this method can be effectively used in many cases to separate proteins.

Many types of hydrophobic interaction functional groups are available. Some of the more common ones include propyl-, butyl-, phenyl-, octyl-, hexyl-, and aminohexyl groups. One of the more popular substances used in calpain/calpastatin purification techniques is phenyl-Sepharose. This matrix is an agarose based matrix to which a benzene ring is attached by a glycerol-ether linkage.

When using hydrophobic interaction chromatography, just as in ion exchange chromatography, the volume of the sample loaded is less important than the concentration of the sample. When considering the elution method, often a decrease in salt concentration, increase in pH, or an organic solvent such as ethylene glycol is used in either a continuous gradient or in a stepwise manner. Often the protein is loaded at an ammonium sulfate level just less than that required to salt out the protein, and elution is then done by using a decreasing ammonium sulfate gradient.

Dye binding chromatography is a form of nonspecific affinity chromatography. In this family of methods, the stationary phase has a covalently linked textile dye attached to it. This dye contains the functional group that attaches to some proteins with a fairly high degree of specificity. The dyes most commonly used are polysulfonated chromophores that are linked to a chlorotriazine group by way of an aminoether linkage. Often the stationary phase used is an agarose matrix as it can provide a relatively large number of hydroxyl groups to couple to the dye (Scopes, 1987). The exact reason many dyes bind to specific proteins is not known, so often proper selection of the most efficient dye can be difficult. An electrostatic attraction may play a partial role in the binding of proteins to certain dyes, as elution of the bound protein is often accomplished by using an increasing salt gradient (usually NaCl, as KCl may cause dyes to precipitate). In the case of calpains, in which often a Reactive Red 120 (Sigma Chemical Company) column is used, the interaction involved may be hydrophobic as the calpains bind to the column at high ionic strengths and elute at low ionic strengths (Edmunds et al., 1991).

Calpain/Calpastatin Purification Techniques

There are several published methods for purifying the μ and m-calpain enzymes and their inhibitor, calpastatin. Most of them follow very similar techniques, and in most cases the differences between them are quite subtle (Wolfe et al., 1989; Koohmaraie, 1990; Edmunds et al., 1991; Wheeler and Koohmaraie, 1991). The techniques for the most part utilize a homogenization/centrifugation procedure followed by various chromatography steps. Some of the procedures involve the use of a dialysis step prior to the first column (Wheeler and Koohmaraie, 1991). The methods outlined below (Edmunds et al., 1991) use many of the principles previously discussed and involve ion exchange, hydrophobic interaction, and dye binding chromatography as well as gel filtration techniques.

Initially, bovine skeletal muscle sample is collected and trimmed free of any visible fat and/or connective tissue and is ground or finely minced. Care must be taken throughout the procedure to keep the sample temperature at 2–4°C. The meat is then homogenized in a blender in 6 volumes of a buffer containing 20 mM Tris-HCl (pH 8.0), 5 mM EDTA-Tris (pH 7.2), 0.1% 2-mercaptoethanol (MCE), 100 mg/L ovomucoid (a trypsin inhibitor), 2.5 μM E-64 (*trans*-epoxysuccinyl L-leucylamido (4-Guanidino)-butane, a cysteine enzyme inhibitor), and 2 mM PMSF (phenylmethylsulfonyl fluoride, an inhibitor of serine proteases). The homogenate is centrifuged at 17,700 × g for 15 min. The resulting supernatant is filtered through glass wool, and the pH is adjusted to 7.5 using solid Tris. The supernatant is mixed and stirred for approximately 2 hr with DEAE cellulose that has been previously equilibrated in 20 mM Tris-HCl (pH 7.5), 1 mM EDTA, and 0.1% MCE. Next, the slurry is filtered and washed with the equilibration buffer. After washing, the DEAE cellulose is suspended in equilibration buffer and packed into a column. A gradient of 0–500 mM KCl is then begun through the column. Calpastatin is eluted off the column at about 80–110 mM KCl, μ-calpain is eluted from 90–200 mM KCl, and m-calpain is eluted from 200–400 mM KCl. These fractions are collected and screened for either inhibitor or enzymatic activity by measuring the degradation of casein in solution. Casein degradation is used to screen all of the collected fractions from the columns throughout the entire procedure.

In the procedure for the purification of calpastatin, the fractions that show calpastatin activity are pooled and the sample is adjusted to 1 M ammonium sulfate by addition of solid ammonium sulfate. The fraction is then centrifuged at 12,000 × g for 10 min to remove any protein that may have salted out of solution after the addition of ammonium sulfate. The supernatant is then loaded onto a hydrophobic interaction column (phenyl Sepharose) that has been equilibrated in 1 M ammonium sulfate, 20 mM Tris-HCl (pH 7.5), 1 mM EDTA, and 0.1% MCE. Calpastatin is eluted from this column by using a linear gradient of 1 M ammonium sulfate to 0 M ammonium sulfate solution in 20 mM Tris-HCl (pH 7.5), 1 mM EDTA, and 0.1% MCE. This is followed by flushing the phenyl sepharose column with 1 mM EDTA and 0.1% MCE. The calpastatin is expected to elute between 0.7 and 0.5 M ammonium sulfate, while any μ-calpain that may have inadvertently been collected along with the calpastatin is eluted in the EDTA wash.

Fractions from the phenyl Sepharose column with calpastatin activity are pooled and the calpastatin is salted out at 65–70% ammonium sulfate. The sample is centrifuged at 12,000 × g for 10 min. The sediment is dissolved in 20 mM Tris-HCl (pH 7.5), 1 mM EDTA, and 0.1% MCE. The suspension is clarified by centrifuging at 48,000 × g for 10 min. The supernatant is then loaded onto a Sephacryl S-300 gel filtration column and is eluted using a buffer containing 20 mM Tris-HCl (pH 7.5), 1 mM EDTA, 0.1% MCE, and 1 mM sodium azide. Sephacryl

S-300 is a cross-linked co-polymer of allyl dextran and N,N'-methylene-bisacrylamide that has a fractionation range for globular proteins of 10,000–150,000.

The active fractions are collected from the S-300 gel filtration column and are loaded onto a DEAE-TSK column that has been equilibrated with 20 mM Tris-HCl (pH 7.5), 1 mM EDTA, and 0.1% MCE. Elution is done by using a 0–125 mM KCl gradient in the equilibration buffer. At this point, the calpastatin coming off the column should have little or no absorbance as detected by UV spectrometer. This is due to the fact that there is very little tryptophan in calpastatin.

For μ-calpain purification, the μ-calpain fractions that were eluted off of the DEAE-cellulose column are pooled and are then loaded onto a phenyl Sepharose column that has been equilibrated with 125 mM KCl in 20 mM Tris-HCl (pH 7.5), 1 mM EDTA, and 0.1% MCE. The column is then flushed with 50 mM KCl in 20 mM Tris-HCl (pH 7.5), 1 mM EDTA, and 0.1% MCE until the conductivity drops. The column is then flushed with a 1 mM EDTA, 0.1% MCE solution. The μ-calpain should elute in the EDTA flush.

The active μ-calpain fractions are identified, pooled, and the ammonium sulfate concentration is adjusted to 0.8 M. This pooled sample is then loaded onto a butyl Sepharose column that has been previously equilibrated with 0.8 M ammonium sulfate in 20 mM Tris-HCl (pH 7.5), 1 mM EDTA, and 0.1% MCE. The column is flushed with the equilibration buffer (which is 0.8 M with respect to ammonium sulfate) until the absorbance coming off the column reads zero. This indicates the major, unbound impurities have been removed. This step is followed by a 0.8–0 M ammonium sulfate gradient in 20 mM Tris-HCl (pH 7.5), 1 mM EDTA, and 0.1% MCE. μ-calpain will come off the column between 0.7 and 0.4 M ammonium sulfate. The active fractions are collected and enough solid ammonium sulfate is added to make the final concentration of the sample 60–65% ammonium sulfate. The sample is centrifuged at $12,000 \times g$ for 10 min. The sediment containing μ-calpain is then dissolved in 20 mM Tris-HCl (pH 7.5), 1 mM EDTA, and 0.1% MCE. The dissolved μ-calpain is clarified at $48,000 \times g$ for 10 min. The supernatant is then loaded onto a Sephacryl S-300 gel filtration column that has been equilibrated in 20 mM Tris-HCl (pH 7.5), 1 mM EDTA, 0.1% MCE, and 1 mM sodium azide.

The μ-calpain is eluted off the gel filtration column with the same buffer the column was equilibrated with. The active fractions from this column are pooled, and the KCl concentration is adjusted to 85 mM and the solution is loaded onto a DEAE-TSK column that has been equilibrated with 85 mM KCl, 20 mM Tris-MES (pH 6.5), 1 M EDTA, and 0.1% MCE. The protein is eluted using a linear 85–125 mM KCl gradient in Tris-MES (pH 6.5), 1.0 mM EDTA, 0.1% MCE followed by a 2 M KCl, 20 mM Tris-MES (pH 6.5), 1.0 mM EDTA, and 0.1% MCE. Tris-MES pH 6.5 is used in this step to bring the pH of the column environment nearer to the pI of calpain (pIs of calpains are 5.1–5.5). This lowering of the pH aids in improving the resolution of closely related proteins.

In order to purify m-calpain, the m-calpain fractions that were eluted from the DEAE cellulose column are loaded directly onto a phenyl Sepharose column that has been previously equilibrated with 100 mM KCl in 20 mM Tris-HCl (pH 7.5), 1 mM EDTA, and 0.1% MCE. After the protein has been loaded, the column is flushed with 50 mM KCl in 20 mM Tris-HCl (pH 7.5), 1 mM EDTA, and 0.1% MCE until a drop in conductivity is noted. The column is then flushed with 1 mM EDTA and 0.1% MCE. The m-calpain should come off the column in the EDTA wash. Active m-calpain fractions are collected and 4 M NaCl is added to make the solution 0.5 M with respect to NaCl. This solution is applied to a Reactive Red 120 (Sigma Chemical Company) column that has been equilibrated in 0.5 M NaCl in 20 mM Tris-HCl (pH 7.5), 1 mM EDTA, and 0.1% MCE. This column is then flushed with the same solution used to equilibrate it until the absorbance of the solution coming off the column reaches zero, an indication that a large portion of the unbound protein has been removed. The m-calpain is then eluted by flushing the column with 20 mM Tris-HCl (pH 8.5), 1 mM EDTA, and 0.1% MCE until the absorbance again returns to the baseline value. The active fractions are pooled and the m-calpain is salted out at 65–70% ammonium sulfate. This is centrifuged at 12,000 × g for 10 min to sediment the m-calpain. The sediment is dissolved in 20 mM Tris-HCl (pH 7.5), 1 mM EDTA, and 0.1% MCE and is loaded onto a Sephacryl S-300 gel filtration column.

The m-calpain is eluted off the S-300 column using 20 mM Tris-HCl (pH 7.5), 1 mM EDTA, and 0.1% MCE with 1 mM sodium azide. The major impurities should elute ahead of the m-calpain. The active fractions from this column are pooled and the KCl concentration of the pooled sample is adjusted to 135 mM, and the solution is loaded onto a DEAE-TSK column that has been equilibrated with 135 mM KCl in 20 mM Tris-MES (pH 6.5), 1 mM EDTA, and 0.1% MCE. The protein is eluted by using a linear 135–275 mM KCl gradient in the same buffer used as the base for equilibrating the column. This is followed by flushing with a 2 M KCl solution in the same buffer.

Evaluation of the purity of the recovered μ- and m-calpain and calpastatin fractions is most frequently monitored by SDS-PAGE.

Principles and Techniques of Sodium Dodecyl Sulfate Polyacrylamide Gel Electrophoresis (SDS-PAGE)

Electrophoresis refers to the transport of particles through a solvent by an electric field. If the molecule is in a matrix, the force of the electrical field is opposed by the friction that occurs between the accelerating molecule and the solution. The degree of frictional force on this molecule is therefore dependent on the size and shape of the molecule as well as the viscosity of the medium through which it moves. Consequently, the movement of the molecule is proportional to the field strength and charge on the molecule, but it is inversely proportional to its size and solution viscosity. All of this seems rather simple, but accomplishing the actual technique of electrophoresis is complex. That is, movement of complex macromolecules in electrophoretic fields and in the presence of rigid matrices, such as polyacrylamide, makes the technique much more complicated. The optimal matrix for carrying out protein separation by electrophoresis is one that is stable, that decreases or eliminates convection, and that does not react with the sample or retard its movement. One of the most widely used matrices is polyacrylamide.

Formation of a polyacrylamide gel occurs by the polymerization and cross-linking of the monomer, acrylamide, over the cross-linking co-monomer, N, N^1-methylene-bis-acrylamide (commonly called Bis). The weight ratios of acrylamide to Bis are critical. If the ratio is too small (<10), the gels tend to be brittle and opaque. Conversely, if the ratio exceeds 100, gels are hard to handle and very easily broken.

The polymerization reaction requires initiators, usually added as a combination of ammonium persulfate and TEMED (N, N, N^1, N^1-tetra-methylethylenediamine). Another combination, riboflavin-TEMED, is a photocatalytic system; however, it requires special equipment for photopolymerization as well as several hours for completion. Both of these initiator systems furnish free radicals for polymerization, but most commonly polymerization is initiated by using ammonium persulfate and TEMED. The physical properties (viscosity, elasticity, density, and mechanical strength) of the gel are determined by the concentration of monomer in the original solution, the degree of polymerization (chain length), and the amount of cross-linking. The final gel can be made to have pore radius of about 0.5 to 3 or 4 nm by adjusting the total concentration of acrylamide (the lower the acrylamide concentration, the greater is the pore dimension) and the amount of cross-linking agent. Hence, polyacrylamide gels can be used to study a wide range of molecular weights (MW < 1000 to MW > 10^6).

The ratio of acrylamide to Bis is critical to the formation of proper pore size. That is, 3–7%, 7%, 7–30% of acrylamide concentration are used to separate $1–2 \times 10^6$, $10^4–10^5$, and $2 \times 10^3–10^4$ molecular weights, respectively.

Gels can be formed in either a tube or a slab (currently, slab gels are preferred) and are positioned between upper and lower buffer reservoirs of the electrophoresis apparatus. The buffer system may be dissociating or nondissociating, with the most commonly used dissociating agent being SDS.

The extensive use and versatility of SDS-PAGE is due largely to the ability of SDS to solubilize many insoluble cellular organelles, including myofibrils and myofibrillar/cytoskeletal proteins. About 1.4 g SDS are bound per gram of protein, rendering all molecules negatively charged and dissociating them to monomeric chains in the presence of MCE. Hydrodynamic characteristics indicate that each polypeptide chain with bound SDS has an extended, rodlike configuration. Uniquely, the length varies with the molecular weight of the protein moiety. Moreover, the electrophoretic mobility of each polypeptide is inversely proportional to the log of its molecular weight.

The system using the same buffer ions in the sample, gel, and electrode vessel reservoirs is termed a continuous buffer system (e.g., the method of Weber and Osborn, 1969). In this system, the protein sample is loaded directly onto the gel in which separation will take place. Conversely, discontinuous buffer systems (e.g., the method of Laemmli, 1970) use different buffer ions in the gel and electrode reservoirs. Also, both buffer composition and pH may differ. Gel samples are loaded onto a large-pore stacking gel polyermized on top of a small-pore resolving gel.

The use of stacking gels in PAGE is popular because of increased resolution of protein separation. The stacking gel is a smaller layer of acrylamide (2–4%), which gives larger pore size. The buffer system of this gel will cause proteins to migrate rapidly and pile up in the stacking layer before entering the running layer, the lower running gel. The lower running gel is the gel in which the proteins separate (usually 5–20% acrylamide or acrylamide gradient).

A sample of protein(s) denatured by heating in a medium containing SDS (0.1–1.0%), either MCE or DTT to reduce and prevent disulfide linkages, and 10–50% sucrose or glycerol to increase density is applied to the top of the gel surface or into wells formed in the gels. An electrical field is applied across the gel, and the negatively charged protein-SDS complexes migrate from the cathode to the anode. A visible tracking dye, premixed in the sample, such as bromphenol blue or pyronin Y, acts as a marker during migration. Once the protein(s) move into the resolving gel, separation occurs dependent upon their size as they migrate through the gel pores (molecular sieving). The migrating units in SDS-PAGE are the monomeric chains, with the amount of migration dependent on a combination of the mobility in the buffer and the constraint imposed by gel filtration. Another factor affecting particle mobility is ionic strength. Mobility is approximately inversely proportional to the square root of the ionic strength. Hence, buffer solutions of low ionic strength allow high migration velocities with little heat production, whereas high ionic strength buffers result in low migration velocities and high heat production.

SDS-electrophoretic mobilities plotted against the logarithm of the molecular weights within an appropriate range of known polypeptide chains result in a linear curve from which a molecular weight of the subunit of an unknown protein can be determined. After the run, rods or slabs are removed from the gel apparatus and then stained, usually with Coomassie blue, to reveal the position of the protein bands. Subsequently the gels are destained to provide clear backgrounds. Thereafter, gels can be stored, photographed, or scanned by densitometry. For greater insight and understanding about the use of SDS-PAGE for analyzing myofibrillar proteins, reviews by Robson et al. (1974) and Greaser et al. (1983) are recommended. Also, five papers commonly referred to in the literature for methods used in electrophoresis, namely, those of Fairbanks et al. (1971), Laemmli (1970), Porzio and Pearson (1977), Studier (1973), and Weber and Osborn (1969), are recommended.

Our (Olson et al., 1977; Parrish, 1977) first use of SDS-PAGE to study proteolysis of myofibrillar proteins during the postmortem storage of beef muscles was with the procedure of Weber and Osborn (1969). Meat scientists are very interested in the use of SDS-PAGE because of its detection of changes in myofibrillar proteins related to postmortem tenderization thought to be caused by proteolysis. Both 7½% acrylamide and bis-acrylamide in a 75:1 weight ratio and 10% acrylamide and bis-acrylamide in a 37:1 weight ratio were used in gels, which were run in 5 mm (inner diameter) × 120 mm tubes. The gels and the upper and lower reservoirs contained 100 mM Na phosphate (pH 7.1), 0.1% SDS. As can be observed in the electrophoretogram of Fig. 3.1, the major change detectable by SDS-PAGE that occurred during conventional postmortem aging of bovine longissimus and semitendinosus muscle was a simultaneous disappearance of troponin-T and the appearance of a 30,000-dalton component. This 30,000-dalton component was subsequently found to be a degradation product of troponin-T. Furthermore, this 30,000-dalton component occurred because calpain, the natural protease in skeletal muscle, caused the degradation of tropinin-T (Olson et

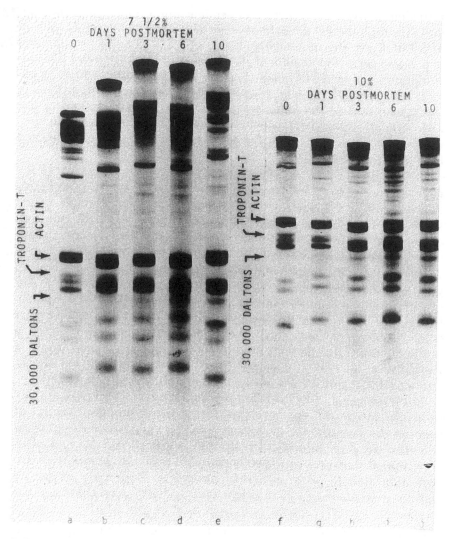

FIG. 3.1   SDS-PAGE of 7½ and 10% gels of myofibrils prpared from bovine longissimus muscle at death and at different times of postmortem storage at 2°C. (a–e): Note the gradual decrease of the troponin-T band and the gradual increase of the 30,000-dalton band from 0 to 10 days postmortem. (f–j): Note the decrease of the troponin-T band and the increase of the 30,000 dalton band from 0 to 10 days postmortem. (*From Olson et al., 1977, and reproduced by permission of the Institute of Food Technologists and the authors.*)

al., 1977). In addition to our work on beef muscle showing the presence of a 30,000-dalton component, the results of Hay et al. (1973), Samejima and Wolfe (1976), and Yamamoto et al. (1979) on chicken muscle, and that of Penny (1976) on pork muscle and Penny and Ferguson-Bryce (1979) on beef muscle support the presence of a 30,000-dalton component in postmortem muscle. Based on our observations, i.e., the strong relationship between the 30,000-dalton component, degradation of the myofibril at or near the Z-lines, and beef steak tenderness (Olson and Parrish, 1977), and the presence of the 30,000-dalton component in tender and not tough beef steaks, we termed this kind of tenderness "myofibril fragmentation tenderness" (MacBride and Parrish, 1977) (Fig. 3.2).

SDS-PAGE has also been used to demonstrate the degradation of desmin during postmortem storage of muscle. Desmin is the major protein component of muscle 10-nm filaments, commonly referred to as intermediate filaments because their diameter is between the 6-nm actin filaments and the 14- to 16-nm myosin filaments. O'Shea et al. (1979, 1981) successfully purified desmin from porcine skeletal muscle and found that desmin forms 10-nm filaments in vitro. Also, immunofluorescence studies show that desmin is located at or near the Z-disk. Postmortem studies show that desmin disappears with an increase in postmortem time (Robson et al., 1991; Young et al., 1980). Coincidentally, the decrease in desmin content parallels a decrease in troponin-T and an increase in the 30,000-dalton component. Furthermore, purified desmin makes an excellent substrate for calpain (O'Shea et al., 1979). Because desmin and the 10-nm filaments appear to function in connecting and organizing myofibrils in muscle cells, postmortem degradation of desmin may also be important in postmortem tenderization and water-holding capacity.

Studies by Lusby et al. (1983) on the effects of postmortem storage on the large molecular weight myofibrillar/cytoskeletal proteins, titin and nebulin, show that these proteins were degraded during postmortem storage of bovine longissimus muscle. To study changes in these large molecular weight proteins during postmortem storage a sample of bovine longissimus muscle was excised at death (within 45 min postmortem), and portions were stored at 2°, 25°, or 37°C (Lusby et al., 1983) and samples removed 1, 3, and 7 days postmortem. SDS-PAGE was performed according to the method of Studier (1973) with 3.2% acrylamide [bis-acrylamide/acrylamide, 1:30 (w/v)] slab gels (14 cm × 10 cm × 0.12 cm).

The results of this study show that titin and nebulin are degraded in bovine longissimus muscle postmortem, with the degree of degradation

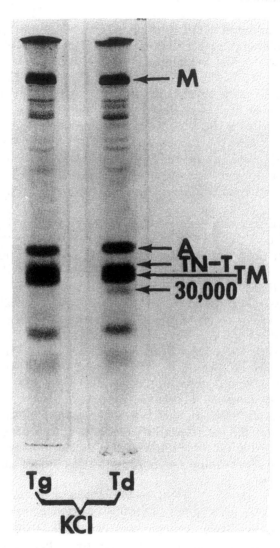

FIG. 3.2   SDS-PAGE of 7½% gels of myofibrillar proteins extracted with high ionic-strength salt solution (0.6 M KCl, 0.1 M K phosphate, pH 7.4) from 1 day postmortem (2°C storage) bovine tough (Tg) and tender (Td) longissimus muscle. Myosin (M), actin (A), troponin T (TN-T), tropomyosin (TM) and 30,000 (30,000-dalton component). Note the presence of the 30,000-dalton component in the tender sample and the absence of the 30,000-dalton component in the tough sample. Protein load: 25 μg. (*From MacBride and Parrish, 1977, and reproduced by permission of the Institute of Food Technologists and the authors.*)

dependent upon storage time and temperature. These results suggest that proteolysis is the cause for the degradation and are consistent with those shown by others that titin (Wang and Ramirez-Mitchell, 1979, 1983) and the high molecular weight components of connectin (titin) (King et al., 1981; Maruyama et al., 1981a) are highly susceptible to proteolytic degradation. Results by Anderson and Parrish (1989) using SDS-PAGE showed that titin and nebulin of myofibrils from tender loin steaks are more extensively degraded than from less tender loin steaks.

Our most recent studies by Huff et al. (1993), using SDS-PAGE, have shown that the degradation of titin and nebulin are affected by postmortem aging time of beef longissimus muscle and maturity and sex of the beef animals. For this study, slightly different methods were used for preparing purified myofibrils for electrophoresis. Following protein determination of a purified myofibril solution, a tracking dye solution [30 mM Tris-HCl (pH 8.0), 3 mM EDTA, 3% (w/v) SDS, 30% (v/v) glycerol, and 0.3% (w/v) pyronin Y] was added to the protein suspension at a ratio of 2:1 (protein solution: tracking dye). MCE was also added at a 1:1 ratio (protein solution volume before tracking dye addition: MCE). Samples were heated at a lower temperature for a longer period of time (50°C for 20 min) than was used previously (100°C for 3 min). Recent studies (Granzier and Wang, 1993) have shown that heating samples at temperatures above 60°C can cause extensive degradation of titin.

Increasing the time of postmortem aging of longissimus muscle was accomplished by a loss of the T_1 band of titin (upper band) and by a loss of the nebulin band at 7 days postmortem. Among those samples that initially had relatively high shear force values and lower sensory scores, the nebulin band was present and the T_1 band of titin was more intense at 3 days postmortem. Conversely, the T_1 band tended to disappear earlier in the more tender samples. This indicates that titin and nebulin were degraded faster in those samples classified as more tender (Fig. 3.3).

Sex of the animal also had an impact on the degradation of titin and nebulin (Fig. 3.4). Nebulin was consistently present in samples of longissimus muscle from bullocks (lanes 4, 5, 6) and cows (lanes 7, 8, 9), whereas it was not always present on SDS-PAGE electrophoretograms at 3 days postmortem in longissimus muscle from steers (lanes 1, 2, 3) (Fig. 3.4). The myofibrils isolated from the longissimus muscle samples of the most tender steer (lane 1) had no nebulin band whereas nebulin was detectable in all other samples at 3 days postmortem. The intensity of nebulin was greatest in the toughest sample as measured by both shear force values and sensory scores for fiber fragmentation (lanes 5 and 9). The T_1 band of titin, while present in all samples at 3 days postmortem, tended to be more intense in those samples that also showed the more

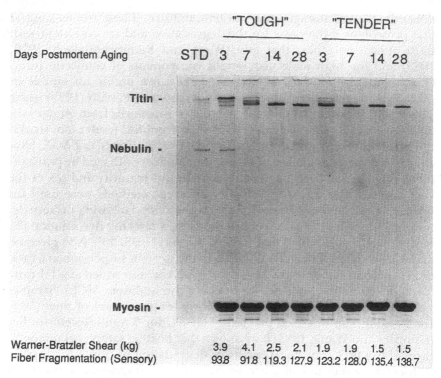

FIG. 3.3 5% SDS-PAGE gel of purified myofibrils from "tough" and "tender" longissimus samples from two steer carcasses over all postmortem aging times. Compare lanes 3 and 7 day postmortem "tough" with 3- and 7-day postmortem "tender" longissimus samples, and note more titin bands for tough than for tender samples. Warner-Bratzler shear and sensory fiber fragmentation values for tender are lower and higher, respectively, than for tough samples. (*From Huff, 1991.*)

intense nebulin band. This indicates the possibility that the intensity of the bands of these two proteins are reflecting differences among animals in the activity of a protease that acts on both titin and nebulin.

Figure 3.5 is a 5% SDS-PAGE electrophoretogram of purified myofibrils isolated from the longissimus samples of steers, bulls, and cows at 14 days postmortem. Those samples having lowest sensory scores for fiber fragmentation (lanes 5 and 9) exhibited the most prominent T_1 band at the 14-day aging period. In addition, the T_1 band of titin, while often faint, was more frequently seen at 14 days postmortem in samples from older animals (cows, lanes 7, 8, 9) than in samples from steers and

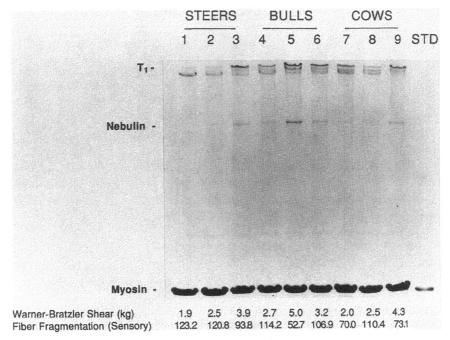

FIG 3.4 5% SDS-PAGE gel of purified myofibrils from longissimus samples from carcasses of bulls, steers and older animals (cows) at 3 days postmortem. Note the increased number of T_1 and nebulin bands for tougher (higher shear and lower fragmentation values) longissimus samples. These increased number of bands are especially noteworthy for samples from bulls and cows. Note the T_1 and nebulin bands in tough bull and cow samples (4–9). (*From Huff, 1991.*)

bulls (lanes 1 through 6). The samples from older animals (cows, lanes 7, 8, 9) also had the heaviest T_2 bands (lower bands). No evidence of the T_1 band of titin was observed in any of the samples from steer carcasses (lanes 1, 2, 3) at 14 days postmortem, possibly indicating that more proteolytic degradation was occurring in those samples. The T_1 band of titin was also not consistently seen in 14-day postmortem samples isolated from longissimus samples from bulls (lanes 4, 5, 6). The evidence presented seems to suggest the possibility that a decreased activity of a protease such as calpain or possibly increased activity of an inhibitor such as calpastatin may be responsible for the decreased tenderness seen in some samples, especially in the older animals. Likewise, the more tender samples may be reflective of an increased activity of a protease such as

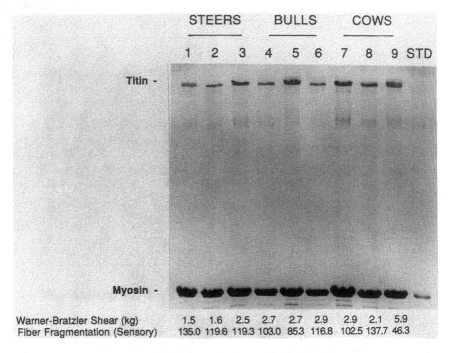

FIG. 3.5 5% SDS-PAGE gel of purified myofibrils from longissimus samples from carcasses of bulls, steers, and older animals (cows) at 14 days postmortem. Note the prominent T_1 band (lanes 5 and 9) occupied by the tough (lowest fiber fragmentation score) samples. (*From Huff, 1991.*)

calpain or the decreased activity of an inhibitor such as calpastatin. Whipple et al. (1990) reported that *Bos indicus* cattle, which are traditionally less tender than *Bos tarus* cattle, exhibited a higher activity of calcium-dependent protease inhibitor activity (calpastatin).

Principles and Techniques of Immunoblot (Western Blotting) Method

The principle of this method is that specific proteins are recognized by antibodies. Although the degradation of titin can be observed with SDS-PAGE, Western blotting is an important technique to use to identify proteolytic breakdown products of titin. This method involves the transfer of proteins by electrophoresis usually to an immunoblotting matrix such as nitrocellulose paper. Subsequently, molecules are identified with a specific antibody. Bandman and Zdanis (1988) have successfully used

Western blotting to detect proteolytic fragments of titin, desmin, and α-actinin. Also, Fritz and Greaser (1991) have used this technique to study changes in titin and nebulin during postmortem refrigerated storage of beef. In our laboratory (Zeece, 1984) a blotting technique was done in a trans blot electrophoresis apparatus (Model 160) from Bio-Rad Corporation. In this study, immediately after slab gel electrophoresis the protein bands were transferred to nitrocellulose paper by electrophoresis. They were transferred for 16 hr at a potential setting of 140 V and 0.6 amps, with a cooling coil used to dissipate heat. The transfer buffer contained 20% methanol and 0.15 M Tris-glycine at pH 8.5. At the end of the transfer time, the paper was removed and placed in a solution of 2% BSA in phosphate buffered saline (PBS), pH 7.0, for 2 hr. The blot was washed three times for 30 min each in PBS (no BSA), then incubated with antibody to the desired antigen. The antibodies were raised in New Zealand white rabbits by the methods of Richardson et al. (1981). The blot was again washed three times in PBS to remove excess antibody. Detection of antigen-antibody complexes was done by using horseradish peroxidase (HRP)–labeled goat-antirabbit (GAR) antibody. The indirect method with GAR-HRP was detected by incubating 0.6% 4-chloro-1-naphthol in 0.01% H_2O_2, which gives a dark blue reaction product.

A study using an immunoblot technique was done to localize the protein and/or proteolytic fragments in electrophoretic separations of myofibril fractions after crude calpain treatment. Figure 3.6 contains results of an immunoblot of a slab gel run with supernatants and sediments from myofibrils that had been incubated with calpain (CAF) at 25°C and pH 7.5. In the immunoblot method, the samples were electrophoresed and then transferred to nitrocellulose paper. The blot was incubated with polyclonal antibodies to titin. In the supernatant lanes (b through f) in Fig. 3.6, many bands are recognized by the titin antibody. No proteins were recognized by the antibody in the control supernatant (lane a). In lanes b through f (supernatant fractions of myofibrils are treated with calpain (CAF) as incubation times increased), the number of bands recognized by the titin antibody increased with increase in incubation time. Furthermore, the population of bands shifts from a few heavily labeled bands present in samples from the short incubations to a greater number of bands of lower molecular weight present in samples from longer incubations (c.f., lane b with lanes e and f). This is a typical pattern of proteolytic digestion as larger polypeptides are degraded to smaller ones. Lanes g through l contain the sediment fractions of myofibrils after corresponding incubation times with calpain. Lane g shows the position of titin in the control (0-time). This band decreases in intensity and migrates slightly faster with increasing incubation time. (The

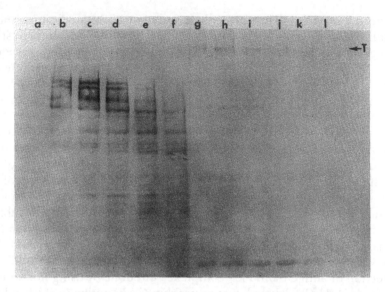

FIG. 3.6 Identification of titin and/or titin fragments by an immunoblot procedure. Supernatants and sediments from myofibrils that had been treated with CAF at 25°C and pH 7.5 were subjected to SDS-PAGE on a 10% acrylamide slab gel, containing a 5% stacking gel. The dark bands in the blot have been recognized by polyclonal antibodies to titin. Lanes a through f in the slab gel were loaded with 50 µL of supernatant fractions corresponding to 0, 2, 4, 8, 15, and 30-min incubations, respectively. Lanes g through l were loaded with 48 µg of corresponding sediment fractions from 0, 2, 4, 8, 15, and 30-min incubations, respectively. The letter T at the right edge of the gel indicates the position of titin in the control sediment (lane g). (*From Zeece, 1986, and reproduced by permission of the author.*)

nature of the two or three other bands labeled in the lanes with the sediments is unknown).

Principles and Techniques of Indirect Immunofluorescence Used in Titin Localization and Organization

The ability to observe cellular and subcellular structures containing various proteins has long been of interest to meat scientists. Where the protein is located within a given subcellular component or how the protein reacts under certain conditions would be just two examples of the useful information supplied by immunofluorescent procedures. Since titin was discovered in 1979 (Wang and Ramirez-Mitchell, 1979) a number of studies has used immunofluorescent techniques to observe the location, structure, and organization of titin (connectin) within the sarcomere (Wang and Greaser, 1985; Furst et al., 1988, 1989; Pierobon-Bormioli et al., 1989). Indirect immunofluorescence techniques rely on the specificity of the antibody-antigen binding relationship. When isolated myofibrils are exposed to a polyclonal antititin with a fluorescein stain almost the entire sarcomere is fluorescent, therefore, it is difficult to specifically determine exactly where titin is located within the sarcomere. To determine specificity Wang and Greaser (1985) successfully developed a monoclonal antibody to titin. A monoclonal antibody will bind only to a single antigenic site on the titin molecule, whereas polyclonal antibodies will bind to every antigenic determinant site on titin. Once the monoclonal antititin was developed it allowed for a single titin epitope to be bound, indeed, its specificity is so strong that a single amino acid switch in the binding areas will make the binding relationship inactive. Currently, several titin antibodies can be purchased from chemical supply companies, and many research laboratories are now making their own titin monoclonal antibodies. As a larger number of monoclonal antibodies becomes available, the epitopes they recognize will allow researchers to effectively "map" the position of titin in the myofibril. Most of the monoclonal antititins bind in the A-I junction area of the myofibril, however, there have been several epitopes located near the M-line area (Furst et al., 1988, 1989). Because fragmentation of the myofibril occurs at or near the Z-line, the goal for most researchers is to develop antititin antibodies in or very near the Z-lines.

The first step in using any indirect immunofluorescent technique is the isolation of myofibrils from muscle samples. This procedure is essential in viewing staining patterns of myofibrils and follows a typical isolating procedure (Beekman and Ringkob, 1992). The final myofibril suspension may be combined with 50% glycerol and frozen for later analysis, or the isolated myofibrils can be directly advanced to the staining procedure.

Isolated myofibrils will have approximately 10–20 sarcomeres, and structurally they are of excellent quality for microscopic viewing. If the myofibrils were glycerinated, 1 mL of the mixture is combined with 10 mL of buffer and centrifuged at $1000 \times g$ for 10 min. The pellet should be resuspended in 10 mL of buffer. The first step of the staining procedure for titin detection is to place 6–8 drops of the myofibril suspension on a glass coverslip. The myofibril suspension is allowed to sit on the coverslips for a 2-min period, and then is fixed to the coverslips by bathing (6–8 drops of the fixative bead over the coverslips) them in a 3.7% formaldehyde solution for 10 min followed by a buffer rinse. Once the myofibrils are fixed, the coverslips are incubated with the antititin antibody by placing 6–8 drops of the antititin over the coverslip and allowing it to sit covered for 1 hr at 5°C. Subsequently, the coverslips are rinsed and bathed with buffer for 10 min. The secondary stain, a fluorescein labeled goat antimouse IgG, is applied by bathing the coverslips for 2 hr at 5°C. After the incubation period the coverslips are rinsed with buffer two or three times. The coverslips are then mounted onto slides using a mounting medium (70% glycerol, 75 mM potassium chloride, 10 mM Tris HCl (pH 8.9), 2 mM EGTA, 2 mM sodium azide, and 1.5 mg/mL p-phenylenediamine). The coverslips are sealed to the slides using fingernail polish.

The laser scanning confocal microscope yields excellent resolution at high magnifications, and it also has the ability to create three-dimensional images. This type of microscope depends on the passage of light through a tiny slit, thus eliminating the out-of-focus images. Laser technology allows sectioning through the depth of the myofibril, and depending on the microscope software, a recreation of the images superimposed on one another in exact register is created. This image can then be displayed in a three-dimensional format and rotated at any angle in order to see the stained bands. Many times multiple myofibrils will be stacked on top of one another, but with the laser scanning confocal system these myofibrils can be sectioned at intervals into the tenths of microns, thus allowing each myofibril to be viewed. A key part of the microscopic evaluation is the adjustments of the image collected. Many times the brightness of the sample may be quite high, thus creating a blurred effect. This condition is easily modified once the image has been collected by lowering the brightness, changing background colors, or assigning pseudocolors to the samples. The brightness could also be decreased by narrowing the slit or pinhole that allows the light into the system. The use of this microscopy system allows the viewing of the fluorescent banding patterns of titin, or any other protein; sarcomere lengths can also be measured quite accurately by a confocal microscope with the required computer software.

If the above procedures are followed, very crisp and clear titin bands can be seen and the images collected on computer disc. If dual staining with anti-alpha actinin is used, the exact location of certain titin epitopes in relation to the Z-line can be identified, thus allowing a "map" of titin to be developed and the location of titin in the myofibril to be determined.

SUMMARY

Many techniques based on sound experimental principles are available to facilitate separation and analysis of muscle/meat proteins. Centrifugation, chromatography, gel electrophoresis, Western blots, and confocal microscopy are discussed. Use of these techniques requires hard work and long hours. The implementation of these techniques holds the potential of providing answers for questions concerning quality meat production. An example of the successful use of some of these techniques and principles is the discovery and molecular explanation of myofibril fragmentation tenderness.

REFERENCES

Anderson, T. J. and Parrish, F. C., Jr. 1989. Postmortem degradation of titin and nebulin of beef steaks varying in tenderness. *J. Food Sci.* 54: 748.

Bandman, E. and Zandis, D. 1988. An immunological method to assess protein degradation in post-mortem muscle. *Meat Sci.* 22: 1.

Bechtel, P. J. and Parrish, F. C., Jr. 1983. Effects of postmortem storage and temperature on muscle protein degradation. Analysis by SDS gel electrophoresis. *J. Food Sci.* 48: 294.

Beekman, D. D. and Ringkob, T. P. 1992. Confocal fluorescence microscopy of immunolabled titin and alpha-actinin in beef myofibrils. *Proc. Western Sec. Am. Soc. Anim. Sci.* 43: 284.

Busch, W. A., Stromer, M. A., Goll, D. E., and Suzuki, A., 1972. Ca^{2+}-specific removal of Z-lines from rabbit skeletal muscle. *J. Cell Biol.* 52: 367.

Cheftel, J. C., and Cuq, J. L. 1985. Amino acids, peptides and proteins. Ch. 5. In *Food Chemistry*, O. R. Fennema (Ed.), p. 245. Marcel Dekker, New York.

Dayton, W. R., Goll, D. E., Stromer, M. H., Reville, W. J., Zeece, M. G., and Robson, R. M. 1975. Some properties of a Ca^{2+}-activated protease that may be involved in myofibrillar protein turnover. In *Cold Springs Harbor Conf. Cell Proliferation*, Vol. 2, *Proteases and Biological Control*, E. Reich, D. B. Rifkin, and E. Shaw (Ed.), p. 551. Cold Spring Harbor Laboratory, Cold Spring Harbor, NY.

Dayton, W. R., Goll, D. E., Zeece, M. G., Robson, R. M., and Reville, W. J. 1976a. A Ca^{2+}-activated protease possibly involved in myofibrillar protein turnover. Purification from porcine muscle. *Biochemistry* 15: 2150.

Dayton, W. R., Reville, W. J., Goll, D. E., and Stromer, M. H. 1976b. A Ca^{2+}-activated protease possibly involved in myofibrillar protein turnover. Partial characterization of the purified enzyme. *Biochemistry* 15: 2159.

Edmunds, T., Nagainis, P. A., Sathe, S. K., Thompson, V. F., and Goll, D. E. 1991. Comparison of the autolyzed and unautolyzed forms of μ and m-calpain from bovine skeletal muscle. *Biochem. Biophys. Acta* 1077: 197.

Fairbanks, G., Steck, T. L., and Wallace D. F. H. 1971. Electrophoretic analysis of the major polypeptides of the human erythrocyte membrane. *Biochemistry* 10: 2606.

Fritz, J. D. and Greaser, M. L. 1991. Changes in titin and nebulin in postmortem bovine muscle revealed by gel electrophoresis, Western blotting and immunofluorescence microscopy. *J. Food Sci.* 56: 607.

Fritz, J. D., Mitchell, M. C., Marsh, B. B., and Greaser, M. L. 1992. Titin content of beef in relation to tenderness. *Meat Sci.* 33: 41.

Furst, D. O., Nave, R. Osborn, M., and Weber, K. 1989. Repetitive titin epitopes with a 42 nm spacing coincide in relative position with known A band striations also identified by major myosin-associated proteins. *J. Cell Sci.* 94: 119.

Furst, D. O., Osborn, M., Nave, R., and Weber, K. 1988. The organization of titin filaments in the half-sarcomere revealed by monoclonal antibodies in immunoelectron microscopy: A map of ten nonrepetitive epitopes starting at the Z line extends close to the M line. *J. Cell Biol.* 106: 1563.

Goll, D. E., Young, R. B., and Stromer, M. H. 1974. Separation of subcellular organelles by differential and density gradient centrifugation. In *Reciprocal Meat Conference Proceedings*, Vol. 27, pp. 250–290. American Meat Science Assn., Chicago, IL.

Goll, D. E., Thompson, V. F., Taylor, R. G., and Zalewska, T. 1992. Is calpain activity regulated by membranes and autolysis or by calcium and calpastatin? *BioEssays* 14(8):549.

Granzier, H. L. M. and Wang, K. 1993. Gel electrophoresis of giant proteins: Solubilization and silver-staining of titin and nebulin from single muscle fiber segments. *Electrophoresis* 14: 56.

Greaser, M. L., Yates, L. D., Krzywicki, K., and Roelke, D. L. 1983. Electrophoretic methods for the separation and identification of muscle proteins. In *Reciprocal Meat Conference Proceedings*, Vol., 35, pp. 87–91. American Meat Science Assn., Chicago, IL.

Hay, J. D., Currie, R. W., and Wolfe, F. H. 1973. Polyacrylamide disc gel electrophoresis of fresh and aged chicken muscle proteins in sodium dodecylsulfate. *J. Food Sci.* 38: 987.

Huff, E. J. 1991. The effects of postmortem aging time, animal age and sex on selected characteristics of bovine longissimus muscle. M.S. thesis, Iowa State University, Ames, IA.

Huff, E. J., Parrish, F. C., Jr., and Robson, R. M. 1993. The effects of postmortem aging time, animal age and sex on the degradation of myofibrillar/cytoskeletal proteins in bovine longissimus muscle. *J. Anim. Sci.* (Submitted).

Hwan, S. F. and Bandman, E. 1989. Studies of desmin and α-actinin degradation in bovine semitendinosus muscle. *J. Food Sci.* 54: 1426.

King, N. L., Kurth, L., and Shorthose, W. R. 1981. Proteolytic degradation of connectin, a high molecular weight myofibrillar protein, during heating of meat. *Meat Sci.* 5: 389.

Koohmaraie, M. 1990. Quantification of Ca^{2+}-dependent protease activity by hydrophobic and ion-exchange chromatography. *J. Anim. Sci.* 68: 659.

Koohmaraie, M. 1992. Role of the neutral proteinases in postmortem muscle protein degradation and meat tenderness. In *Reciprocal Meat Conference Proceedings*, Vol., 45, pp. 63–71. American Meat Science Assn., Chicago, IL.

Koohmaraie, M., Babiker, A. S., Schroeder, A. L., Merkel, R. A., and Dutson, T. R. 1988. Acceleration of postmortem tenderization in ovine carcasses through activation of Ca^{2+}-dependent proteases. *J. Food Sci.* 53: 1638.

Koohmaraie, M., Crouse, J. D., and Mersmann, H. J. 1989. Acceleration of postmortem tenderization in ovine carcasses through infusion of calcium chloride: effect of concentration and ionic strength. *J. Anim. Sci.* 67: 934.

Kurzban, G. P. and Wang, K. 1988. Giant polypeptides of skeletal muscle titin: Sedimentation equilibrium in guanidine hydrochloride. *Biochem. Biophys. Res. Commun.* 155: 1155.

Laemmli, U. K. 1970. Cleavage of structural proteins during assembly of the head of bacteriophage T4. *Nature* 227: 680.

LaSalle, F., Robson, R. M., Lusby, M. L., and Parrish, F. C., Jr. 1983. Localization of titin in bovine skeletal muscle by immunofluorescence and immunoelectron microscopy. *J. Cell Biol.* 97: 285.

Locker, R. H. 1982. A new theory of tenderness in meat, based on gap filaments. In *Reciprocal Meat Conference Proceedings,* Vol., 35, p. 92. American Meat Science Assn., Chicago, Il.

Lusby, M. L., Ridpath, J. F., Parrish, F. C., Jr., and Robson, R. M. 1983. Effect of postmortem storage on degradation of the myofibrillar protein titin on bovine longissimus muscle. *J. Food Sci.* 48: 1627.

MacBride, M. A. and Parrish, F. C., Jr. 1977. The 30,000-dalton component of tender bovine longissimus muscle. *J. Food Sci.* 42: 1627.

Maki, M., Hatanaka, M., Takano, E., and Murachi, T. 1990. Structure-function relationships of the calpastatins. Ch. 3. In *Intracellular Calcium Dependent Proteolysis*, R. L. Mellgren and T. Murachi (Ed.), p. 37. CRC Press Inc., Boca Raton, FL.

Maruyama, K., Kimura, M., Kimura, S., Ohaski, K., Suzuki, K., and Katunuma, N. 1981a. Connectin, an elastic protein of muscle. Effects of proteolysis *in situ*. *J. Biochem.* 89: 711.

Maruyama, K., Kimura, S., Ohashi, K., and Kuwano, Y. 1981b. Connectin, an elastic protein of muscle. Identification of "titin" with connectin. *J. Biochem.* 89: 701.

Morgan, J. B., Miller, R. K., Mendez, F. M., Hale, D. S., and Savell, J. W. 1991. Using calcium chloride injection to improve tenderness of beef from mature cows. *J. Anim. Sci.* 69: 4469.

Nave, R., Fürst, D. O., and Weber, K. 1989. Visualization of the polarity of isolated titin molecules: A single globular head on a long thin rod as the M-band anchoring domain. *J. Cell Biol.* 109: 2177.

O'Shea, J. M., Robson, R. M., Hartzer, M. K., Huiatt, T. W., Rathbun, W. E., and Stromer, M. H. 1981. Purification of desmin from adult mammalian skeletal muscle. *Biochem J.* 195: 345.

O'Shea, J. M., Robson, R. M., Huiatt, T. W., Hartzer, M. K., and Stromer, M. H. 1979. Purified desmin from adult mammalian skeletal muscle: A peptide mapping comparison with desmins from adult mammalian and avian smooth muscle. *Biochem. Biophys. Res. Commun.* 89: 972.

Offer, G. and Trinick, J. 1983. On the mechanism of water holding in meat: The swelling and shrinking of myofibrils. *Meat Sci.* 8: 245.

Olson, D. G. and Parrish, F. C., Jr. 1977. Relationship of myofibril fragmentation index to measures of beefsteak tenderness. *J. Food Sci.* 42: 506.

Olson, D. G., Parrish, F. C., Jr., Dayton, W. R., and Goll, D. E. 1977. Effect of postmortem storage and calcium activated factor on the myofibrillar proteins of bovine skeletal muscle. *J. Food Sci.* 42: 117.

Parrish, F. C., Jr. 1977. Skeletal muscle tissue disruption. In *Reciprocal Meat Conference Proceedings*, Vol., 30, pp. 87–98. American Meat Science Assn., Chicago, IL.

Paterson, B. C. and Parrish, F. C., Jr. 1987. SDS-PAGE conditions for detection of titin and nebulin in tender and tough bovine muscles. *J. Food Sci.* 52: 509.

Paterson, B. C., Parrish, F. C., Jr., and Stromer, M. H. 1988. Effects of salt and pyrophosphate on the physical and chemical properties of beef muscle. *J. Food Sci.* 53: 1258.

Paxhia, J. M. and Parrish, F. C., Jr. 1988. Effect of postmortem storage on titin and nebulin in pork and poultry light and dark muscles. *J. Food Sci.* 53: 1599.

Penny, I. F. 1976. The effect of conditioning on the myofibrillar proteins of pork muscle. *J. Sci. Food Agric.* 27: 1147.

Penny, I. F. and Ferguson-Bryce, R. 1979. Measurement of autolysis in beef muscle homogenates. *Meat Sci.* 3: 121.

Pierobon-Bormioli, S., Betto, R., and Salviati, G. 1989. The organization of titin (connectin) and nebulin in the sarcomere: An immunocytolocalization study. *J. Musc. Res. Cell Motil.* 10: 446.

Porzio, M. A. and Pearson, A. M. 1977. Improved resolution of myofibrillar proteins with sodium dodecyl sulfate polyacrylamide gel electrophoresis. *Biochem. Biophys. Acta* 490: 27.

Reville, W. J., Goll, D. E., Stromer, M. H., Robson, R. M., and Dayton, W. R. 1976. A Ca^{2+}-activated protease possibly involved in myofibrillar protein turnover. Subcellular localization of the protease in porcine skeletal muscle. *J. Cell Biol.* 70: 1.

Richardson, F. L., Stromer, M. H., Huiatt, T. W., and Robson, R. M. 1981. Immunoelectron and immunofluorescence localization of desmin in mature avian muscle. *Eur. J. Cell Biol.* 26: 91.

Ringkob, T. P., Marsh, B. B., and Greaser, M. L. 1988. Change in titin position in postmortem bovine muscle. *J. Food Sci.* 53: 276.

Robson, R. M., Goll, D. E., and Temple, M. J. 1968. Determination of protein in "tris" buffer by the biuret reaction. *Anal. Biochem.* 24: 339.

Robson, R. M., Huiatt, T. W., and Parrish, F. C., Jr. 1991. Biochemical and structural properties of titin, nebulin and intermediate filaments in muscle. In *Reciprocal Meat Conference Proceedings*, Vol. 44, pp. 7–20. American Meat Science Assn., Chicago, IL.

Robson, R. M., Tabatabai, L. B., Dayton, W. R., Zeece, M. G., Goll, D. E., and Stromer, M. H. 1974. Polyacrylamide gel electrophoresis in the presence and absence of denaturing solvents. In *Reciprocal Meat Conference Proceedings*, Vol. 27, pp. 199–225. American Meat Science Assn., Chicago, IL.

Roe, S. 1989. Separation based on structure. Ch. 4. In *Protein Purification*

Methods: A Practical Approach, E. L. V. Harris and S. Angal (Ed.), p. 175. IRL Press at Oxford University Press, Oxford, England.

Samejima, K. and Wolfe, F. H. 1976. Degradation of myofibrillar protein components during postmortem aging of chicken muscle. *J. Food Sci.* 41: 250.

Scopes, R. K. 1987. *Protein Purification: Principles and Practice*, 2nd ed. Springer-Verlag, New York.

Siegel, D. G. and Schmidt, G. R. 1979. Crude myosin fractions as meat binders. *J. Food Sci.* 44: 1129.

Sorimachi, H., Imajoh-Ohmi, S., Emori, Y., Kawasaki, H., Ohno, S., Minami, Y., and Suzuki, K. 1989. Molecular cloning of a novel mammalian calcium-dependent protease distinct from both m and μ types. Specific expression of the mRNA in skeletal muscle. *J. Biol. Chem.* 264: 20106.

Soteriou, A., Gamage, M., and Trinick, J. 1993. A survey of interactions made by the giant protein titin. *J. Cell Sci.* 104: 119.

Studier, F. W. 1973. Analysis of bacteriophage T7 early RNA's and protein on slab gels. *J. Mol. Biol.* 79: 237.

Takano, E. and Murachi, T. 1982. Purification and some properties of human erythrocyte calpastatin. *J. Biochem.* 92: 2021.

Trinick, J., Knight, P., and Whiting, A. 1984. Purification and properties of native titin. *J. Mol. Biol.* 180: 331.

Wang, K. 1982. Purification of titin and nebulin. *Methods Enzymol.* 85: 264.

Wang, K. and Ramirez-Mitchell, R. 1979. Titin: Possible candidate as putative longitudinal filaments in striated muscle. *J. Cell Biol.* 83: 389.

Wang, K. and Ramirez-Mitchell, R. 1983. A network of transverse and longitudinal intermediate filaments is associated with sarcomeres of adult vertebrate skeletal muscle. *J. Cell Biol.* 96: 562.

Wang, K. and Wright, J. 1988. Architecture of the sarcomere matrix of skeletal muscle: Immunoelectron microscope evidence that suggests a set of parallel inextensible nebulin filaments anchored at the Z line. *J. Cell Biol.* 107: 2199.

Wang, S. M. and Greaser, M. L. 1985. Immunocytochemical studies using a monoclonal antibody to bovine cardiac titin on intact and extracted myofibrils. *J. Musc. Res. Cell Motil.* 6: 293.

Weber, K. and Osborn, M. 1969. The reliability of molecular weight determinations by dodecyl sulfate-polyacrylamide gel electrophoresis. *J. Biol Chem.* 244: 4406.

Wheeler, T. L. and Koohmaraie, M. 1991. A modified procedure for simultaneous extraction and subsequent assay of calcium-dependent and lysosomal protease systems from a skeletal muscle biopsy. *J. Anim. Sci.* 69: 1559.

Wheeler, T. L., Koohmaraie, M., and Crouse, J. D. 1991. Effects of calcium chloride injection and hot boning on the tenderness of round muscles. *J. Anim. Sci.* 69: 4871.

Whipple, G., Koomaraie, M., Dikeman, M. E., Crouse, J. D., Hunt, M. C., and Klemm, R. D. 1990. Evaluation of attributes that affect longissimus muscle tenderness in *Bos taurus* and *Bos indicus* cattle. *J. Anim. Sci.* 68: 2716.

Whiting, A., Wardale, J., and Trinick, J. 1989. Does titin regulate the length of muscle thick filaments? *J. Mol. Biol.* 205: 163.

Wolfe, F. H., Sathe, S. K., Goll, D. E., Kleese, W. C., Edmunds, T., and Duperret, S. M. 1989. Chicken skeletal muscle has three Ca^{2+}-dependent proteases. *Biochem. Biophys. Acta* 998: 236.

Yamamoto, K., Samejima, K., and Yasui, T. 1979. Changes produced in muscle proteins during incubation of muscle homogenates. *J. Food Sci.* 44: 51.

Young, O. A., Grafhuis, A. E., and Davey, C. L. 1980. Postmortem changes in cytoskeletal proteins of muscle. *Meat Sci.* 5: 41.

Zeece, M. G. 1984. The effect of calcium activated protease (CAF) and cathepsin D on bovine muscle myofibrils under varying conditions of pH and temperature. Ph.D. dissertation, Iowa State Univ., Ames, IA.

Zeece, M. G., Robson, R. M., Lusby, M. L., and Parrish, F. C., Jr. 1986. Effect of calcium activated protease (CAF) on bovine myofibrils under different conditions of pH and temperature. *J. Food Sci.* 51(3): 797.

4

Computer-Aided Techniques for Quantitative Structure Activity Relationships Study of Food Proteins

Shuryo Nakai, Guillermo E. Arteaga, and Eunice C. Y. Li-Chan

The University of British Columbia
Vancouver, British Columbia, Canada

INTRODUCTION

Quantitative structure-activity relationships (QSAR or SAR) may be the most common concept in chemistry to explain the function of chemical compounds based on their structure. To derive a quantitative relation equation, $y = f(x)$, regression analysis—simple or multiple linear regression analysis depending on the number (n) of structure parameters (x_n) that are correlated with activity or function (y)—has been most frequently applied.

Hansch and colleagues may have been the first group to introduce the hydrophobicity concept into QSAR; a good review work was published by Hansch and Klein (1991). According to Stuper et al. (1979), "Hansch analysis" is linear free energy relations using a first- or second-order linear model of parameters in relation to the hydrophobic, steric, and electronic changes caused by modifying the molecular structure. With the advent of modern computers, Stuper et al. (1979) applied, for the first time, pattern-recognition techniques using multivariate analysis to QSAR. However, the application was limited mostly to small molecular compounds, e.g., drug and olfactory stimulants. Extension of the QSAR concept to macromolecules is no doubt extremely difficult, especially in the expression of steric parameters. To be honest, any attempt to apply QSAR to food proteins, including enzymes, is difficult.

In a previous review article (Nakai and Li-Chan, 1993), the most important question asked was whether we could justify QSAR in explaining functional properties of food proteins based on their structure. Accordingly, the approach taken should be in the direction from small molecular compounds, i.e., amino acids, towards peptides and finally proteins.

QSAR of Peptides and Proteins

The function of amino acids as building blocks of proteins is based on charge, hydrophobicity, and structure-forming capacity including covalent bonds and noncovalent interactions. It is possible to classify all amino acids using scales based primarily on charge and hydrophobicity, along with some structure effects. However, in the context of macromolecules, structure-forming capacity such as the propensity of hydrogen bond formation and hydrophobic interaction should be taken into consideration. There are still other characteristic amino acids, which are difficult to classify based on the above scales. A good example is histidine residue with a strong ability to catalyze chemical reactions at physiological pH. These hypothetical scales characterizing amino acids should be used in the QSAR study of peptides and proteins and may be useful in the future for prediction or optimization for acquiring desired functions in protein engineering.

An orthodox approach using hydrophobicity, molecular size, and electronic effects was employed for prediction of functional properties of peptides such as bitterness, biological activities as opioids and hormones, and interfacial properties (Nakai and Li-Chan, 1993).

A QSAR approach was attempted by Klein et al. (1986) to predict 26 protein functions based on compositional or physicochemical properties of amino acid residues in 1603 protein sequences. The result showed that three or four attributes were generally sufficient to distinguish each of the 26 functional categories from the remainder of the database. The attributes used were related to hydrophobicity, charge, and their distribution, e.g., frequency of occurrence, a periodicity of appearance, as well as secondary structure parameters (average propensity to form α-helix, β-sheet, or β-turns).

QSAR study of genetically modified protein molecules is still in its infancy. Charton (1986, 1990) correlated activities of tyrosyl-tRNA synthetase and T4 lysozyme with single-site mutations. Intermolecular force equations were developed as functions of polarizability, electrical effects, hydrogen-bonding parameter, lone pairs, ionic side chains, charge transfer side chains, and steric parameter derived from van der Waals radii. Polarizability, ionic side chains, and steric effects were the major factors in transition state binding of ATP and tyrosine by tyrosyl-tRNA synthetase. Whereas, bacterial cell lysis by T4 lysozyme was affected by hydrogen bonding, dispersion forces, and steric effects, in that order. However, the QSAR information obtained from this study may have been limited, since single-site mutation cannot be expected to induce great steric changes in the molecular structure.

QSAR of Food Proteins

The major difficulty encountered in the case of QSAR of food proteins is the lack of information on three-dimensional structure and knowledge of their amino acid sequence. Furthermore, they are frequently mixtures of individual proteins, making the situation even more difficult. However, it may be worth trying the Hansch analysis to food proteins. In our last review article (Nakai and Li-Chan, 1993), we covered emulsification, foaming, gelation, and breadmaking as the important properties among many other functional properties of food proteins.

It was observed that the emulsifying ability was closely related to hydrophobicity and that the foaming activity required hydrophobicity as well as factors pertaining to the adsorption of proteins at the interface to obtain adequate foam lamella strength. Hydrophobicity as well as other factors relating to the intermolecular interactions, e.g., calcium and sulfhydryl groups, were involved in thermally induced gelation. Although no extensive QSAR work has been conducted, the critical function of hydrophobicity was demonstrated in breadmaking properties. Therefore, the validity of QSAR approach for elucidating functional properties of food proteins has been established, despite difficulty in selecting appropriate attributes as discussed previously (Nakai and Li-Chan, 1993).

The objective of this paper is to discuss another important aspect of QSAR, namely, the mathematical treatments. QSAR is a study of relationships; therefore, the algorithm to be used for computing the relationships between structure and activity (function) may derive important consequences. QSAR equations obtained should be able to be used for prediction of quality parameters of food protein products. A second important algorithm in food research involves pattern-recognition techniques to classify foods into groups. "Good" or "bad" product quality is one of the critical judgments to be made in the food industry. Although QSAR is capable of predicting quality parameters to be used in classification, sometimes noncontinuum cases exist. Relation equations cannot be applied for these cases, such as variety classification. For instance, occasionally there is a need to find whether spoilage is due to proteolytic or lipolytic action, which are not related to each other. Finally, a third important algorithm is to find how to formulate food products to achieve the best quality utilizing QSAR relationships of ingredients. Optimization techniques should be used for this purpose.

In the following discussion, therefore, three groups of algorithms are described: computation for relationships, classification, and formula optimization.

COMPUTATION FOR RELATIONSHIPS

Regression analysis is used for describing the functional dependence of variables on other independent variables. For QSAR purposes, multiple regression analysis (MRA) has been the most popular technique, especially in stepwise modes. However, for avoiding multicolinearity and reducing data dimension, principal component regression (PCR) is recommended. Lately, partial least squares (PLS) regression was reported to be superior to PCR (Martens and Martens, 1986). Also, artificial neural networks (ANN) has been introduced as the most powerful approach for pattern recognition. ANN can be used also for prediction study, although it does not derive relation equations.

Multiple Regression Analysis

MRA has been perhaps the most widely used statistical method for analyzing relationships among variables. Since QSAR usually has to handle more than one independent variable, models or equations involving polynomial terms as well as product terms to investigate interactions are required because most QSAR relations of food proteins are not linear. Multiple regression analysis has been commonly used to establish the QSAR equations as discussed in the last QSAR review article (Nakai and Li-Chan, 1993). Table 4.1 lists the QSAR equations cited in that article; they were all highly significant.

Problems of MRA, when applied to food research, were reviewed by Pike (1986). Major failures in the use of MRA arise when routine analysis is automatically applied to large data sets without understanding the underlying data structure. Regression analysis assumes that the prediction variables ($X\eta$) are independent of each other. The relationship describing dependence of a variable (y, the dependent variable) on the independent $\chi\eta$ variables can then be derived using MRA. Multicolinearity due to significant correlations between predictor variables may bring about unstable regression coefficients that carry a wrong sign or have a magnitude much greater than the true value.

Several procedures for variable selection are available: (a) all possible variable regression, (b) forward selection, (c) backward elimination, and (d) their stepwise procedures. Although the best equation may be derived by combining all variables, it is extremely time consuming when the number of variables is large, especially when nonlinearity has to be taken into consideration. Furthermore, a large number of experimental data are required to obtain reliable QSAR equations. The inclusion of redundant prediction variables in "all possible variable regression" may promote the occurrence of multicolinearity problems. Among the methods suggested, stepwise multiple regression is used most frequently for finding the significant influencing predictors.

QSAR of biological phenomena is likely to be nonlinear, because there are frequently optimal values of structure parameters that yield the best functions, with bell-shaped relationships. Successful results of data transformations, which involve linearization of the data distribution, have been reported in an attempt to improve the fitting of linear equations, because MRA should be applied, in principle, under assumption of the normal distribution and linearity of variables.

Principal Component Regression

In principal component analysis (PCA), the correlation coefficient matrix $\mathbf{R}$ is transformed to an eigenvalue matrix Λ, that is, a diagonal matrix without intercorrelations:

$$\mathbf{e'R} = \Lambda\mathbf{e'} \qquad (1)$$

where $\mathbf{e}$ is the eigenvector. Principal component score $\mathbf{Y}$ is then calculated by:

$$\mathbf{Y} = \mathbf{Xe} \qquad (2)$$

where $\mathbf{X}$ is the standardized original data matrix.

The main purpose of the PCA is to determine parameters, i.e., principal components, in order to explain as much of the total variation in the original data as possible with as few principal components as possible.

The number of variables is reduced while maintaining as much of the original information as is possible by transforming the original set of variables into a smaller set of linear combination in the form of principal components. Thus, PCA is a technique for dimension reduction and avoidance of colinearity between variables. It can be regarded as similar to stepwise MRA, except that complete elimination of multicolinearity cannot be expected from the latter.

Using a number of principal components (PC) fewer than the number of original predictor variables, stepwise MRA is carried out for principal component regression (PCR) analysis. Selection of PCs is customarily done by choosing PC scores with eigenvalues larger than 1.0. This rule should be handled in a flexible manner.

TABLE 4.1 QSAR Equations Cited in the QSAR Review Article of Nakai and Li-Chan (1993)

Regression equation[a]	Ref.
$EAI = 74.26\ \Delta\log K + 59.66$ ($r^2 = 0.846$, $n = 28$, $P < 0.01$)	Kato and Nakai (1980)
$EAI = 0.07\ CPA + 53.9$ ($r^2 = 0.706$, $n = 28$, $P < 0.01$)	Kato and Nakai (1980)
$EAI = 0.932\ s - 0.007\ s^2 + 0.214\ CPA$ ($R^2 = 0.583$, $n = 52$, $P < 0.001$)	Voutsinas et al. (1983)
$EC = 6.46\ s + 0.00132\ CPA.s - 0.0546\ s^2 - 35.3$ ($R^2 = 0.917$, $n = 26$, $P < 0.000001$)	Li-Chan et al. (1984)
$EC = 0.091\ D - 0.024\ CPA - 0.000046\ ANS^2 + 0.0015\ W^2 + 0.00028\ s.CPA + 0.00040\ ANS.CPA + 33.26$ ($R^2 = 0.774$, $n = 230$, $P < 0.001$)	Li-Chan et al. (1987)
$EC = 192.2 + 96.1\ WOAI$ ($r^2 = 0.962$, $n = 9$, $P < 0.01$)	De Kanterewicz et al. (1987)
$EC = 422.4 - 26.7\ WOAI$ ($r^2 = 0.914$, $n = 9$, $P < 0.01$)	De Kanterewicz et al. (1987)
$FBC = 1.38\ CPA - 0.014\ CPA.s + 30.2$ ($R^2 = 0.802$, $n = 11$, $P = 0.0015$)	Voutsinas & Nakai (1983)
$FBC = -16.2\ SEP - 0.00029\ s^2 + 0.00034\ SH^2 + 7.6\ SEP^2 + 0.073\ SEP.s + 8.22$ ($R^2 = 0.556$, $n = 230$, $P < 0.001$)	Li-Chan et al. (1987)
$FE = j.(\log P) + A$ ($r^2 = 0.957$ to 0.968, $n = 4$ to 6)	Britten and Lavoie (1992)

TABLE 4.1 (*continued*)

Regression equation[a]	Ref.
FC = 1493 η + 25.93 ln S_e − 1775 (R^2 = 0.779, n = 16, P < 0.01)	Townsend & Nakai (1983)
GS = −0.000088 Ca + 0.0113 HB + 0.514 (R^2 = 0.93, n = 11, P = 0.0001)	Kohnhorst and Mangino (1985)
GS = 821.1 − 8.91 SH − 0.063 ANS (R^2 = 0.621, n = 26, P < 0.001)	Hayakawa & Nakai (1985)
GS = −0.029 ANS − 0.38 fat − 3.15 pH + 0.17 D + 0.000024 ANS^2 − 2.70 SEP^2 − 0.0018 W^2 + 0.0085 s.SH + 0.44 SEP.P + 27.74 (R^2 = 0.834, n = 114, P < 0.001)	Li-Chan et al. (1987).

[a]EAI, emulsifying activity index of Pearce and Kinsella (1978); EC, emulsifying capacity; FBC, fat-binding capacity; FE, foam expansion; FC, foaming capacity; GS, gel strength; Δlog K, hydrophobic coefficient defined by Shanbhag and Axclsson (1975); CPA, surface hydrophobicity measured using *cis*-parinaric acid; s, solubility index; D, dispersibility; ANS, surface hydrophobicity measured using 8-anilinonaphthalenesulfonic acid; W, moisture; WOAI, water-oil absorption index; SEP, salt-soluble protein; SH, sulfhydryl contcnt; P, protein content; A and j, constants; η, viscosity; S_e, exposed hydrophobicity measured using CPA; Ca, calcium content; HB, heptane-binding capacity.

Partial Least Squares Regression

PLS uses latent variables (**T**) found by an iterative algorithm (Geladi, 1988). Unlike multiple regression, which correlates x variables directly with y-variable(s), PLS uses **T** to correlate with **Y** (Martens and Martens, 1986). To improve the accuracy of obtained models, cross-validation is used to change weighting constants of **T** based on **Y**. Martens and Martens (1986) stated that PCR is the statistically most established method, and PLS regression may be considered as an extension of PCR. The PLS algorithm was discussed by Hoskuldsson (1988). Recently, Banks et al. (1993) applied PLS to GC data of Cheddar cheese and found an excellent correlation with sensory flavor scores of the cheese samples of different ages. For the model including both volatiles and measures of proteolysis, up to 98% of the total variation in the maturity scores was accounted for using PLS, while PCR did not yield reliable predictions.

Artificial Neural Networks

Artificial neural networks consist of computer-simulated layers of processing elements or artificial neurons. The processing unit receives some number of input signals and calculates its own output in a two-step process: first the weighted sum of its inputs is computed and then the resulting sum or activation level is passed through an output or transfer function (e.g., a sigmoid function) to obtain the neuron output. The processing units are usually arranged in three layers. The first layer consists of neurons, which simply take on the input value of a pattern, and is called the "input layer." The input layer performs no processing on its inputs and serves merely to distribute them to the next layer. Following the input layer are one or more "hidden layers," so called because they receive no input from, and produce no output to, the outside world. Hidden layers act as layers of abstraction by extracting features from inputs. Finally, the "output layer" produces the output result of the network to the user. The "weights" on the input lines represent the strength of the connection to a unit, and learning rules (such as the back-propagation algorithm) alter these weights to create a desired input/output response from an ANN.

ANN is, in principle, a technique for pattern recognition, however, it can be used also for relation study. Once a training set is established, it can be used for prediction. QSAR data of 11 food-related proteins were recomputed for prediction of foaming and emulsifying properties using ANN and PCR (Arteaga and Nakai, 1993). ANN was found to be superior to PCR in prediction ability. Figure 4.1 illustrates a comparison of

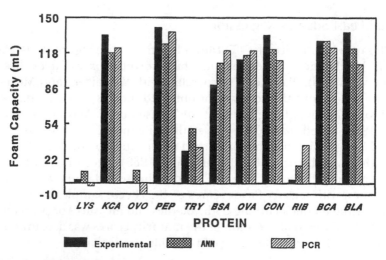

FIG. 4.1 Comparison of experimental foam capacity vs. values predicted by ANN and PCR. ANN consisted of 11 input neurons, 2 hidden neurons. Lys, lysozyme; KCA, κ-casein; OVO, ovomucoid; PEP, pepsin; Try, trypsin; BSA, bovine serum albumin; OVA, ovalbumin; CON, conalbumin; RIB, ribonuclease; BCA, β-casein; BLA, β-lactoglobulin.

ANN and PCR in terms of predictability of foam capacity. Except for some proteins, e.g., κ-casein, pepsin, and trypsin, ANN prediction is closer to experimental values than PCR prediction. For foam stability, PCR could not give a significant model.

Once ANN has "learned" the association between inputs and outputs, it can be used for prediction or for simulation experiments. Figure 4.2 shows the prediction effect, calculated using a trained ANN, of changing the surface hydrophobicity and charge density of lysozyme on its foam capacity class (FCC). FCC = 0 and 4 indicate poor and excellent foam capacity, respectively (Arteaga and Nakai, 1993). Although these results are a simulation, they suggest a new way for improving the functional properties of proteins.

Hasegawa et al. (1992) reported the use of PLS for development of a QSAR model for classification of the antiarrhythmic activity of 14 phenylpyridines. Antiarrhythmic activity was assessed in mice, and a binary scheme was used to classify the antiarrhythmic activity of 14 compounds. A value of 0 or 1 was assigned if the compound was inactive or active, respectively. Three physicochemical parameters, representing polarizability, steric (torsion angle between the amide and benzene group) and charge (proton affinity of the amide group) effects were

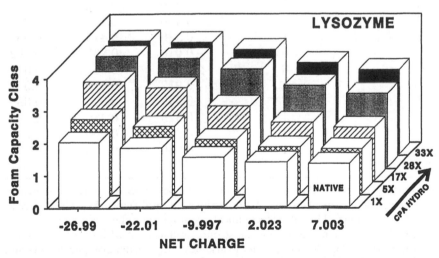

FIG. 4.2   Artificial neural networks (ANN) simulation of the effect of changing the net charge (units/residue) and fluorescence probe hydrophobicity (CPA HYDRO) of lysozyme. The ANN used for the simulation consisted of 12 neurons, 2 hidden neurons, and the foam capacity class as the output neuron. The CPA HYDRO of the native protein was changed by a factor of 5, 17, 28, and 33.

130   NAKAI et al.

used for the development of the models. Although PLS and PCR gave relatively good prediction (78%), we obtained a significant improvement in the predictability (93%) by applying ANN with three hidden neurons to the data reported by Hasegawa et al. (1992) as shown in Table 4.2.

For nonlinear deconvolution, it was reported that ANN was also superior to PLS in terms of the ability to estimate the composition of mixtures of compounds computed from fluorescence spectra (McAvoy et al., 1992).

Table 4.2 Experimental and Predicted Antiarrhythmic Activity Class (0 = inactive, 1 = active) of Phenylpyridines[a]

Compound number	Activity			
	Experimental	PLS	PCR	ANN
1	0	0.215	0.351	0.013
2	1	1.171	0.708	0.999
3	0	0.732[b]	0.810[b]	0.251
4	0	−0.917	0.080	0.000
5	0	0.489	0.502	0.599[b]
6	0	0.436	0.422	0.266
7	1	0.376[b]	0.415[b]	0.883
8	0	−0.016	0.238	0.002
9	0	0.212	0.250	0.019
10	1	0.912	0.720	0.998
11	1	0.542	0.510	0.835
12	1	0.325[b]	0.470[b]	0.799
13	1	0.902	0.900	0.989
14	1	0.946	0.772	0.995
% correct[c]		78	78	93

[a]Prediction of activity class was performed using partial least square (PLS), principal component regression (PCR), or artificial neural networks (ANN). Cross-validation results. The physicochemical parameters molar refractory, torsion angle between amide and benzene group, and proton affinity of the amino group were used as the independent variables. Physicochemical parameters and experimental activity class were taken from Hasegawa et al. (1992).
[b]Misclassified compound.
[c]According to Hasegawa et al. (1992) if predicted activity > 0.5 or predicted activity < 0.5, then compound is classified as being active or inactive, respectively.

COMPUTATION FOR CLASSIFICATION

Classification techniques are divided into two categories. The first category is "unsupervised" techniques, the purpose of which is grouping of samples that may be mutually related. The second category is "supervised," in which samples are classified into known groups.

Supervised classification uses an optimal classification rule based on a training set of data. In this sense, a variety of discriminant analysis techniques is mathematically robust, except for the occurrence of random or chance classification when the number of samples is small compared to the number of variables (Aishima and Nakai, 1991). However, in food research, sample grouping is not always *a priori* defined. For instance, when defective products are discovered, the cause for the defects is frequently unknown, yet it is urgent and necessary to solve the problem by identifying the cause. The most popular unsupervised classification is PCA. Recently, principal component similarity (PCS) analysis was proposed (Vodovotz et al., 1993) for unsupervised classification. Although ANN is probably the most powerful technique for pattern recognition, it is a method mainly for supervised classification. (Kohonen network is used for unsupervised cases).

Stepwise Discriminant Analysis

In discriminant analysis, the between-group variance $(\hat{b}'\mathbf{B}\mathbf{b})$ is maximized relative to the within-group variance $(\hat{b}'\mathbf{W}\mathbf{b})$, thus the following relationship is derived from the maximization:

$$(\mathbf{W}^{-1}\mathbf{B} - \hat{\lambda}\mathbf{I})\hat{\mathbf{b}} = 0 \qquad (3)$$

where $\mathbf{W}$ and $\mathbf{B}$ are the matrix within-group and between-group sums of squares and cross products, respectively. This equation is solved for eigenvalue $\hat{\lambda}$ and corresponding eigenvector $\mathbf{b}$ of the matrix $\mathbf{W}^{-1}\mathbf{B}$. Eigenvectors corresponding to the largest and second-largest eigenvalues are used as a set of discriminant weights in two linear discriminant equations. The separation of group-means and the scattergram of individual samples are demonstrated in a two-dimensional canonical plot.

In practical classification, linear discriminant functions are most frequently used (Jeon, 1991). A linear combination of variables is chosen to maximize the one-way analysis of variance F-test. This method requires that the multivariate normal distribution assumption holds and that the quality of the covariance matrices assumptions also holds. Nonparametric discriminant analysis is recommended when the distribution of the data is unknown. In our laboratory, nonparametric discriminant analysis yielded the highest rate of effectiveness in classification compared to linear and quadratic discriminant analysis in segregating canned salmon of different species, stages of sexual maturity, and quality levels (Girard, 1991).

Principal Component Similarity Analysis

For unsupervised classification, PCA is currently the most popular method. From the plots of combinations of two or three PC scores, the most suitable combination for the classification purposes is selected, although sample grouping may have been loosely defined. This approach is quite useful when only a small number of variables affect quality parameters.

We have introduced the concept of pattern similarity, which is frequently used for comparison of chromatographic patterns (Aishima et al. 1987), in PCA (Vodovotz et al., 1993). Principal component similarity (PCS) analysis is an extension of PCA, as all PC scores (excluding minor PC scores, making only a slight contribution to the total variation) are used in computation of the similarity constants of samples vs. reference. A scattergram is drawn by plotting slope vs. coefficient of determination (r^2) of the linear regression for deviation of sample PC scores from reference PC scores vs. proportion of variation accounted for by the PC scores.

PCS was useful when applied to HPLC data for discriminating irregular samples of Cheddar cheese from regular samples, as shown in Fig. 4.3 (Furtula et al. 1994a). While a similar aging route from mild to medium, then old, and finally extra-old are shown for the two regular aged sample sets, the irregular set aged inadvertently at a higher temperature remains within a small area near the medium zone of normally aged cheeses. Probably, the age difference was minimized due to different mechanisms of degradation of proteins at temperatures higher than those normally used for cheese ripening. Since the flavor of cheese consists of a combination of compounds rather than a small number of characteristic compounds, a PCS approach may be more effective in discriminating abnormal samples from normal samples than PCA. Compared to the scattergram obtained by PCA (Fig. 4.4), as the shift from C (old) to D (extra-old) is intermingled with the shift from B (medium) to C, the shift by aging is smoother in the case of PCS. This method was found useful for objectively judging the effects of accelerated cheese ripening (Furtula et al., 1994b). The same method was applied to the analytical data of different milk clotting enzymes with good classification results being superior to those of PCA (Nakai, 1992).

Therefore, for unsupervised classification or even in supervised cases

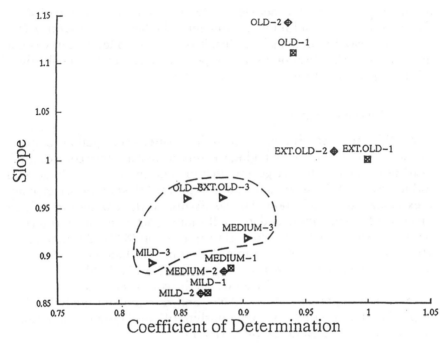

FIG. 4.3   PCS scattergram of three sets of Cheddar cheese samples. Sets 1 and 2 are regular aged samples, and set 3 is cheese samples aged under abnormal conditions (at higher aging temperatures).

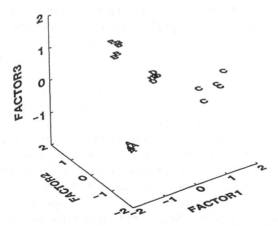

FIG. 4.4   Three-dimensional plot of PC (factor) scores for milk (A), medium (B), old (C), and extra-old (D) cheeses.

when sample grouping is not clearly defined, simultaneous use of PCA and PCS may be a useful tool in obtaining the best classification. After an unsupervised search such as this, it is recommended to carry out a more detailed classification using a supervised technique to finally identify the cause of defects.

Artificial Neural Networks

ANN is the most advanced approach in pattern recognition. Because of excellent flexibility by changing weights, as well as other control functions associated with a large number of neurons included in several hidden layers, ANN is best suited to complicated classification, e.g., complex nonlinear classification. To classify proteolytic enzymes into three groups of high, medium, and low milk-clotting activities, linear discriminant analysis yielded the correct classification of 78.6% (Aishima et al., 1987). When applied to the same data using six hidden layers, ANN improved the correct classification rate to 93% (T. Aishima, personal communication). A similar difference in the correct classification of 94% vs. 81% was reported between ANN and discriminant analysis when applied to the structure-odor relationships of nitrobenzenic compounds (Chastrette and de Saint Laumer, 1991).

Drawbacks of ANN are (1) arbitrary selection of the control factors, e.g., learning rate, smoothing factors, noise level to be introduced if necessary, and number of hidden layers and number of neurons in a layer, and (2) possible landing on local optimum. In general, to improve the prediction accuracy on the training set, the number of neurons in the hidden layers is to be increased. Alternatively, to improve generalization capabilities, thus improving performance on new cases, the size of the hidden layers is to be reduced. If too many hidden neurons are used, the ANN tends to perform excellently in the training set, however, when presented with new cases, its prediction ability may be deteriorated by overfitting. In this case, ANN does not learn to generalize, but it only memorizes the individual patterns. On the other hand, if too few hidden neurons are used, the prediction ability of the ANN may be inferior, making it useless by underfitting (Arteaga and Nakai, 1993).

An attempt was made recently in our laboratory to optimize the ANN protocol by applying random-centroid optimization (RCO) for selecting the control factors in ANN. It is possible that the use of RCO may effectively promote homing-in on the global optimum. A similar approach was reported using simplex optimization to minimize a cost function for determining the threshold values (internal values to activate a neuron triggering the output) of ANN (Richard et al., 1993).

FORMULA OPTIMIZATION

For formula optimization in food processing, the least cost formulation (LCF) based on linear programming is most popular in the food industry, especially the meat industry. However, as already discussed, the ingredient vs. quality parameters relations cannot always be expected to be linear. To circumvent this problem, the Complex (constrained simplex) of Box (1965) was successfully applied to meat formulation (Vazquez-Arteaga, 1990). Meanwhile, random-centroid optimization of Nakai (1990), which is appropriate to use for experimental optimization, rather than computational optimization as in the case of LCF, was modified for formula optimization by introducing a capacity to manipulate constraints (Dou et al., 1993).

Linear Programming

Linear programming is an optimization technique used to find a combination of independent variables, restricted by constraints, that gives the best value for the dependent variables. Linear relationships between the independent and dependent variables are required to be known *a priori*. Because of this property, linear programming has been best utilized for scheduling of transportation, labor control for working plans, etc., mainly for economic reasons. It is popular among nutritionists in menu formulation to find the most economical combination of ingredients without lowering the nutritional quality of prepared meals. The meat industry has been extremely active in utilizing linear programming for product formulation (Pearson and Tauber, 1984). Least cost formula for meat processing seeks to find the best blend of the ingredient meats to minimize costs while satisfying the constraint conditions, e.g., texture, fat content, lower limit of protein and upper limit of moisture, or color range. An important aspect of linear programming is the assumption that the objective functions should be linear, meaning that the relations are always simply additive.

A recent example of LCF is the work of Beausire et al. (1988) on the formulation of a low-cost, fresh turkey bratwurst, a coarse ground–type sausage. An acceptability constraint, in addition to the contents of protein, fat, and moisture, was used to attain the maximum consumer acceptability of the product. An advantage of the LCF is accurate control of ingredient costs while maintaining uniform composition and quality (Pearson and Tauber, 1984). However, the LCF places excessive emphasis on cost reduction and frequently sacrifices the product quality, since the quality parameters are used as constraints and not as objective functions. As Fishken (1983) stated, the product acceptability is nonlinearly related to ingredient levels, and, therefore, linear programming cannot properly accommodate these relations required for achieving good product acceptability.

Constrained Simplex Optimization

After the introduction of one of the most successful, unconstrained direct-search algorithms by Spendley et al. (1962), Box (1965) modified this simplex optimization algorithm to accommodate constraints and the method was called "Complex" (constrained simplex). Unlike linear programming, working equations, both linear and nonlinear, to compute response values can be included in subprograms separated from the main programs for optimization algorithm. The major strategy of Complex for avoiding the violation of implicit constraints is to contract the vertex by moving halfway towards the centroid. This contraction is repeated until all of the implicit constraints are satisfied. Moskowitz and Jacobs (1987) used the Complex to find pie crust formulas with specific sensory profiles.

The original Complex program of Box (1965) was modified to "Forplex" to find a formulation with the best quality while meeting constraints for composition and cost of ingredients (Vazquez-Arteaga, 1990). This is contrary to the LCF for minimizing the ingredient cost while maintaining bind constants representing quality within allowable ranges. As a model for assessing the optimization capacity, equations to predict quality parameters based on the ingredient meats as derived from frankfurter-formulation experiments was used. This approach is a practical replacement of the prediction equations derived from QSAR study. The QSAR equations, if available, should be more reliable and versatile as working equations for formula optimization.

A model optimization was conducted to formulate a frankfurter from pork fat, mechanically deboned chicken, and beef. Three to four quality parameters were used as a multiobjective function with respective target values, and one quality parameter was also used as a constraint. The predicted quality values for the formulation optimized by the Forplex program were in reasonable agreement with the preset target values. Four LCFs based on linear programming (LP) were computed to meet the request for increasingly higher total bind values for the product, which increased the ingredient costs. Although the ingredient cost was substantially lower in LP1 ($1.89/kg) and LP2 ($1.95/kg), those of LP3 ($2.07/kg) and LP4 ($2.23/kg) are not much different from that of For-

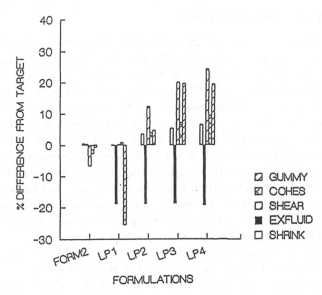

FIG. 4.5 Differences between specified target quality values and the predicted quality values of FORM2 and the least-cost formulations. The quality parameters were gumminess, cohesiveness, and shear values obtained by Texture Profile Analysis of Instron data and % expressible fluid and % shrink measured by weight loss after processing.

plex (FORM2, $2.13/kg). As shown in Fig. 4.5, the deviations from the target quality values are much greater in LCF (LP1 to LP4) than in Forplex (FORM2).

An advantage of Forplex is its capacity to accommodate any form of prediction equations either as objective functions or constraints. Unlike LP, Forplex can accommodate the antagonistic or synergistic interactions of ingredients which are expressed as quadratic and product terms in regression equations (Arteaga et al., 1993). Forplex can optimize many quality parameters simultaneously in the form of multiobjective function searching for the best compromise, which linear programming cannot manipulate. Based on the above comparison, it may be true that Forplex is preferable to LCF for the purpose of routine process control for food formulation. It was found that Forplex could avoid producing products of low quality due to pursuing the least costs for ingredients using linear programming.

Random-Centroid Optimization

Random-centroid optimization (RCO) was recommended for experimental optimization (Nakai, 1990). The main reason for introducing RCO into food research and development, thereby replacing the simplex optimization of Nakai et al. (1984), was the difficulty of simplex optimization in achieving the global optimum. Once inappropriate search spaces are chosen for the initial simplex, the global optimum will never be reached by simplex optimization. When it is not evident that the global optimum has been reached, it is recommended to restart optimization from several different, widely distributed initial points, and if all or most lead to a single solution, then that point is considered to be the global optimum (Saguy, 1983). However, this replication of simplex search immediately increases the number of experiments to reach the optimum, thus deteriorating the optimization efficiency.

According to Schwefel (1981), the most reliable global search method is the grid method using a network search plan. However, this approach is extremely inefficient, because the number of experiments for optimization increases exponentially as the number of factors increases. Random search is a risky alternative of the grid search and may also be inefficient, as it is totally reliant on luck in selection of experimental conditions. We decided to use a regulated random search, which is an intermediate between full random and grid search (Nakai, 1990). Incorporation of the centroid search (Aishima and Nakai, 1986) and a mapping process (Nakai et al., 1984) into the random search plan significantly improved the optimization efficiency.

The random-centroid optimization of Nakai (1990) was further modified by introducing a penalty function to accommodate constraints in formula optimization (Dou et al., 1993). Quality parameters not significantly different ($p > 0.05$) from those obtained by the Forplex were computed by RCO and yielded similar ingredient compositions to those computed by Forplex as seen in Table 4.3. RCO is therefore appropriate for use in experimental optimization such as cases of research and development, while Forplex is suitable for computational optimization such as routine formulation due to simplicity in the computer operation.

DISCUSSION

The application of QSAR to protein seems to have gained general recognition. However, much remains to be pursued before rules or guidelines to define useful parameters and mathematical algorithms for use in QSAR of food proteins can be established. Because of large molecular

TABLE 4.3 Comparison of FORPLEX and Random Centroid Optimization for Frankfurter Formula Optimization Trials Where a Quality Parameter Was Considered as a Constraint

Quality parameter	Target value	FORPLEX Optimum ingredients[a]	Response value[b]	Predicted quality	Random-centroid Optimum ingredients[a]	Response value[b]	Predicted quality
SHRINK	8.7	$X_1 = 0.223$	372.8	9.1	$X_1 = 0.238$	287.0	9.03
SHEAR	4.8	$X_2 = 0.198$		4.78	$X_2 = 0.180$		4.68
COHES	0.255	$X_3 = 0.579$		0.255	$X_3 = 0.581$		0.253
GUMMY	40.0			39.8			39.15
EXWATER	6.0	$X_1 = 0.233$	127.8	5.49	$X_1 = 0.231$	125.3	5.7
EXFAT	4.5	$X_2 = 0.164$		4.53	$X_2 = 0.149$		4.49
HARD1	160	$X_3 = 0.602$		160.6	$X_3 = 0.620$		167.2
COHES	0.255			0.255			0.255
SHRINK	9.0	$X_1 = 0.230$	33.35	9.06	$X_1 = 0.238$	64.6	9.14
EXFAT	4.5	$X_2 = 0.194$		4.5	$X_2 = 0.155$		4.59
COHES	0.255	$X_3 = 0.576$		0.254	$X_3 = 0.606$		0.254

Quality parameters used as constraints are: model 1: EXFLUID 9.50–10.50; model 2: TWLOSS 22.00–24.00; model 3: HARD1: 150.0–180.0.

[a]X_1, skinless pork fat; X_2, frozen mechanically deboned poultry meat; X_3, lean beef.

[b]Response value calculated as the product of the absolute values of standardized differences between target and predicted quality.

Source: Dou et al., 1993.

size in addition to the general structural properties, the distributions of charge, hydrophobicity, and other characteristic functional groups must be taken into consideration. Application of QSAR to food proteins is complicated by the unavailability of full information on molecular structure. However, from a practical point of view, a simpler approach is possible, as in the case of LCF and Forplex, using regression equations to correlate ingredient proteins directly to their functionality without considering basic physical/chemical properties of the proteins.

Despite the lack of experience due to insufficient opportunities of trial and comparison with other techniques, ANN has already been shown to outperform almost all other multivariate analytical techniques both for computation of relation and for classification. For quality control purposes, the prediction of quality using ANN may be again the most powerful approach. However, for classification or investigation of unknown cause of problems, i.e., cases in which supervised training is not possible since available information is critically limited, a variety of principal component analyses or factor analyses are preferable.

For formula optimization, the methods to search for the highest quality within allowable cost constraint are recommended to replace the currently popular least cost formulation based on linear programming. In the case of limited information on product quality and processing conditions, a fuzzy logic–based modeling and optimization technique was recently proposed for routine quality control (Qian et al., 1993). This approach is especially effective when empirical knowledge of processes has been accumulated through years of practical operation and represented by a set of imprecise and empirical equations. However, a new sequential optimization technique, i.e., random-centroid optimization, should be useful also in this case, especially for research and development when the experience on formula and processing conditions are restricted.

Few papers have been published on the application of ANN to optimization, probably because ANN is a data-processing technique. Most of the papers published so far have used data to construct accurate training sets for ANN, thus being eventually the optimization of prediction capacity of ANN. Therefore, more reliable ANN prediction can replace working equations being employed in the optimization of routine processing using computational techniques. However, in the case of sequential (or experimental) optimization for research and development with almost no or only few data available, it may not be realistic to expect ANN to replace the entire process of optimization. Application of computational neural networks to draw a 3-D response surface designed to model QSAR predictions was discussed (Maggiora et al., 1992). It may be possible to improve the accuracy of the mapping process in RCO, which is an approximation of the response surface, thereby improving the potential optimization efficiency using ANN.

A newly emerging technique for optimization, namely biological algorithms, that claims a superior capacity in finding the global optimum, requires special attention (Lucasius and Kateman, 1993).

CONCLUSION

For prediction, classification, and optimization as the final objectives of QSAR study of food proteins, the rapid developments in multivariate analysis techniques should be carefully watched. It may be interesting to note the recent explosive increase in the number of ANN papers, especially in QSAR study; many of these papers have reported that ANN outperforms the traditional statistical methods. Selection of a technique for any specific purpose should be made by careful comparison of all techniques available at the moment. Some compromise may be required in selecting a technique, weighing the benefits of increasingly complicated computer-aided techniques with the need for simplification of the techniques to apply to food processing.

ACKNOWLEDGEMENT

This work was supported by a Research Grant from the Natural Science and Engineering Research Council of Canada.

REFERENCES

Aishima, T. and Nakai, S. 1986. Centroid mapping optimization: A new efficient optimization for food research and processing. *J. Food Sci.* 51: 1297–1300, 1320.

Aishima, T. and Nakai, S. 1991. Chemometrics in flavor research. *Food Res. Int.* 7: 33–101.

Aishima, T., Wilson, D. L., and Nakai, S. 1987. Application of simplex algorithm to flavour optimization based on pattern similarity of GC profile. In *Flavour Science and Technology*, M. Martens, G. A. Dallen, and H. Russwurm (Ed.), pp. 501–508. John Wiley & Sons, New York.

Arteaga, G. E. and Nakai, S. 1993. Prediction of protein functionality using artificial neural networks: Foaming and emulsifying properties. *J. Food Sci.* 58: 1152–1156.

Arteaga, G. E., Li-Chan, E., Nakai, S., Cofrades, S., and Jimenez-Colmenero, F. 1993. Ingredient interaction effects on protein functionality: Mixture design approach. *J. Food Sci.* 58: 656–662.

Banks, J. M., Brechany, E. Y., Christie, W. W., Hunter, E. A., and Muir, D. D. 1993. Volatile components in steam distillates of Cheddar cheese as indicator indices of cheese maturity, flavour and odour. *Food Res. Int.* 25: 365–373.

Beausire, R. L. W., Norback, J. P., and Maurer, A. J. 1988. Development of an acceptability constraint for a linear programming model in food formulation. *J. Sensory Stud.* 3: 137–149.

Box, M. J. 1965. A new method of constrained optimization and a comparison with other methods. *Computer J.* 8: 42–52.

Britten, M. and Lavoie, L. 1992. Foaming properties of proteins as affected by concentration. *J. Food Sci.* 57: 1219–1222, 1241.

Charton, M. 1986. Quantitative description of side chain effects on binding to protein. *Int. J. Peptide Protein Res.* 28: 201–207.

Charton, M. 1990. The quantitation of protein bioactivity. Phage T4 lysozymes substituted at residue 86. *Collect. Czech. Chem. Commun.* 55: 273–281.

Chastrette, M. and de Saint Laumer, J. Y. 1991. Structure-odor relationships using neural networks. *Eur. J. Med. Chem.* 26: 829–833.

De Kanterewicz, R. J., Elizalde, B. E., Pilosof, A. M. R., and Bartholomai, G. B. 1987. Water-oil absorption index (WOAI): A simple method for predicting the emulsifying capacity of food proteins. *J. Food Sci.* 52: 1381–1383.

Dou, J., Toma, S., and Nakai, S. 1993. Random-centroid optimization for food formulation. *Food Res. Int.* 26: 27–37.

Fishken, D. 1983. Consumer-oriented product optimization. *Food Technol.* 37(11): 49–52.

Furtula, V., Nakai, S., Amantea, G. F., and Laleye, L. 1994a. Assessment of accelerated cheese ripening by reverse-phase HPLC. I. Analysis of reference Cheddar cheese samples. *J. Food Sci.* 59: accepted.

Furtula, V., Nakai, S., Amantea, G. F., and Laleye, L. 1994b. II. Analysis of Cheddar cheese samples aged by a fast-ripening process. *J. Food Sci.* 59: accepted.

Geladi, P. 1988. Notes on the history and nature of partial least squares (PLS) modelling. *J. Chemometr.* 2: 231–246.

Girard, B. 1991. Headspace gas chromatography for quality assessment of canned Pacific salmon. Ph.D. thesis, University of British Columbia.

Hansch, C. and Klein, T. E. 1991. Quantitative structure-activity relationships and molecular graphics in evaluation of enzyme-ligand interactions. *Methods Enzym.* 202: 512–543.

Hasegawa, K., Miyashita, Y., and Sasaki, S. 1992. Quantitative structure-activity relationship study of antiarrhythmic phenylpyridines using multivariate partial least squares modelling. *Chemometr. Intel. Lab. Systems* 16: 69–75.

Hayakawa, S. and Nakai, S. 1985. Contribution of hydrophobicity, net charge and sulfhydryl groups to thermal properties of ovalbumin. *Can. Inst. Food Sci. Technol. J.* 18: 290–295.

Hoskuldsson, A. 1988. PLS regression methods. *J. Chemometr.* 2: 211–228.

Jeon, I. J. 1991. Pattern recognition techniques for food research and quality assurance. Ch. 12. In *Instrumental Methods for Quality Assurance in Foods*, D. Y. C. Fung and R. F. Matthews (Ed.), pp. 271–298. Marcel Dekker, New York.

Kato, A. and Nakai, S. 1980. Hydrophobicity determined by a fluorescence probe method and its correlation with surface properties of proteins. *Biochim. Biophys. Acta* 624: 13–20.

Klein, P., Jacquez, I. A., and Delis, C. 1986. Prediction of protein function by discriminant analysis. *Math. Biosci.* 81: 177–189.

Kohnhorst, A. L. and Mangino, M. E. 1985. Prediction of the strength of whey protein gels based on composition. *J. Food Sci.* 50: 1403–1405.

Li-Chan, E., Nakai, S., and Wood, D. F. 1984. Hydrophobicity and solubility of meat proteins and their relationship to emulsifying properties. *J. Food Sci.* 49: 345–350.

Li-Chan, E., Nakai, S., and Wood, D. F. 1987. Muscle protein structure-function relationships and discrimination of functionality by multivariate analysis. *J. Food Sci.* 52: 31–41.

Lucasius, C. B. and Kateman, G. 1993. Understanding and using genetic algorithms. part 1. Concepts, properties and context. *Chemometr. Intel. Lab. Systems* 19: 1–33.

Maggiora, G. M., Eirod, D. W., and Trenary, R. G. 1992. Computational neural networks as model-free mapping devices. *J. Chem. Inf. Comput. Sci.* 32: 732–741.

Martens, M. and Martens, H. 1986. Partial least squares regression. Ch. 13. In *Statistical Procedures in Food Research*, J. R. Piggott (Ed.), pp. 293–359. Elsevier Applied Science, New York.

McAvoy, T. J., Su, H. T., Wang, N. S., He, M., Horvath, J., and Semerjian, H. 1992. A comparison of neural networks and partial least squares for deconvoluting fluorescence spectra. *Biotech. Bioeng.* 40: 53–62.

Moskowitz, H. R. and Jacobs, B. E. 1987. Consumer evaluation and optimization of food texture. Ch. 12. In *Food Texture. Instrumental and Sensory Measurement*, H. R. Moskowitz (Ed.), pp. 293–328. Marcel Dekker, New York.

Nakai, S. 1990. Computer-aided optimization with potential application in biorheology. *J. Jap. Soc. Biorheology* 4: 143–152.

Nakai, S. 1992. Importance of protein functionality in improving food quality: Role of hydrophobic interaction. *Comments Agric. Food Chem.* 2: 339–387.

Nakai, S. and Li-Chan, E. 1993. Recent advances in structure and function of food proteins: QSAR approach. *Crit. Rev. Food Sci. Nutr.* 33: 477–499.

Nakai, S., Koide. K., and Euguster, K. 1984. A new mapping super-simplex optimization for food product and process development. *J. Food Sci.* 49: 1143–1148.

Pearce, K. H. and Kinsella, J. E. 1978. Emulsifying properties of proteins: Evaluation of a turbidimetric technique. *J. Agric. Food Chem.* 26: 716–723.

Pearson, A. M. and Tauber, F. W. (Ed.) 1984. Least-cost formulation and preblending of sausage. Ch. 8. In *Processed Meats*, 2nd ed., pp. 158–186. AVI Publishing, Westport, CT.

Pike, D. J. 1986. A practical approach to regression. Ch. 3. In *Statistical Procedures in Food Research*, J. R. Piggott (Ed.), pp. 101–123. Elsevier Applied Science, New York.

Qian, Y., Tessier, P., and Dumont, G. A. 1993. Process modeling and optimization of systems with imprecise and conflicting equations. *Eng. Appl. Artif. Intell.* 6: 39–47.

Richard, D., Cachet, C., and Cabrol-Bass, D. 1993. Neural network approach to structural feature recognition from infrared spectra. *J. Chem. Inf. Comput. Sci.* 33: 202–210.

Saguy, I. 1983. Optimization methods and application. Ch. 10. In *Computer-Aided Techniques in Food Technology*, I. Saguy (Ed.) pp. 263–320. Marcel Dekker, New York.

Schwefel, H.-P. 1981. Random strategies. Ch. 4, In *Neumerical Optimization of Computer Models*, H.-P. Schwefel (Ed.), pp. 87–103. John Wiley & Sons, New York.

Shanbhag, V. P. and Axelsson, G. -G. 1975. Hydrophobic interaction determined by partition in aqueous two-phase systems. *Eur. J. Biochem.* 60: 17–22.

Spendley, W., Hext, C. R., and Himsworth, F. R. 1962. Sequential application of simplex design in optimization and evolutionary operation. *Technometrics* 4: 441–461.

Stuper, A. J., Brugger, W. E., and Jurs, P. C. (Ed.) 1979. *Computer Assisted Studies of Chemical Structure and Biological Function.* John Wiley & Sons, New York.

Townsend, A. -A. and Nakai, S. 1983. Relationships between hydrophobicity and foaming characteristics of food proteins. *J. Food Sci.* 48: 588–594.

Vazquez-Arteaga, M. C. 1990. Computer-aided formula optimization. M.Sc. thesis, University of British Columbia.

Vodovotz, Y., Arteaga, G. E., and Nakai, S. 1993. Principal component similarity analysis for classification and its application to GC data of mango. *Food Res. Int.* 26: 355–363.

Voutsinas, L. P. and Nakai, S. 1983. A simple turbidimetric method for determining the fat binding capacity of proteins. *J. Agric. Food Chem.* 48: 58–63.

Voutsinas, L. P., Cheung, E., and Nakai, S. 1983. Relationships of hydrophobicity to emulsifying properties of heat denatured proteins. *J. Food Sci.* 48: 26–32.

5
Protein Interactions in Emulsions: Protein–Lipid Interactions

Michael E. Mangino

The Ohio State University
Columbus, Ohio

INTRODUCTION

Butter, cream, ice cream, milk, mayonnaise, and salad dressing are representative of food products that contain both water and fat in the form of emulsions. Some formulated meat products have from time to time been referred to as emulsions, but are more properly excluded for this classification. Some true emulsions tend to be solid or semi-solid and are quite different from products that are liquid. Such solid emulsions will not be discussed in this chapter. Fluid emulsions are thermodynamically unstable mixtures of immiscible liquids such as vegetable oil and water. Their formation requires the application of energy. When energy is applied to a mixture of water and oil, the phases may be dispersed, but increased surface energy causes rapid coalescence unless an energy barrier is established. A molecule that contains a moiety that is soluble in water and another that is soluble in nonpolar solvents is termed amphiphilic. Immiscible liquids may be stabilized against coalescence by the addition of these types of molecules, generally referred to as emulsifiers. Proteins are capable of coating lipid droplets and can provide an energy barrier to both particle association and phase separation. The number of component interactions possible makes the study of emulsion systems complex.

Conditions that are important in dilute solutions, where the systems are easier to study, may not apply to conditions likely to be found in foods, increasing the difficulty of relevant research. Nevertheless, it is possible to explain much of how proteins function in emulsions from a knowledge of the forces that are operational during emulsion formation and that govern protein structure.

Some food emulsions, such as liquid infant formula, are generally expected to remain stable for months or even years. It is important, therefore, that the mechanisms and the nature of the forces responsible for emulsion stability or breakdown be understood. This chapter will review the forces involved in emulsion formation and relate these to the forces that govern protein-lipid interactions. The mechanism of emulsion destabilization and the forces involved are discussed. The recent text by Larson and Friberg (1990) contains a number of excellent reviews of emulsions and is highly recommended for those seeking more detail on a number of the topics covered in this chapter.

INTERACTIONS

A discussion of the interactions of water molecules with themselves and with other molecules is necessary before a discussion of emulsions can be attempted. Water is characterized by a large permanent dipole moment. The structure of a single water molecule is shown in Fig. 5.1A. The angle of the hydrogen-oxygen bonds is such that each of the two lone-pair electrons associated with the oxygen is available to interact with a hydrogen associated with another water molecule. In the normal state of ice, each water is involved in four hydrogen bonds, as shown in Fig. 5.1B. The structure in liquid water is not known with certainty, but appears to contain approximately half the number of hydrogen bonds found in ice. Many of these bonds are strained, and the structure undergoes continual rearrangement (Eisenberg and Kauzman, 1969). The formation of water-water hydrogen bonds should result in a favorable negative enthalpy. The formation of such interactions, however, imposes order on the system and results in an unfavorable decrease in entropy. Lumry and Rajendar (1970) have suggested that these effects are approximately equal and compensate for each other. The interactions of

A

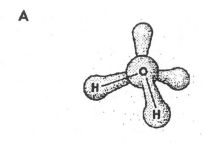

B

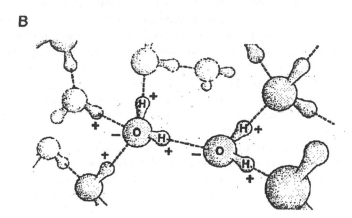

FIG. 5.1   Structure of water. (A) Asymmetric bond angles between hydrogen and oxygen and the lobes of negative charge due to the lone pair oxygen electrons. (B) Tetrahedral arrangement of water in ice. Dashed lines indicate hydrogen bonds. (*Adapted from Starr and Taggart, 1989.*)

water with charged molecules result in hydration energies from -50 to -200 kcal/mole and are strongly favored.

It has been hypothesized that for a molecule to dissolve in water, a cavity in the structure of water must be created. The energy required to create such a cavity is approximately equal to the strength of the water-water hydrogen bonds that must be broken to accommodate the molecule. This can only occur if the interactions between water and the inserted molecule are more favorable than the interactions of water molecules with themselves. The hydration of charged particles is an example of such a case. For nonpolar molecules, the free energy of transfer from a nonpolar solvent to water is positive. Such a transfer has been shown to result in a negative enthalpy. This is likely the result of an increase in the hydrogen bonding of water with other water molecules. The increase in hydrogen bonding imparts more structure to the system and results in a decrease in entropy. The decrease in entropy is greater than the decrease in enthalpy, resulting in a positive free energy.

Ordered structures of water called clathrates have been observed to surround inert gases dispersed in water. These water "cages" serve to minimize the contact of water with the nonpolar molecules and thus minimize the number of water molecules with perturbed hydrogen bonding. Water associated with most nonpolar molecules is probably less ordered than that found in the clathrates observed with noble gasses. The highly unfavorable interactions of nonpolar molecules with water causes them to associate with themselves in an attempt to minimize their contact with water. Such associations are known as hydrophobic interactions. These molecules will interact with one another through transient dipole-induced dipole interactions. These van der Waals forces lower the energy of the system. The accompanying gain in energy is inadequate to compensate for the loss of hydrogen bonding of water required for the creation of the initial cavity. The driving force for these interactions is not, however, derived from favorable van der Waals associations. Rather, it is due to an attempt by the water molecules to minimize the unfavorable interactions with the nonpolar substance. Thus, the addition of nonpolar molecules to water results in a situation where the area of contact between water and the nonpolar molecules is minimized (Tanford, 1980).

Energy considerations require that nonpolar molecules in water associate with each other as much as possible to minimize their contact with water. The formation of these hydrophobic associations is favored by an increase in entropy of the associated state when compared to the nonassociated state. Such interactions should become more favorable as the temperature of the mixture increases. This is true to a point, and the energy of hydrophobic associations is generally considered to be maximal at about 60°C (Tanford, 1980). Further increase in temperature causes a weakening of hydrophobic associations. Changes in the structure of water and the nature of its hydrogen bonding at higher temperatures are the likely cause for the decrease in the strength of hydrophobic associations at elevated temperatures.

PROTEINS

Proteins contain both polar and nonpolar amino acids. One of the ways that proteins minimize their energy is by folding into structures of low free energy. These structures generally result when the interactions of polar groups with water are maximized and the interactions of nonpolar groups with water are minimized. Native proteins exist in structures that represent the lowest kinetically attainable state of free energy (Anfinsen, 1973). This generally results in removal of hydrophobic groups from the aqueous environment. While individual hydrophobic associations are weak and their strength is strongly dependent on the nature of the solvent (Tanford, 1961), they are present in proteins in large numbers. The presence of molecules or conditions that affect the strength of hydrophobic associations can have a large effect on protein functionality.

The conformation of lowest free energy is strongly dependent upon the composition of the solvent. Different conformations are assumed under different environmental conditions. Factors that can cause changes in protein conformation include pH, temperature, dielectric constant, ionic strength, and the presence of other components including gas, liquid, and other proteins. (Mangino, 1984).

The native conformations of proteins are dynamic structures (Karplus and McCammon, 1983). There is rotational freedom about many of the bonds within the protein molecule, and the entropy gain of this freedom lowers the total free energy of the native structure. Many portions of the molecule are stabilized by relatively weak secondary interactions, and only small inputs of energy are necessary for the molecule to assume slightly different conformations. These alternate conformations generally lead to structures of higher free energy and thus are not stable or long lived. If the energy differences are slight, a number of similar conformations will coexist. A protein may be envisioned as a dynamic entity that is constantly sampling a variety of structures. These new structures are usually only slightly different from the native conformation and almost always lead to a situation where the free energy of the system increases. The increase in free energy causes the protein to spontaneously refold into the state of lowest free energy. Thus, the native structure of a protein is not the only structure it can assume, but rather the one of lowest free energy and hence of greatest probability. Slight changes in the environment can cause alternate structures to be of lowest free energy and thus lead to conformational changes.

EMULSION FORMATION

Water and oil are immiscible because of the unfavorable increase in energy that occurs when the structure of water is perturbed by contact with nonpolar substances. The energy of this system is minimal when the area of contact between the two immiscible liquids is minimized. This is usually manifested by phase separation. The two liquids may be brought into intimate contact with the application of a significant amount of energy. The energy applied must be greater than the surface energy of contact resulting from their mixing. The interfacial area of the emulsion depends on the amount of work done on the system. Smaller droplets result in a larger surface area than do larger droplets and will require a greater input of work. As the work applied increases, the droplet size becomes smaller. Creation of new interfacial area results in a high energy state and the system attempts to reach a lower energy state by coalescence of fat globules. The rate of droplet collision and the presence or absence of an energy barrier will determine the rate of fat globule coalescence. For uncoated fat globules the energy barrier to coalescence is so small that it can be ignored.

The critical factor in emulsion technology depends upon the application of emulsifying agents, which function to overcome the unfavorable energy situation associated with finely dispersed oil droplets in water. Emulsifying agents are amphiphilic molecules for which a portion of the molecule is soluble in water while the other portion is soluble in nonpolar solvents. Phospholipids are examples of naturally occurring amphiphilic molecules. Such molecules can serve as emulsifiers. When phospholipids are added to water, a micelle forms with the fatty acid tail groups oriented away from and the charged head groups in contact with the aqueous phase. This structure tends to maximize contact of the polar groups and minimize the contact of the hydrophobic groups with water. The approach of a phospholipid micelle to a lipid droplet causes the structure to reorient in an attempt to prevent the dehydration of the charged phospholipid head groups. The removal of the charged moiety of the molecule from the aqueous phase would require either that water-water hydrogen bonds or water-ion interactions be broken. This is a high-energy state and results in the charged portion of the molecules being repelled from the interface. This can cause the micelle to reorient, resulting in exposure of the fatty acid tail portions of the molecules. As these come into contact with the lipid phase, the fatty acid portions of the molecules orient into the lipid, while the charged head groups remain in contact with the water. This results in the formation of a monolayer of phospholipid molecules at the surface of the droplet and a considerable decrease in surface energy. The structure of a phospholipid micelle is shown schematically in Fig. 5.2. For lipid in water with no emulsifier present, a rapid coalescence of fat globules occurs and phase separation follows. If emulsifier molecules are present, they diffuse to the fat lipid interface as coalescence is occurring (Mangino, 1984).

Instantaneous coating of the lipid droplets with emulsifier would re-

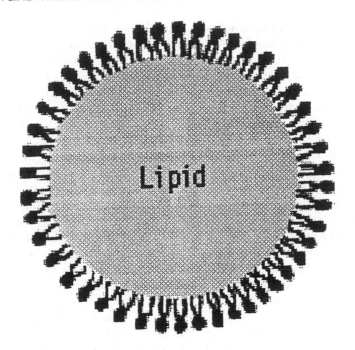

FIG. 5.2 Arrangement of phospholipids around a lipid droplet in an aqueous medium.

sult in an emulsion containing particles having the same size distribution as they did at the moment of emulsification. Emulsifier molecules do not instantaneously coat the droplet surface, but require a finite time to diffuse to the interface and be adsorbed. The rate of droplet coating by phospholipid is a complex function of the rate of fat droplet coalescence, the rate at which the phospholipid molecules reach the lipid surface and the rate at which the micelles are able to reorient. Small emulsifier molecules are able to diffuse rapidly compared to proteins and the amount of reorientation required to interact in the surface is small. In general, the use of small emulsifier molecules results in a relatively narrow particle size distribution and in the formation of emulsions with relatively small fat globules (Darling and Birkett, 1987).

PROTEINS AT INTERFACES

Many proteins can be added to emulsions to aid in their formation and to increase their stability. Being amphiphilic, proteins will tend to orient at the polar-nonpolar interface. Proteins are large when compared to

molecules such as phospholipids, and the forces involved in their unfolding are more complex. As with any emulsifier, the protein molecule
must reach the interface to exhibit functionality. In native proteins most
of the nonpolar amino acid side chains are located in the interior of
the molecules. It has been estimated that the removal of one mole of
hydrophobic groups from the surface results in an energy gain of 12 KJ
(Kinsella, 1982). Any hydrophobic groups that remain at the surface
increase the total energy of the system. Thus, in the typical case, the
surface of soluble proteins is represented by charged groups, which
remain in contact with water molecules. The favorable interaction of
water with surface charge lowers the total energy of the protein molecule.
In some respects the protein may be envisioned as resembling phospholipid micelles with their hydrophobic groups removed from contact with
water and their charged amino acids located on the surface to maximize
contact with the aqueous phase. The charged groups on the surface of
a protein resist removal from the aqueous phase. As insertion of hydrated ions into the nonpolar medium is energetically unfavorable, these
groups are repelled from the interface. If the groups nearing the interface are in a region of the protein molecule that is flexible, the molecule
may begin to unfold with exposure of hydrophobic groups to the surface.
Exposure of the groups to the aqueous environment results in a high
energy state and are resisted. Limited exposure of hydrophobic groups
is possible and results in an increase in total energy of the system. In
time, random fluctuations in protein structure cause these groups to
return to the interior of the molecule. When protein unfolding occurs
at a water-lipid interface, the hydrophobic groups will be inserted into
the lipid phase. Insertion of hydrophobic amino acids into the lipid has
a low energy of activation and proceeds spontaneously (Tanford, 1970).
However, a hydrophobic residue can only enter the lipid phase if there
are no charged groups in close proximity. It has been reported that a
minimum of from six to eight adjacent hydrophobic amino acids are
required for insertion of a hydrophobic region into a nonpolar environment (MacRitchie, 1987). Removal of the hydrophobic water of hydration results in a positive enthalpy (MacRitchie, 1978), Therefore, the
driving force for the protein-lipid interaction must be an increase in the
entropy of the system. This increased entropy consists of two components, one due to the conformational entropy of the protein and one
due to the change in structure of water near hydrophobic groups. There
are more ways that the hydrophobic groups of the protein can be arranged in a lipid environment than were possible in the native protein
structure. This results in an increase in the conformational entropy of
the protein. The solvent molecules at the interface are arranged in highly

ordered structures. The approach of the protein with the insertion of hydrophobic groups into the oil phase will, in essence, coat the nonpolar material and will allow for the release of solvent from the surface. The release of this water is responsible for a significant increase in the entropy of the system.

MacRitchie (1987) has studied the adsorption of proteins at interfaces in considerable detail and has determined that the energy of activation for insertion of a hydrophobic area into the lipid phase is minimal. The reaction is not, however, readily reversible. Proteins have a number of charged groups at the surface and in time other sections of the protein molecule will approach the surface. As long as these charged groups are in areas of the protein sufficiently flexible to unfold, additional hydrophobic regions will become inserted into the lipid phase. As this continues there is a steady progression in protein unfolding at the interface.

The removal of a hydrophobic group from the lipid phase would bring it into contact with water and result in an increase in total energy. In practice, this high energy state could be fairly easily accommodated with a slight rearrangement of the protein molecule. For proteins attached at more than one hydrophobic site, desorption occurs very slowly, if at all. With multiple points of attachment, it becomes difficult to have proper protein refolding upon desorption. Even if a given site can desorb and be refolded, there are still many points of attachment. By the time another site is able to desorb, it is likely that the first site will become reattached. Thus, the probability of reversibility becomes a function of the number of points of attachment and thus of protein flexibility.

Significant desorption can only occur if there are enough additionally surface-active, hydrophobic moieties present to coat the surface exposed upon desorption of the protein. This will inhibit resorption of the protein sites. As will be seen later, this can be significant with small emulsifier molecules. In the case of proteins, it has been shown that relatively inflexible proteins such as gelatin can be displaced by more hydrophobic and flexible casein molecules from the water-lipid interface (Mussellwhite, 1966). Once a layer of protein has been adsorbed additional protein cannot be added in the same manner because an energy barrier to adsorption exists (Phillips, 1981)). In order for more protein to be adsorbed, the protein already at the surface must be compressed to make room. The amount of compression that is possible depends on the surface charge and rigidity of the protein. Extensive compression leads to the situation where the adsorption of more protein will require more energy than can be gained by the insertion of hydrophobic groups into the lipid layer. Further interaction involves the interaction of protein molecules in the bulk phase with those already adsorbed to the lipid and the formation of multilayers.

The insertion of the first hydrophobic group at the interface represents adsorption, while the continued unfolding is known as spreading. The extent and rate of protein spreading will depend on both the hydrophobicity and flexibility of the protein. Graham and Phillips (1979b) have studied the time course of protein adsorption and the increase in surface pressure that accompanies spreading. A typical result for a flexible molecule such as β-casein is shown in Fig. 5.3. It should be noted that the time course of protein adsorption and increased surface pressure are essentially identical. There is a rapid initial increase in surface concentration. This is followed by a more gradual increase and finally a plateau. The initial rapid increase in surface protein concentration results from the adsorption of the first molecules to arrive when there is essentially no barrier to adsorption. As the surface is nearly covered, the adsorption of new molecules is slower. The changes in surface pressure parallel those for surface concentration, suggesting that as the molecules are adsorbed at the interface they are capable of exhibiting their full effect on interfacial tension. In the case of β-casein, this very flexible molecule (Swaisgood, 1982) is probably adsorbed in the unfolded state. A similar curve for β-lactoglobulin is presented in Fig. 5.4. In this case, it should be noted that there is a considerable lag between the adsorption of the protein and the increase in surface pressure. In fact, most of the increase in surface pressure occurs only after protein adsorption has reached a maximum level. While a rapid increase in surface concentration with time is observed (similar to that noted with β-casein), the rate of change in interfacial pressure shows a lag and continues to decrease with time long after adsorption of new protein molecules has ceased. This may be interpreted as being due to the unfolding of structured areas of the proteins already adsorbed at the interface (Stainsby, 1985). It is likely that β-lactoglobulin is adsorbed as a compact globular molecule possessing considerable tertiary structure. Given time, more hydrophobic groups within the molecule are able to come into contact with the interface and become adsorbed. Thus, there is a further increase in surface pressure with no change in surface concentration of the protein.

The adsorption of pure proteins to lipid-water interfaces has been studied extensively. Data for the surface concentration for a number of food proteins often gives values with units of milligrams of protein adsorbed per meter of interfacial area (Dickinson and Iveson, 1993; Phillips, 1981; Graham and Phillips, 1979a). Thus, complete surface coverage requires comparatively small amounts of protein. In typical

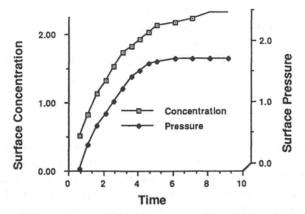

FIG. 5.3 Plot of surface concentration (arbitrary units) and surface pressure (arbitrary units) vs. time (arbitrary units) for β-casein. (*Adapted from Phillips, 1981.*)

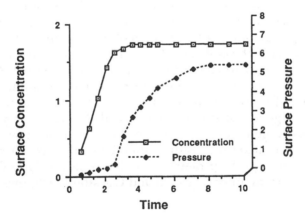

FIG. 5.4 Plot of surface concentration (arbitrary units) and surface pressure (arbitrary units) vs. time (arbitrary units) for β-lactoglobulin. (*Adapted from Phillips, 1981.*)

food applications, there is considerably more protein present than would be required for surface coverage. This additional protein can be adsorbed to the interface in the form of multilayers or it may remain in true solution in the aqueous phase.

Once a layer of protein has been adsorbed additional protein cannot be added in the same manner because an energy barrier to adsorption exists (Phillips, 1981). Further interaction involves the interaction of protein molecules in the bulk phase with those already adsorbed to the lipid and the formation of multilayers. While these interactions will not lead to further decreases in the interfacial tension of the emulsion, they can have significant effects on the stability of the emulsion.

EMULSION STABILITY

The causes of emulsion instability can generally be divided into four main types. These include coalescence, flocculation, gravitational creaming, and Ostwald ripening. Coalescence leads to complete failure of the emulsion and is therefore of greatest concern. The DLVO theory has been utilized to rationalize the forces favoring and inhibiting coalescence of lipid droplets (Petrowski, 1976). According to this theory, residual charges on the surface of the lipid droplet will repel one another and lead to stability. This will be counteracted by the transit dipole–induced dipole interactions that are induced as the lipid droplets approach one another. These are relatively weak forces, the strength of which is proportional to the distance of separation of the lipid droplets to the power of minus 6. While the individual forces are weak, they are strong enough to cause coalescence when the distance of separation is small. Calculations of the charge repulsion required to overcome the attractive van der Waals forces for food emulsions have demonstrated that there is rarely enough charge at the surface of the lipid droplets to result in stable emulsions (Bergenstahl and Claesson, 1990). The fact that stable emulsions can be formed suggests that other forces also provide stability. For emulsions stabilized by proteins, there is a repulsive force due to other factors. According to Bergenstahl and Claesson (1990) these factors include:

1. An ideal entropy of mixing between segments
2. Changes in solution of polymer segments
3. Volume exclusion effects that decrease the number of possible conformations and give rise to true steric repulsion
4. Electrostatic effects

Consideration must also be given to the energy required to desorb a protein from the surface of a fat globule. Thus, the attractive forces must be strong enough not only to overcome the charge repulsion that is present, but also to cause rearrangement of the protein as it desorbs from the interface.

Friberg et al. (1990) have also suggested that under certain conditions the formation of liquid crystalline regions of emulsifier molecules may add to stability. The conditions required for the formation of liquid crystalline structures depend upon the ratios of oil, emulsifier, and water, the temperature of the system, and the nature of the emulsifier molecules. Generally, it has been estimated that the formation of such crystalline structures is favored when the concentration of emulsifier is about 3–6% of the total lipid concentration. The formation of liquid crystals effectively increases the distance of separation of the hydrophobic groups and greatly decreases the strength of van der Waals attractive forces. In most food emulsions, instability due to coalescence can be avoided through proper formulation and processing.

Flocculation refers to the aggregation of fat globules through interactions of the adsorbed macromolecules. Enough of the continuous phase remains associated with the globules to prevent coalescence from occurring. The flocculated particles acquire an increased effective diameter and tend to cream at a more rapid rate than do individual fat globules. If the forces causing flocculation are weak, the floc may be redispersed by shaking. Flocculation is often ion mediated, especially when the stabilizing proteins are sensitive to specific ions. The addition of excess calcium to a casein-stabilized emulsion, for example, will lead to flocculation. Generally, flocculation is less serious than coalescence, but it may present a significant problem when the attractive forces between the particles are strong enough to prevent easy redispersion of the system.

Ostwald ripening refers to the diffusional transport of material from smaller droplets into larger ones. The chemical potential is higher in droplets with a large radius of curvature. This difference in chemical potential can serve as the driving force for the transport of material. Recent work by Dungan and McClements (1993) has demonstrated that such transport can occur in emulsions formed from lipids having different melting points. When mixtures of the two emulsions were made and aged in the presence of excess emulsifier, it was possible to demonstrate the existence of lipid droplets having melting points intermediate between those of the original lipids. The rate of transport was directly related to the concentration of free emulsifier molecules, which presumably served as carriers for the transported lipids. Interestingly, during the time frame of the experiments, the size distribution of the emulsions did not change. If significant Ostwald ripening were to occur, the average droplet diameter would be expected to increase. In emulsions stabilized by proteins, Ostwald ripening probably has little impact upon emulsion stability.

Creaming refers to separation due to the upward rise of fat globules due to differences in density between the dispersed and the continuous phase. The rate of creaming is given by Stoke's law:

$$v = \frac{2r^2g\Delta p}{9\mu}$$

Where v equals the velocity of the fat globule, r is the radius of the fat globule, g is the force of gravity, Δp is the density difference between the two phases, and μ is the viscosity of the continuous phase. The calculated rate of fat globule creaming suggests that for fat globules of diameters less than about 0.1 μm, the rate of creaming is so slow that it can be ignored. For large globules, the extent of the problem will depend upon the duration of storage of the product. For all fluid products with a mean diameter very much in excess of 1 μm, creaming will be a definite problem.

Stoke's law shows clearly that the rate of creaming can be minimized by decreasing the average particle diameter, decreasing the density difference between the two phases or increasing the viscosity of the continuous phase. The average fat globule diameter is a function of both the type of emulsifier present and the method of emulsion formation (Mangino, 1984). Present formulations and equipment yield products with average diameters a little less than 1 μm. It is doubtful that for most food products much improvement in this diameter is possible.

Increasing product viscosity is possible in some applications. Products like salad dressings are, by design, sufficiently viscous to inhibit gravitational creaming. For many products, increasing the viscosity of the continuous phase to a level effective for inhibition of creaming would render an unacceptable product. An infant formula with a high enough viscosity to inhibit creaming, for example, would not pass freely through a nipple and would be of little use in infant nutrition.

Decreasing the density differences between the two phases can only be used in a limited number of applications. There is too great a range in average fat globule diameter obtained during emulsion formation to allow the processor to predictably adsorb the required amounts of protein to negate the density differences. In some applications with essential flavorings, it is possible to increase the density of the discontinuous phase with substances such as brominated vegetable oil (Hwang, 1990). The number of substances suitable for increasing the density of the lipid phase is limited, and for the vast majority of emulsions processors will have to accept problems arising from density differences.

Despite the fact that the previously described approaches are of limited value in increasing stability towards gravitational creaming, practical experience tells us that there are commercially available products for which the defect has been minimized. While increasing viscosity of the continuous phase has only limited applications, rheological considerations are still important. Emulsions are pseudoplastic systems, and the viscosity decreases rapidly with shear. If a weak gel with a low yield stress can be established, the stability of the emulsion towards creaming can be greatly enhanced. Walstra (1987) has suggested that a yield stress of greater than 0.1 Pa would be adequate to prevent creaming. For a yield stress less than about 10 Pa, the gel would be readily reversible and would be fluid when poured. Even if there is not a yield value of more than 0.1 Pa, most emulsions are sufficiently pseudoplastic to exhibit higher than expected viscosities at very low shear rates, and thus creaming is often slower than predicted.

The nature of the proteins associated with the fat globule surface may exert a large effect on the viscosity of the emulsion. We (Fligner et al., 1990) have observed that for model infant formulas, the ratio of casein to whey was significantly correlated to the extent of creaming in the product. These systems were thermally processed to commercial sterility, thus heat-induced changes in protein structure were possible. We observed that as the ratio of casein to whey proteins in the formulation increased, the amount of creaming also increased. The increased stability in the formulations containing increased levels of whey protein was attributed to the formation of a weak protein gel. Complete replacement of casein with whey was not possible in this system, as a certain level of casein was required to prevent total gelation of the product. It has been hypothesized that carrageenan increases the stability of these types of products by a similar mechanism (Hansen, 1982; Sharma, 1981).

FACTORS IMPORTANT TO PROTEIN-STABILIZED EMULSIONS

Tornberg and Hermansson (1977) have reported that the method of emulsion formation is the most important factor in determining the nature of the emulsion. These authors have suggested that the development of a standardized procedure for emulsion formation must precede any meaningful investigation of emulsion characteristics. The methods commonly available for emulsion formation and the relative merits of each procedure are thoroughly reviewed by Tornberg et al. (1990). Walstra (1983) has presented an extensive discussion of the forces involved in the disruption of lipids necessary for the formation of emulsions. Discussion in this section will be limited to a description of the factors that have a major impact on the formation of protein-stabilized emulsions.

The temperatures associated with emulsion formation and subsequent storage are important considerations. Food emulsions are generally formed at temperatures around 60°C. The interactions of hydrophobic groups are optimal at this temperature (Tanford, 1980), and interactions between emulsifiers and lipids as well as the hydrophobic groups of proteins and lipids will be optimal. Emulsion formation at this temperature will favor the interactions necessary for stable emulsions. The ability of a protein or small emulsifier to diffuse to the interface increases as the viscosity of the solution decreases. Thus, emulsion formation at elevated temperatures will increase the opportunities for emulsifier molecules to interact with lipids before significant coalescence can occur. In principle, the beneficial effects of decreased solution viscosity should continue as the temperature is raised beyond 60°C. In practice, the weakening of the hydrophobic associations between emulsifier molecules and the lipid phase that result from increased temperature negate any gains from increase in temperature beyond 60°C. The temperature of storage can also have a significant effect on emulsion stability. As the temperature is lowered, water attains more and more structure. As the water becomes more ordered, there is less of an energy difference between hydrophobic groups exposed to the aqueous phase and those buried in the oil phase. Low temperature alone does not usually cause an emulsion to break, but it can be the deciding factor in the stability of an otherwise poorly emulsified system (Mangino, 1990). The largest temperature-induced changes to emulsions occur upon freezing and subsequent thawing. Not only is the energy difference between the associated and free state minimized by the low temperature, but the formation of ice crystals causes physical damage to the emulsion. When the system is thawed, coalescence occurs if the physical damage has been extensive. Appropriate ways to minimize this type of damage include the addition of substances that will modify the size and extent of water crystal formation (Courts, 1980).

Predictably, the pH of emulsion formation is an important criterion with respect to the amount of protein adsorbed at the interface and also affects the stability of the emulsion that is formed. Halling (1981) has extensively reviewed the literature and suggested that the effects of pH on emulsification can be explained by one of four mechanisms:

1. The isoelectric point is the pH of minimal protein solubility, and for many proteins the reduced solubility may affect emulsification functionality.

2. The cohesiveness of protein films tend to be maximal near the isoelectric point. Cohesive films tend to be more stable.

3. Charge repulsion of emulsion droplets is minimal near the isoelectric point, resulting in decreased stability.

4. Near the isoelectric point, repulsion of charged segments of protein chains is minimized, allowing for the adsorption of compact protein structures.

Associated pH-dependent effects on protein conformation have been reported, which affect emulsion formation through changes in protein hydrophobicity (Yamauchi et al., 1980; Shimizu et al., 1985). Generally the net effect of these mechanisms is to maximize protein adsorption near the isoelectric point. For example, Waniska and Kinsella (1988) demonstrated that maximum surface viscosity of β-lactoglobulin solutions occurred near the isoelectric point. At this pH, electrostatic repulsion is minimized allowing hydrophobic residues to stabilize a more compact tertiary structure. Near the isoelectric point, proteins are able to form close-packed, condensed films. Thus, increased protein concentration at the interface will potentially cause the formation of a more viscous interfacial film, beneficial for emulsion stability. In some cases (Shimizu et al., 1981; Lu, 1987), emulsion stability has increased at pH values away from the isoelectric point. This is especially true when the emulsion was formed with shear devices such as a polytron. The increased energy input, compared to that of a valve homogenizer, may allow for increased adsorption of more highly charged molecules. Once adsorbed, the charge repulsion due to the residual charge should increase emulsion stability. Many food emulsions are formed and consumed at pH values removed from the isoelectric point of the proteins used in emulsion formation. For these products, pH adjustment is not an option to increase emulsion stability.

The interactions of proteins with lipids are hydrophobic in nature, and an examination of the hydrophobic nature of protein emulsifiers may be helpful. Attempts to correlate average hydrophobicity of individual proteins as calculated by the Bigelow method (Bigelow, 1967) to the functionality of proteins in emulsions have generally failed. Mangino (1984) has suggested that this results from the failure of this method to account for the availability of the hydrophobic groups to interact with the interface. Thus, a protein may have a large number of hydrophobic groups held tightly in its interior and give a high value for average hydrophobicity. Such a protein might not function well as an emulsifier because the hydrophobic groups are not able to unfold and interact with the lipid molecules at the interface. A number of studies have demonstrated that a measure of effective hydrophobicity can be related to the function of proteins as emulsifiers.

Kato and Nakai (1980) were the first to report a correlation between surface hydrophobicity and emulsifying activity. This observation has been confirmed in other studies (Kato et al., 1983; Nakai, 1983; Li-Chan et al., 1984). Investigations by Lee and Kim (1987) of several methods for determining hydrophobicity have shown a relationship between emulsifying index and surface hydrophobicity as determined by the *cis*-parinaric acid technique. Results for partition and hydrophobic chromatography were also correlated with surfactant activity, but the 8-anilo-1-napthalenesulfonic acid (ANS) probe technique was not. In general, proteins that possess high levels of surface hydrophobicity exhibit favorable emulsification characteristics. Shimizu et al. (1981) have proposed that average hydrophobicity is not meaningful for studying the adsorption of proteins, whereas conformational flexibility is more informative. Kinsella and Whitehead (1988) and Dickinson and Stainsby (1988) also concluded that the emulsifying behavior of β-lactoglobulin correlated better with molecular flexibility than with hydrophobicity. Thus, emulsification is more strongly related to the accessibility of hydrophobic groups at the surface rather than to the total number of hydrophobic groups present.

The importance of solubility has been reviewed by Halling (1981). Protein solubility has been described as the most important factor governing the functionality of proteins in food systems (Kinsella, 1976; deWit et al., 1986). While it is difficult to reconcile functional characteristics with insoluble proteins, such as result from excessive denaturation, there is evidence that 100% solubility is not required. For many applications it has been observed that over reasonable ranges of solubility (35–95%), protein solubility is not the primary factor in the determination of protein functionality. We (Liao and Mangino, 1987) examined 10 commercially available acid whey protein concentrates and found that the protein solubility ranged from 25.4 to 82.4%. Models were generated to determine factors important to specific functional performance in a number of systems (whipped topping overrun, foaming, foam stability, and emulsion capacity). In these models, solubility was not the most important factor for any of these attributes. It should be noted that three of these model systems measured factors that related to either emulsion capacity or emulsion stability. In another study, we (Peltonen-Shalaby and Mangino, 1986) found that the pH 4.6 insoluble fraction of a number of commercially available whey protein concentrates to be a more effective emulsifying agent than either the pH 4.6 soluble material or the unfractionated whey protein concentrates. However, this same insoluble fraction was less functionally effective in all other applications. We suggested that the increased functionality in emulsion formation could be attributed to the presence of residual phospholipid material. These data are consistent with the concept for emulsification-complete protein solubility is not required for the complex adsorption onto the fat globule surface. However, it is a minimum requirement that a protein must be able to interact with other molecules and, thus, must be able to be dispersed in the continuous phase.

The formation of emulsions with the addition of proteins and small emulsifiers makes the situation more complex. While information on the interactions of proteins with lipids is necessary and valuable, consideration must also be given to the interactions that occur between small emulsifier molecules and proteins. In a series of papers, Dickinson and coworkers (Dickinson et al., 1989; Courthaudon et al., 1991; Dickinson and Tanai, 1992; Dickinson and Iveson, 1993) have studied the effects of various emulsifier molecules on the fat globule size distribution, surface rheology, and protein load of a number of milk protein–stabilized emulsions. The effects observed are often different depending upon the lipid source, protein, and emulsifier molecule, but some general trends can be noted.

At emulsifier-to-protein ratios above 16, increases in the concentration of emulsifier result in (1) a decrease in protein load, (2) a slight decrease to no change in the average fat globule diameter, and (3) a decrease in the surface viscosity of the film formed. Such data are pertinent to the long-term stability of emulsions. The amount of protein adsorbed is affected by not only the amount of emulsifier present, but also by the type of emulsifier molecule. Yamauchi et al. (1980), for instance, have demonstrated a pronounced effect of type of emulsifier on the amount of whey protein adsorbed at the lipid interface. For studies that aim to assess the stability of emulsions under conditions that reflect processing, such interactions must not be ignored. The data in Fig. 5.5 is from Fligner et al. (1991) and demonstrates the effect of added lecithin on the creaming stability of a model infant formula emulsion. Lecithin content is expressed as percent of the lipid in the product. These findings demonstrate that the addition of lecithin up to about 0.5% of the total fat increases the stability of the emulsion toward gravitational creaming. The further addition of lecithin causes a decrease in the stability of the emulsion under these conditions. An explanation for the physical cause of this observation may be deduced from Fig. 5.6. This data correlates the protein load in a model infant formula to level of added lecithin. This plot indicates that as lecithin content increases above 0.5%, the protein load on the fat globules decreases. This observation is in qualita-

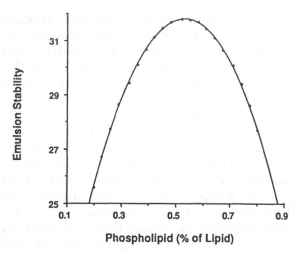

FIG. 5.5 Plot of emulsion stability (EVI) vs. added phospholipid (percent of total lipid) for model infant formula emulsions. (*Adapted from Fligner et al., 1991.*)

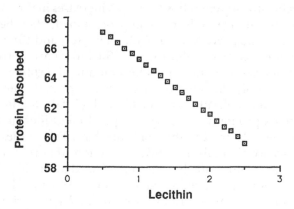

FIG. 5.6 Plot of protein absorbed (percent of total) vs. added phospholipid (percent of total lipid) for model infant formula emulsions. (*Adapted from Fligner et al., 1991.*)

tive agreement with the data of Dickinson's group. We suggested that the displacement of protein from the interface had a negative effect upon the yield value of the system and made the fat globules more susceptible to creaming.

MEASUREMENT OF EMULSION QUALITY

The study of emulsions is complicated by the wide variety of food emulsions that exist and by the diverse information that is sought. Many studies to determine the nature of the interactions between proteins and hydrophobic interfaces have been conducted by observing the interactions between molecules at quiescent interfaces. The rate of adsorption, changes in interfacial tension, changes in surface rheology, and interactions of various emulsifier molecules have been studied extensively by these techniques. Much useful information regarding the nature of the molecular interactions have been obtained from such studies (Graham and Phillips, 1976; Tornberg, 1978a; Kim and Kinsella, 1987). Extrapolation of data obtained from these studies to the situation that exists in food products must be done with care as these studies are often performed at protein-to-lipid ratios and at shear rates that are quite different from what occurs in food products (Halling, 1981). In this section, we describe the methodology utilized to obtain data relevant to food processors. The underlying principles for each test are described, and the relative advantages and disadvantages of each approach are discussed.

Much of the early work with food emulsions involved the measurement of the emulsion capacity of food proteins (Swift et al., 1961). Emulsion capacity (EC) is generally defined as the maximum amount of lipid emulsified by a protein dispersion. Oil is added at a given rate to a defined quantity of a protein dispersion containing a fixed amount of protein. Oil addition is continued until the viscosity decreases or inversion occurs. This test has been especially popular for selecting proteins to be utilized in meat emulsions. The main advantages of this method are that it is relatively simple to perform and that several trials can be conducted within a relatively short period of time without consumption of large quantities of protein. Regenstein (1988) has described the problems encountered with this method and has concluded that the test is not very useful for selection of proteins for use in emulsion formation. One problem with the method is that it measures the ability of proteins to form emulsions at protein-to-lipid ratios that are very different from those that will be encountered in the finished product. Values for emulsion capacity by this method are often in the range of several tens to a few hundred milliliters of oil emulsified per gram of protein, which is far in excess of practical protein-to-lipid levels, which are usually closer to one to one. Thus, data obtained from these studies may indicate differences that have no relationship to the functionality of the proteins being evaluated under typical conditions of use. Another shortcoming of EC tests is that little or no information regarding the stability of the emulsion can be obtained from this data. Halling (1981) has thoroughly reviewed this method and attempted to rationalize the data obtained with a theoretical evaluation of the forces involved. He has concluded that the surface properties of proteins are responsible for differences observed with this test. He has further concluded that the identification of the critical surface property or properties has yet to be made.

No single method has been accepted as a measure of emulsion stability. Horie et al. (1978) have classified the methods of emulsion stability testing into three groups:

1. Droplet size distribution. These methods attempt to determine the extent of coalescence by determining the size distribution of the emulsions. Common procedures include microscopy, electronic counters, and light scattering.

2. Estimates of phase separation. The purpose of these methods is to determine the extent of oil separation at a specified time and temperature. These methods determine the extent of separation by using dielectric measurements, conductivity, colorimetry, and changes in lipid or moisture content.

3. Accelerated stability tests. These methods attempt to estimate the creaming rate at some time in the future by applying a stress. Ultracentrifugation is the most commonly applied stress, but heating or freeze thaw cycles are also popular.

The size distribution of the fat globules in an emulsion will be a function of the forces tending to cause coalescence of the fat globules and those that tend to disrupt or stabilize them. In principle, more efficient means of emulsion formation or fat globule stabilization will result in a smaller average fat globule diameter. A number of procedures are available to estimate the average fat globule diameter, the surface area of the emulsion formed, or the size distribution of the fat globules in the emulsion. Both light scattering (Bergenstahl, 1988; Horie et al., 1978; Pearce and Kinsella, 1978) and Coulter counter procedures (Dickinson et al., 1984; Ray et al., 1983; Walstra et al., 1969) have been applied to simple emulsions. Both of these procedures are time consuming. It has been suggested (Walstra et al., 1969; Walstra and Oortwijn, 1969, 1975) that the electronic counter procedures are less tedious and tend to provide more reproducible data than light microscopy.

Probably the most widely utilized of these techniques is the Emulsion Activity Index (EAI) of Pearce and Kinsella (1978). EAI is expressed as the area of interface (determined from turbidity measurements) stabilized per unit weight of protein. The data obtained from the determination can also be utilized to obtain an average fat globule diameter. Changes in the average fat globule diameter with either time or heating have been related to emulsion stability. The method is based on the assumption that a reasonable estimate of fat globule surface area may be obtained from a measurement of the turbidity of the emulsion. While the assumptions behind this method have been debated (Halling, 1981; Howe et al., 1986a), recent data suggests that the values obtained by light scattering are in reasonable agreement with those obtained by more cumbersome techniques (Das and Kinsella, 1989). Major deviations are encountered when the protein suspension makes a significant contribution to the suspension turbidity such as occurs with emulsions stabilized by calcium caseinate or when heat processing induces turbidity in the sample that is not related to the size of the average fat globules. Despite these difficulties, this method has many advantages that make it a valuable method for the study of emulsions. These include the relative simplicity of the test, the small amounts of sample required, and dynamic data acquisition relevant to the new surface area formed during emulsion formation.

Much of the available information on the surface area created per unit of protein for a large number of food proteins has been obtained by this method. One assumption often made by workers who employ this method is that emulsions with inherently smaller average fat globule diameters will be more stable than those composed of larger fat globules. This would appear to be a reasonable assumption from the application of Stoke's law. However, the assumption ignores any stabilization due to steric repulsive forces or due to the formation of a yield value resulting from protein-protein or protein-stabilizer interactions in the emulsion. We (Fligner et al., 1990, 1991) have shown, for example, that in sterilized model infant formula emulsions, the samples with the smallest average fat globule diameter are not necessarily the ones that undergo the least amount of creaming during 4 months of storage. Determination of EAI provides considerable data regarding the efficiency of emulsion formation and allows for interpretation of data regarding the effects of changes in protein conformation on the creation of emulsions. Workers who desire information on the long-term stability toward creaming of emulsions must be aware that this procedure often does not measure factors that are relevant to such long-term stability.

A number of methods exist that attempt to measure emulsion stability (ES) by determining the extent of phase separation. ES can be defined as the maintenance of a homogeneous structure and texture of the system.

Several techniques are available to determine if an emulsion has under-gone any observable changes and commonly involve, but are not limited to, measurement of the amount of oil or cream separation from an emulsion during a given time period at a particular temperature and gravitational field. The determination of the amount of lipid arriving at the surface or leaving the bottom of an emulsion in a given amount of time is an absolute indicator of the stability of that emulsion towards gravitational creaming for that period of time. Often, however, there is need for information that will predict the stability of the emulsion at some time in the future. Periods of months or even years are often in question.

To obtain an estimate of the future state of the emulsion, measurements may be made under conditions of imposed stress. Heating represents one form of stress that may be applied (Chen et al., 1975; Tsai et al., 1972; Wang and Kinsella, 1976). The newly formed emulsion will be subjected to a defined heat treatment and then allowed to cream for a defined period of time. By such an approach, various proteins or treatments can be compared and judgments rendered regarding the stability of emulsions. Analytically the amount of oil separated is measured by any number of methods such as the Mojonnier technique (Titus et al., 1968). The change in fat distribution may be determined based on the change in lipid content in the lower aqueous phase (Tornberg, 1978b; Tornberg and Hermansson, 1977; Mohanty et al., 1988) or the increase in water in the aqueous phase (Acton and Saffle, 1970). Howe et al. (1986b) used ultrasound to monitor the distribution of oil with height and time in a creaming emulsion over a one-month period of storage, the assumption being that the imposed stress caused by heating of the emulsion would allow for almost immediate determination of the stability achievable under normal stress conditions. Clearly, these assumptions must be rigorously tested before they are accepted, and so far this has usually not been the case. It is likely, for example, that the instability induced in emulsions subjected to heating are of a different nature than what the emulsion will experience under conditions of long-term storage at ambient temperature. If the mechanism of emulsion failure is different under these sets of conditions, the relevance of the data obtained is questionable.

Common means of accelerated testing of emulsions relate to the use of high centrifugal forces. It is assumed, for example, that an emulsion exposed to a force of $10,000 \times g$ will cream at a rate 10,000 times faster than will the same emulsion stored for extended periods of time $1 \times g$. This assumption has been criticized (Vold and Mittal, 1972; Vold and Groot, 1964). It has been observed that exposure of emulsions to high gravitational fields may also expose them to forces that are different than those observed in normal storage. Distortion of the surface of the fat globules may result in a different mechanism of emulsion failure (Becher, 1965). It is also likely that such high gravitational fields will be capable of disrupting networks of proteins and or stabilizers that may impart a yield value to the emulsion. Vold and Acevedo (1977) compared stability measured by ultracentrifugation with the absolute creaming rates for emulsions aged for up to 3 years. They found that ultracentrifugal parameters of fresh emulsions did not predict actual stability in all cases. Some of the most unstable emulsions were predicted to be the most stable and vice versa. Thus, centrifugal studies must be correlated with absolute indicators of emulsion stability before they can be generally accepted as indicators of emulsion stability.

In a study of salad dressing emulsions, McDermott et al. (1981) have introduced a parameter called the emulsion volume index (EVI). This parameter is derived from an accelerated aging procedure in which the centrifugation is carried out in a microhematocrit centrifuge. The authors have suggested that this method relates better to long-term emulsion stability than does centrifugation in larger centrifuge tubes. In a study of the factors important to the functionality of proteins in this test, Chmura (1982) reported that the centrifugal effects on the emulsion were different in the capillary system than they were in larger vessels and concluded that factors that tended to increase protein hydration also decreased the tendency for phase separation and increased stability towards creaming under normal gravity conditions. We (Fligner et al., 1990, 1991) compared this method to EAI as a predictor of the stability of model infant formulas to creaming. We found in both the short term (1 week) and long term (4 months) that EVI was significantly correlated to the amount of creaming that occurred under normal storage conditions. We also observed that, under the conditions of this test, EAI did not significantly correlate with emulsion stability. It is possible that EVI will prove to be a predictor of stability towards creaming in a number of systems. Again, rigorous correlation and calibration of the technique with each system studied will be required before its validity can be generally accepted.

Often the properties of emulsions are studied in relatively simple systems consisting of a buffer, a lipid, and the protein to be studied. Such simple systems make interpretation of data much simpler. The use of such simple systems can generally be rationalized when the purpose of the investigation is to obtain information regarding the nature of the component interactions. When the purpose of the study is to apply the information to the formulation of food products, simple systems are more difficult to justify. The arbitrary nature of the selection of protein and lipid level in such systems is open to criticism. Harper (1984) has discussed some of the complex interactions that occur in real food products and has provided formulations for a number of model systems that are based on currently manufactured food products. He has suggested that the use of these systems in research aimed at development of optimal formulations or reformulation should be considered. In the area of emulsions he gives formulations for a model sausage, infant formula, whipped topping, and coffee whitener. This author has used the last three model systems and has found that the data obtained using these formulations is very different from that obtained with simpler models and can often provide information that might not be available in the simple systems (Harper, 1991; Harper et al., 1980). Harper (1991) listed the factors that were significantly related to the minimization of serum separation in a coffee whitener. The amount of emulsifier used was the most important factor. The type of protein utilized was the second most important factor followed by level of polyphosphate, level of gum, and a protein-phosphate interaction term. The type of protein providing the most stable system was different for formulations that did not contain emulsifiers or stabilizer. Thus the conclusions obtained from such a simple model system would have provided erroneous information to the manufacturers of such a product. While the use of these model systems often requires that larger amounts of protein be utilized and that more time be spent on the preparation of the emulsions, the results obtained are probably more relevant to food processors and justify the additional effort. Darling and Birkett (1987) have recently suggested that one of the greatest challenges facing those who study food emulsions is to make their model systems and the data collected relevant to real food systems. They stated, in part, that "too much time is spent burying our heads in the proverbial 'model system' rather than facing up to the complex world of real food emulsions. . . . Perhaps it is time for basic scientists to step into the dirty world of food colloids in practice." We would like to echo this sentiment and suggest that the complexity and uncertainty of working with systems that resemble food may make interpretation of the data more difficult and ambiguous, but the data obtained from such studies are necessary. One goal of research into emulsion systems is to improve our understanding of the physical chemistry involved; another goal is to translate such understanding into meaningful recommendations for the food industry. If this second goal is to be achieved, more work with complex systems will be necessary.

The composition of emulsions as well as the nature of emulsion instability can be quite different in various food products (Dickinson, 1989;

Dickinson and Stainsby, 1987; Fayerman and Harper, 1983; Schmidt et al., 1984). The practical implications of applying information obtained regarding the emulsifying ability of proteins to applications in real food formulations are considerable and must be viewed in the context of their effects on emulsion formation and subsequent stability. In some applications, the emulsions are only marginally stable and failure may result in complete phase separation. In other cases, instability results in the clustering of fat globules that leads to the formation of a cream layer due to differences in density. For many products, this may be only a minor problem, while for others, it may be the factor that limits shelf life. Thus, any study that attempts to define factors or conditions that are important to emulsion formation or stability must carefully define the system under study and the type of instability being observed. The complexity of the systems that contain proteins, small emulsifier molecules, and carbohydrate stabilizers often makes a clear understanding of the relative importance of various components and processes to emulsion formation and stabilization difficult to interpret. Much progress has been made in relating protein structure to functionality in emulsions, and several contributing factors have been established. It is becoming increasingly evident that the selection of a protein to use in an emulsion as well as conditions of emulsion formation must be made with due consideration to other ingredients present, such as surfactants and hydrophilic colloids.

ACKNOWLEDGMENTS

Salaries and research support were provided by state and federal funds appropriated to The Ohio Agricultural Research and Development Center, The Ohio State University. The work was supported in part by The National Dairy Promotion and Research Board.

REFERENCES

Acton, J. D. and Saffle, R. L. 1970. Stability of oil-in-water emulsions. 1. Effects of surface tension, level of oil, viscosity and type of meat protein. *J. Food. Sci.* 35: 852.

Anfinsen, C. B. 1973. Principles that govern the folding of protein chains. *Science* 181: 223.

Becher, P. 1965. *Emulsions: Theory and Practice*, 2nd. ed. Reinhold Publishing Co., New York.

Bergenstahl, B. 1988. Gums as stabilizers for emulsifier covered emulsion droplets. In *Gums and Stabilizers for the Food Industry 4*, Phillips, G. O., Williams, P. A., and Wedlock, D. J. (Ed.), p. 363. IRL Press, Oxford.

Bergenstål, B. A. and Claesson, P. M. 1990. Surface forces in emulsions. In *Food Emulsions*, 2nd ed., Larson, K. and Friberg, S. T. (Ed.), p. 41. Marcel Dekker, New York.

Bigelow, C. C. 1967. On the average hydrophobicity of proteins and the relationship between it and protein structure. *J. Theor. Biol.* 16: 187.

Chen, L.-F., Richardson, T., and Amundson, C. H. 1975. Some functional properties of succinylated proteins from fish protein concentrate. *J. Milk Food Technol.* 38: 89.

Chmura, J. N. 1982. The effect of protein hydration on emulsion stability. Ph.D. thesis, The Ohio State University, Columbus.

Courthaudon, J-L., Dickinson, E., and Christie, W. W. 1991. Competitive adsorption of lecithin and β-caseins in oil in water emulsions. *J. Agric. Food Chem.* 39: 1365.

Courts, A. 1987. Properties and uses of gelatin. In *Applied Protein Chemistry*, Grant, R. A. (Ed.), p. 1. Applied Science, London.

Darling, D. F. and Birkett, R. J. 1987. Food colloids in practice. In *Food Emulsions and Foams*, Dickinson, E. (Ed.), p. 1. The Royal Society of Chemistry, London.

Das, K. P. and Kinsella, J. E. 1989. pH dependent emulsifying properties of β-lactoglobulin. *J. Disp. Sci. Tech.* 10: 77.

deWit, J. N., Klarenbeek, G., and Adamse, M. 1986. Evaluation of functional properties of whey protein concentrates and whey protein isolates. 2. Effects of processing history and composition. *Neth. Milk Dairy J.* 40: 41.

Dickinson, E. 1989. Surface and emulsifying properties of caseins. *J. Dairy Res.* 56: 471.

Dickinson, E. and Iveson, G. 1993. Absorbed films of β-lactoglobulin + lecithin at the hydrocarbon-water and triglyceride-water interfaces. *Food Hydrocolloids* 6: 533.

Dickinson, E., Roberts, T., Robson, E. W., and Stainsby, G. 1984. Effect of salt on stability of casein stabilised butter oil-in-water emulsions. *Lebensm.-Wiss. U.-Technol.* 17: 107.

Dickinson, E., Rolfe, S. E., and Dalglish, D. G. 1989. Competitive absorption in oil-in-water emulsions containing α-lactalbumin and β-lactoglobulin. *Food Hydrocolloids* 3: 193.

Dickinson, E. and Stainsby, G. 1987. Progress in the formulation of food emulsions and foams. *Food Technol.* 41: 74.

Dickinson, E. and Stainsby, G. 1988 Emulsion stability. In *Advances in Food Emulsions and Foams*, Dickinson, E. and Stainsby, G. (Ed.), p. 1. Elsevier Applied Science, New York.

Dickinson, E. and Tanai, S. 1992. Temperature dependence of the competitive displacement of protein from the emulsion droplet surface by surfactants. *Food Hydrocolloids* 6: 163.

Dungan, S. and McClements, D. 1993. Oil transport between oil-in water emulsion droplets stabilized by non-ionic surfactant. Presented at Conference of Food Engineering, Chicago, IL.

Eisenberg, D. and Kauzmann, W. 1969. *The Structure and Properties of Water*. Oxford University Press, Oxford.

Fayerman, A. M. and Harper, W. J. 1983. Whey protein concentrates in sausages. Paper presented at the Australia, New Zealand, United States collaborative conference on whey protein utilization. Melbourne, Australia.

Fligner, K. L., Fligner, M. A., and Mangino, M. E. 1990. The effects of compositional factors on the physical stability of fluid emulsions. *Food Hydrocolloids* 4: 95.

Fligner, K. L., Fligner, M. A., and Mangino, M. E. 1991. The effects of compositional factors on the long-term physical stability of concentrated infant formula. *Food Hydrocolloids* 5: 269.

Friberg, S. E., Goubran, R. F., and Hayai, I. H. 1990. Emulsion stability. In *Food Emulsions*, 2nd ed., Larson. K. and Friberg, S. T. (Ed.), p. 1. Marcel Dekker, Inc., New York.

Graham, D. E. and Phillips, M. C. 1976 The conformation of proteins at interfaces and their role in stabilizing emulsions. In *Theory and Practice of Emulsion Technology*, p. 75. Academic Press, New York.

Graham, D. E. and Phillips, M. C. 1979a. Proteins at liquid interfaces. 1. Kinetics of adsorption and surface denaturation. *J. Coll. Interface Sci.* 70: 427.

Graham, D. E. and Phillips, M. C. 1979b. Proteins at liquid interfaces. *J. Colloid Interface Sci.* 70: 403.

Halling, P. J. 1981. Protein-stabilized foams and emulsions. *CRC Crit. Rev. Food Sci. Nutr.* 15: 155.

Hansen, P. M. T. 1982. Hydrocolloid-protein interactions: Relationship to stabilization of fluid milk products. A review. *Prog. Food Nutr. Sci.* 6: 127.

Harper, W. J. 1984. Model food system approaches for evaluating whey protein functionality. *J. Dairy Sci.* 67: 2745

Harper, W. J. 1991. Evaluation of the functionality of whey proteins in model systems. Presented at The CDR/ADPI Whey Protein Workshop. Madison, WI.

Harper, W. J., Peltonen, R., and Hayes, J. 1980. Model food systems yield clearer utility of whey proteins. *Food Prod. Dev.* 17: 25.

Horie, K., Tanaka, S., and Akabori, T. 1978. Determination of emulsion stabilization by spectral absorption. 1. Relationship between surfactant type, concentration and stability index. *Cosmet. Toil.* 93: 53.

Howe, A. M., Mackie, A. R., Richmond, P., and Robins, M. M. 1986a. Creaming of oil-in-water emulsions containing polysaccharides. In *Gums and Stabilizers for the Food Industry. 3*, Phillips, G. O., Wedlock, D. J., and Williams, P. A. (Ed.), p. 295. Elsevier Applied Science Publishers, New York.

Howe, A. M., Mackie, A., R., and Robins, M. M. 1986b. Technique to measure emulsion creaming by velocity of ultrasound. *J. Dispersion Sci. Technol.* 7: 231.

Hwang, C.-S. 1990. Emulsions as clouding agents for soft drinks. M. S. thesis, The Ohio State University, Columbus.

Karplus, M. and McCammon, J. A. 1983. Dynamics of proteins: Elements and function. *Ann. Rev. Biochem.* 53: 263.

Kato, A. and Nakai, S. 1980. Hydrophobicity determined by a fluorescence probe method and its correlation with surface properties of proteins. *Biochim. Biophys. Acta* 624: 13.

Kato, A., Osako, Y., Matsudomi, N., and Kobayashi, K. 1983. Changes in the emulsifying and foaming properties of proteins during heat denaturation. *Agric. Biol. Chem.* 47: 33.

Kim, S. H. and Kinsella, J. E. 1987. Surface active properties of food proteins. Effects of reduction of disulfide bonds on film properties and foam stability of glycerin. *J. Food Sci.* 52: 128.

Kinsella, J. E. 1976. Functional properties of proteins in foods: A survey. *CRC Crit. Rev. Food Sci. Nutr.* 7: 219.

Kinsella, J. E. 1982. Relationship between structure and functional properties of food proteins. In *Food Proteins*, Fox, P. F. and Cowden, J. J. (Ed.), p. 51. Applied Science, London.

Kinsella, J. E. and Whitehead, D. M. 1988. Emulsifying and foaming properties of chemically modified proteins. In *Advances in Food Emulsions and Foams*, Dickinson, E. and Stainsby, G. (Ed.), p. 163. Elsevier Applied Science, New York.

Larson, K. and Friberg, S. T. 1990. *Food Emulsions,* 2nd ed. Marcel Dekker, New York.

Lee, C-H. and Kim, S-K. 1987. Effects of protein hydrophobicity on the surfactant properties of food proteins. *Food Hydrocolloids* 1:283.

Li-Chan, E., Nakai, S., and Wood, D. F. 1984. Hydrophobicity and solubility of meat proteins and their relationship to emulsifying properties. *J. Food Sci.* 49: 345.

Liao, S. Y. and Mangino, M. E. 1987. Characterization of the composition, physicochemical and functional properties of acid whey protein concentrates. *J. Food Sci.* 52: 1033.

Lu, T. W. 1987. The effect of xanthan gum on protein stabilized emulsions. M. S. thesis, The Ohio State University, Columbus.

Lumry, R. and Rajendar, S. 1970. Enthalpy-entropy compensation phenomena in water solutions of proteins and small molecules: A ubiquitous property of water. *Biopolymers* 9: 1125.

MacRitchie, F. 1978. Proteins at interfaces. *Adv. Prot. Chem.* 32: 283.

MacRitchie, F. 1987. Consequences of protein adsorption at fluid interfaces. In *Protein at Interfaces,* Brash, J. L. and Horbett, T. A. (Ed.), p. 165. American Chem. Society, Washington, DC.

Mangino, M. E. 1984. Physico-chemical basis of whey protein functionality. *J. Dairy Sci.* 67: 2711–2722.

Mangino, M. E. 1990. Molecular properties and functionality of proteins in food emulsions: Liquid food systems. In *Food Proteins: Structure and Functional Relationships,* Kinsella, J. E. and Soucie, W. (Ed.), p. 159. American Oil Chemists Soc., Champaign, IL.

McDermott, R., Harper, W. J., and Whitley, R. 1981. A rapid centrifugal method for characterization of salad dressing emulsions. *Food Technol.* 35: 81.

Mohanty, B., Mulvihill, D. M., and Fox, P. F. 1988. Emulsification and foaming properties of acidic caseins and sodium caseinate. *Food Chem.* 28: 17.

Mussellwhite, P. R. 1966. The surface properties of an oil-water emulsion stabilized by mixtures of casein and gelatin. *J. Colloid Interface Sci.* 21: 99.

Nakai, S. 1983. Structure-function relationships of food proteins with an emphasis on the importance of protein hydrophobicity. *J. Agric. Food Chem.* 31: 676.

Pearce, K. N. and Kinsella, J. E. 1978. Emulsifying properties of proteins: Evaluation of a turbidimetric technique. *J. Agric Food Chem.* 26: 716.

Peltonen-Shalaby, R. and Mangino, M. E.. 1986. Factors that affect the emulsifying and foaming properties of whey protein concentrates. *J. Food Sci.* 51: 103.

Petrowski, G. E. 1976. Emulsion stability and its relation to food. *Adv. Food Res.* 22: 110.

Phillips, M. C. 1981. Protein conformation at liquid interfaces and its role in stabilizing emulsions and foams. *Food Technol.* 35: 50.

Ray, A. K., Johnson, J. K., and Sullivan, R. J. 1983. Refractive index of the dispersed phase in oil-in-water emulsions: Its dependence on droplet size and aging. *J. Food Sci.* 48: 513.

Regenstein, J. M. 1988. Are comminuted meat products emulsions or a gel matrix? Presented at the American Oil Chemist's Society Mtng., Phoenix, AZ.

Schmidt, R. H., Packard, V. S., and Morris, H. A. 1984. Effect of processing on whey protein functionality. *J. Dairy Sci.* 67: 2723.

Sharma, S. C. 1981. Gums and hydrocolloids in oil-water emulsions. *Food Technol.* 35: 59.

Shimizu, M., Kamiya, T., and Yamauchi, K. 1981. The adsorption of whey proteins on the surface of emulsified fat. *Agric. Biol. Chem.* 45: 2491.

Shimizu, M., Saito, M., and Yamauchi, K. 1985. Emulsifying and structural properties of B-Lactoglobulin at different pHs. *Agric. Biol. Chem.* 49: 189.

Stainsby, G. 1985. Foaming and emulsification. In *Food Macromolecules*, Mitchell, J. R. and Ledward, D. A. (Ed.), p. 315. Elsevier, London.

Starr, C. and Taggart, R. 1989. *Biology: The Unity and Diversity of Life.* Wadsworth, Belmont, CA.

Swaisgood, H. E. 1982. Chemistry of milk proteins. In *Developments in Dairy Chemistry. 1-Proteins*, Fox, P. F. (Ed.), p.1, Applied Science, London.

Swift, C. E., Lockett, C., and Frayar, A. J. 1961. Comminuted meat emulsions—capacity of meats for emulsifying fat. *Food Technol.* 15: 486.

Tanford, C. 1961. *Physical Chemistry of Macromolecules.* Wiley, New York.

Tanford, C. 1970. Protein denaturation, Part C. Theoretical models for the mechanism of denaturation. *Adv. Prot. Chem.* 24: 1.

Tanford, C. 1980 *The Hydrophobic Effect: Formation of Micelles and Biological Membranes*, 2nd ed. John Wiley & Sons, New York.

Titus, T. C., Wiancki, N. N., Barbour, H. F., and Mickle, J. B. 1968. Emulsifier efficiency in model systems of milk fat or soybean oil and water. *Food Technol.* 22: 115.

Tornberg, E. 1978a. The interfacial behavior of three food proteins studied by the drop volume technique. *J. Sci. Food Agr.* 29: 904.

Tornberg, E. 1978b. Functional characterization of protein stabilized emulsions. *J. Food Sci.* 43: 1559.

Tornberg, E. and Hermansson, A. -M. 1977. Functional characterization of protein stabilized emulsions: Effect of processing. *J. Food Sci.* 42: 468.

Tornberg, E., Olsson, A., and Persson, K. 1990. The structural and interfacial properties of food proteins in relation to their function in emulsions. In *Food Emulsions*, 2nd ed., Larson. K. and Friberg, S. T. (Ed.), p. 247. Marcel Dekker, New York.

Tsai, R., Cassens, R. G., and Brisky, E. J. 1972, The emulsifying properties of purified muscle proteins. *J. Food Sci.* 37: 286.

Vold, R. D. and Acevedo, M. V. 1977. Comparison of ultracentrifugal stability parameters with long-term shelf stability of emulsions. *J. Am. Oil Chem. Soc.* 54: 84.

Vold, R. D. and Groot, R. C. 1964. The effects of electrolytes on the ultracentrifugal stability of emulsions. *J. Coll. Sci.* 19: 384.

Vold, R. D. and Mittal, K. L. 1972. Differences in ultracentrifugal stability of various oil-in-water emulsions. *J. Soc. Cosmet. Chem.* 23: 171.

Walstra, P. 1983. Formation of emulsions. In *Encyclopedia of Emulsion Technology*, Beucher, P. (Ed.), p. 57. Marcel Dekker, New York.

Walstra, P. 1987. Overview of emulsion and foam stability of protein-stabilized emulsions. In *Food Emulsions and Foams*, Dickinson, E. (Ed.), p. 242. The Royal Society of Chemistry, London.

Walstra, P. and Oortwijn, H. 1969. Estimating globule-size distribution of oil-in-water emulsions by Coulter counter. *J. Colloid Interface Sci.* 29: 424.

Walstra, P. and Oortwijn, H. 1975. Effect of globule size and concentration on creaming in pasteurized milk. *Neth. Milk Dairy J.* 29: 263.

Walstra, P., Oortwijn, H., and de Graff, J. J. 1969. Studies on milk fat dispersions. 1. Methods for determining globule size distribution. *Neth. Milk Dairy J.* 23: 12.

Wang, J. C. and Kinsella, J. E. 1976. Functional properties of novel proteins: Alfalfa protein. *J. Food Sci.* 41: 286.

Waniska, R. D. and Kinsella, J. E. 1988. Surface viscosity of glycosylated derivatives of beta-lactoglobulin at the air-water interface. *Food Hydrocolloids* 2: 59.

Yamauchi, K., Shimizu, M., and Kamiya, T. 1980. Emulsifying properties of whey protein. *J. Food Sci.* 45: 1237.

6
Protein Interactions in Foams: Protein–Gas Phase Interactions

J. Bruce German

University of California
Davis, California

Lance Phillips

Cornell University
Ithaca, New York

INTRODUCTION

The capacity of food processing and formulation is rapidly increasing. As a result, the demands on protein functional properties in food systems has become acute, including the ability to form stable foams. Not all proteins foam, and those that do vary widely in their foaming capacity (Graham and Phillips, 1976; MacRitchie, 1978; Halling, 1981; Kinsella, 1981; Kinsella and Phillips, 1989). In foams, the dispersed phase volume is large in comparison to the continuous phase. High surface energy created by large air-water interfacial surface area and substantial density differences between phases render these dispersions thermodynamically unstable (Table 6.1). For a protein to be a successful foaming agent, it must be able to stabilize the new surface area continuously being created during foaming (MacRitchie, 1978). The migration of proteins to the interface is energetically favorable because some of the conformational

TABLE 6.1 Some Factors Affecting Stability of Protein-Based Foams

Enhanced stability	Reduced stability
Protein's flexible domains increase viscosity	Drainage (gravitational) of aqueous phase
Protein concentration and film thickness	Disproportionation
Film mechanical strength and surface viscoelasticity	Mechanical shock
Gibbs-Marangoni effect	Capillary pressure
Film net surface charge	Permeable film
Heterogeneous proteins with residual tertiary structure	Surface-active lipids

Source: Kinsella and Whitehead, 1989.

TABLE 6.2 Molecular Properties of Proteins That Relate to Foaming Properties

Solubility	Rapid diffusion to the interface
Amphipathicity	Distribution of charged, polar, and nonpolar residues for enhanced interfacial interactions
Segmental flexibility	Facilitate unfolding at the interface
Interactive segments	The disposition of different functional segments facilitaties secondary interactions in the air, aqueous, and interfacial phases
Disposition of charged groups	Charge repulsion between contiguous bubbles
Disposition of polar groups	Prevents close approach of bubbles; hydration, osmotic, and steric effects

energy and some of the energy of hydration of the protein is lost at the interface (Phillips, 1981). The adsorption of proteins to an interface is not necessarily irreversible, however, and for many proteins the net energy is not sufficient to maintain the protein adsorbed. This has led to a reformulation of the concept of protein surface adsorption to contain both a diffusion coefficient term and a probability of adsorption (Damodaran, 1990). Thus, a protein effective in foaming must rapidly diffuse to the interface, adsorb, and then reorient to form a viscous film to maintain discrete bubbles until stabilizing interactions develop (Prins,

1988; Dickinson et al., 1988; Kinsella and Phillips, 1989). This ability is dependent on intrinsic factors including the structure and conformation of the protein, which in turn depends on environmental factors such as pH, ionic strength, and protein interactions (Damodaran, 1990; Halling, 1981; Kinsella and Phillips, 1989). In most food foams, we must place an additional requirement on the film properties of proteins in that stabilizing interactions must be maintained through the gelation and matrix formation phase that ultimately stabilizes the solid food foam (Table 6.2).

FOAM STRUCTURE

Foams are complex two-phase colloidal systems containing at least initially a continuous liquid phase and a gas phase dispersed as bubbles or air cells. The properties of foams depend to a large extent on the methods used to make them. There are three dynamic procedures for determining foaming capacity of proteins: whipping, shaking, or sparging (Cumper, 1953; Yasumatsu et al., 1972; Halling, 1981; Waniska and Kinsella, 1979). One important difference between these methods is the amount of protein required for foam production. The amount of protein ranges from 3 to 40% for whipping, is approximately 1% for shaking, and ranges from 0.01 to 2% for gas sparging (Lawhon and Carter, 1971; Yasumatsu et al., 1972; Buckingham, 1970).

Prins (1988) categorized the methods for making foams into three types: (1) agitation of a given amount of liquid in an unlimited amount of air, (2) agitation of a mixture of gas and liquid in which both volumes are given, or (3) allowing gas to be generated from the liquid in the foam bubbles.

Bubbling

The food industry uses incorporation of air by aeration (bubbling at high pressures) for products such as ice cream and choco-mousse (Prins, 1988). The process is classified as a type 2 process (Prins, 1988) and involves injecting a particular quantity of air through an orifice into a specific amount of liquid. The bubbles leaving the orifice have a diameter that is dependent on the viscous forces exerted on the bubbles by the liquid. The bubble is detached from the orifice when the force exerted on the bubble is greater than the force holding the bubble to the orifice. This can be expressed as $F_g > F_b$:

$$F_g = [4/3]\pi R^3 \rho g$$

where F_g is the buoyancy force, R the bubble radius, ρ the liquid density, g the acceleration of gravity, and F_b the force holding the bubble to the orifice ($F_b = 2\pi r$, r = orifice radius). When $F_g = F_b$ the radius of the bubble (R) leaving the orifice can be approximated by (Prins, 1988):

$$R = [3\gamma r/2\rho g]^{1/3}$$

A major characteristic of the system as described by this equation is that the smaller the surface tension and/or the radius of the orifice, the smaller the bubble size (higher orifice gas flow rates and induced surface viscosity effects can result in larger bubble sizes) (Prins, 1988). When making foods such as ice cream, the bubble size is reduced further by the action of some type of mechanical mixer.

Halling (1981) asserted that the bubbling (sparging) of gas through a porous sparger was the most commonly used method in basic studies on foams. The method yields more uniform bubble sizes, and it allows easy monitoring of the progress of formation. In the bubbling or sparging method, once the foam is formed, the rate of rupture of the bubble is a function of the lamella thickness and interfacial viscoelasticity (Cumper, 1953; Mita et al., 1977).

Waniska and Kinsella (1979) evaluated a modified sparging apparatus. The foam was produced by sparging nitrogen at a known rate through a dilute protein solution (0.01–0.1%). The foam was formed inside a water-jacketed column. Nitrogen gas was sparged into the protein solution until the 70-mL foam chamber was filled. The volume of liquid in the sparging chamber was simultaneously maintained by adding protein solution. The time required to form 70 mL of foam and the volume of buffer added to the sparging chamber were recorded. After 10 min the volume of liquid drained from the foam was recorded. The strength of the foam was determined by measuring the time and distance a brass weight would fall through the foam.

Problems with this method lie largely in its dynamic range. The ability to foam varies quite widely, and it is not always possible to distinguish differences between proteins that perform at the extremes of this range. That is, if a protein is essentially very poorly able to foam, then the low protein concentrations and mild conditions of sparging may not form a foam at all, while extremely surface active proteins, e.g., β-Lg, foam too well to be easily distinguished from each other (Phillips, 1988).

Breadmaking can be considered a reverse bubbling method whereby gas-producing yeast are dispersed throughout such that each yeast cell acts as a miniature sparger (Prins, 1988). Beer and other carbonated beverages are other examples in which the continuous phase is supersaturated with gas such that bubbles are formed upon nucleation and release of the gas from the continuous phase.

Shaking

Shaking fundamentally resembles whipping rather than bubbling, which involves rapidly agitating a fixed volume containing a protein solution producing a foam that can be measured either simply by its volume change (Yasumatsu et al., 1972; Graham and Phillips, 1976) or very sensitively by the difference in internal pressure due to the total curved bubble surface within the closed chamber. The rate at which gas bubbles are introduced into a solution is dependant on a number of parameters including the frequency and amplitude of shaking, the volume and shape of the container, the volume and flow properties of the liquid, and the surface properties of the protein (Halling, 1981).

Whipping Methods

Whipping liquids with a mixer is the final type of foaming, i.e., agitation of a given amount of liquid in an unlimited amount of air (Prins, 1988). In this case the foaming process is related to the speed of the mixer, the geometry of the beaters, the rheological properties of the liquid, as well as the surface properties of the material being foamed (Phillips et al., 1990a; Prins, 1988).

Prins (1988) described the change in bubble surface area (A) when whipping at a particular speed:

$$\frac{\Delta \ln A}{\Delta t} = \frac{\rho v^3}{3Lp}$$

where A is the bubble surface area, t the time in sec, v the circumference velocity of the beaters, L the distance between the beaters, ρ the solution density, and p the atmospheric pressure.

The change or fluctuating surface area of bubbles has important implications for bubble stability because of the concomitant fluctuation in surface tension. Fluctuating surface tension affects the foam stability because an increase in the whipping velocity causes the system to reach a point where the surfactant cannot move to the interface and stabilize the surface tension. Furthermore, the newly formed surfaces may collapse if the surface tension increases to a critical level with bubble coalescence being the result. Prins (1988) observed this behavior by varying the mixer speed and measuring the foam volume of various surfactants.

Whipping protein solutions in a mixer is perhaps closest to the practical production of foamed foods and can readily discern what amounts to practically important differences between proteins (Phillips et al., 1987, 1990a). A variety of devices that vigorously agitate a liquid and its interface with a bulk gas phase can be used for whipping (Halling, 1981).

The process of bubble formation and the history of a single bubble are not well defined. This has made this method unpopular for basic studies (Halling, 1981). Although sparging has been used more for basic studies, one must remember that it is not just a simplified model of the whipping process.

The whipped foam is completely mixed during formation. The stratification sometimes observed during sparging is not observed with whipping. The observed maximum levels of gas incorporation during whipping reflects a much more real dynamic equilibrium between mechanical formation and destruction of bubbles. This in turn gives a more realistic measure of foam stability. Foam stability usually is measured as the amount of time required for a specified amount of liquid to drain from the foam.

Phillips et al. (1990a) through a collaborative study outlined several practical variables that should be considered when using whipping as a method for measuring foam formation and stability. The main source of variability was the mixer itself. A mixer with two beaters was found to be more reproducible than one with a single beater. Other sources of variability, related to the mixer, were nonuniform bowls, beaters, and mixer speeds. Each of these parameters had to be standardized before reproducible results could be obtained from all the contributing laboratories (Phillips et al., 1990a). If protein powders were the starting material, they had to be completely dispersed in a standardized manner (Morr et al., 1985) and a solution volume selected that was adequate for use with the mixer (Phillips et al., 1990a). The protein-protein interactions that occur during foam formation are reflected in the overrun profiles of various proteins.

FOAM STABILITY

The gas dispersion in foams generated by high agitation (whipping and shaking) (Yatsumatsu et al., 1972; Phillips et al., 1990a) or by direct introduction of a gas through a liquid phase (sparging or injecting) (Buckingham, 1970; Waniska and Kinsella, 1979; Kim and Kinsella, 1985) introduces small gas bubbles into the sample yielding a discrete dispersion of air bubbles entrapped by the stabilizing properties of the continuous protein solution (Prins 1988). As more gas is introduced, entrapped bubbles are forced upward carrying with them the continuous aqueous phase. This is a highly unstable state where disintegration mechanisms, mainly drainage of continuous liquid phase and bubble breakage, begin immediately and therefore must be considered a part of the process of foam formation. Most studies evaluating foam stability have neglected this essential component of foaming behavior. Stability of foams has typically been evaluated by measuring foam height and/or drained liquid changes as a function of time beginning after the formation process has taken place (Waniska and Kinsella, 1979; Halling, 1981; German et al., 1985; Kim and Kinsella, 1985). Comparisons of foam stability based on foam drainage without regard to the events of foam formation often assume that the process of foam formation is identical in the systems of comparison. Such an approach, evaluating foam stability by the single event of liquid drainage, severely oversimplifies the multifold processes that comprise a foam's behavior, especially the distinctions between various foaming agents.

Foams vary with the method and equipment used (Halling, 1981; Kinsella, 1984; Phillips et al., 1990a). Foams produced by whipping are fundamentally different from foams produced by sparging and, therefore, are not directly comparable. Studies must develop fundamental principles underlying the process rather than describing the foams themselves. Even sparging at different flow rates produces different foams (Halling, 1981). Furthermore, attempts to describe the foaming process from beginning to end are problematic. Methods tend to be destructive or they confound the events leading to foam disintegration. For example, comparing foams by changes in foam height or amount of drained liquid does not differentiate between foam drainage and bubble rupture. Thus, the effects of composition on the individual events of foam disintegration are not well understood.

One limitation facing researchers in this area has been a lack of highly sensitive techniques that would allow continuous measurement without compromising the complexity of foams. Quantitative techniques for measuring foam density have recently been developed using magnetic resonance imaging (MRI) (Assink et al., 1988; McCarthy et al., 1989). In one study, foam stability in terms of the complete foam history, including formation and breakdown, were determined with well-characterized foaming properties under identical formation conditions (Pilhofer, 1991). Foam formation and disintigration were examined utilizing MRI as the measurement probe with MRI projections providing signal intensity as a function of vertical position. The MRI ^{1}H signal is dependent on proton density and thereby foam density.

A foaming apparatus was designed to fit inside an MRI birdcage coil 0.10 m i.d. and constructed with nonmagnetic materials. Solutions for foaming were dispensed into the foaming column in the imaging coil centered in the bore of the magnet. Foams were generated by metering a constant flow of nitrogen gas into the sparging cylinder.

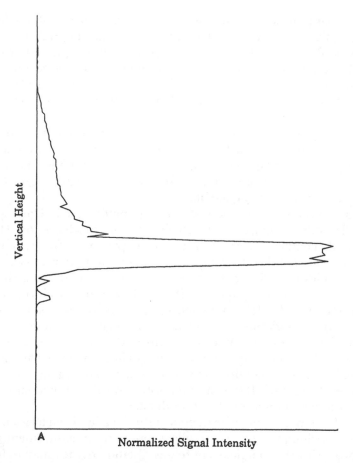

Normalized Signal Intensity

FIG. 6.1   (A) MRI projection of foamed BSA at end of sparging. (B) Initial projections of egg albumin foam.

Measurements of the ¹H spectra of the foams were obtained with a two Tesla CSI-II imaging spectrometer (General Electric Medical Imaging Systems, Freemont, CA). One-dimensional projections of the foamed protein solutions were acquired using a spin-echo Fourier imaging pulse sequence with the phase encode gradient turned off. Relaxation time constants were estimated for the foams by acquiring projections at echo times of 10, 12, and 15 msec. The parameters used for the final spectra were 15 sec predelay time, 10 msec echo time, 128 point resolution, and a 102.3 mm field of view. Foam disintegration was followed as a function of time by one-dimensional projections of the distribution of the continuous phase in the foam.

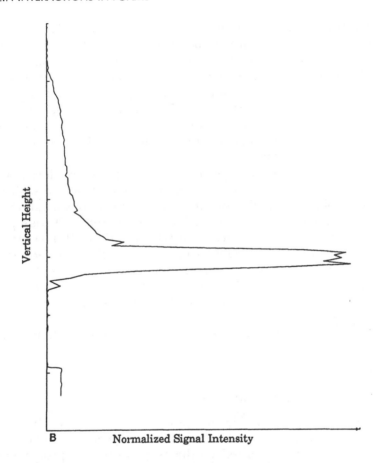

B   Normalized Signal Intensity

Even during foam formation, destabilizing events such as drainage, bubble collapse, and gas diffusion occur. Of the two proteins shown in Fig. 6.1, bovine serum albumin and ovalbumin, ovalbumin exhibited the least stratification of foam density. Complete foam projections detected less stratification of foam density, hence less drainage within the albumin foams before measurement by normal drainage methods could even be conducted.

Sequential projections over time measured the progress of the destabilizing events that contributed to disintigration of the foam. Density at a single position near the top of the foam reflects disintegration. Simple drainage from a system with unchanging bubble size distribution has been suggested to follow close to first-order kinetics (Halling, 1981; Kim and Kinsella, 1985). A first-order equation fitted to these density data, however, was clearly inadequate. If, in fact, simple drainage does follow first-order kinetics, this would suggest that other events are also contributing to foam disintigration, primarily bubble collapse. The magnitude of departure from first order was dependent on the protein. The β-casein foam was least well fit by a first-order equation. Again, assuming that the drainage component was indeed proceeding as a first-order kinetic process, the significantly greater departure would suggest the greatest collapse in the β-casein foam. This would argue that the β-casein bubbles were most fragile to breakage, supporting previous independent studies that found β-casein foams to be weaker than BSA foams (Graham and Phillips, 1976; Phillips, 1981).

Measurement of air incorporation and foam stability does not completely describe important differences between protein foams. Measurement of bubble size distribution and lamellar dimensions are also needed for complete description of foams (German et al., 1985). This is especially important for stability measurements (Halling, 1981). If the breakdown of the foam is caused by bubbles increasing in size and decreasing in number, this will lead to textural differences in the foam without drainage of liquid. If the destabilization is primarily the result of lamellae rupture, this will bring about loss of liquid from the foam (Halling, 1981). Understanding the fundamental process and how to control it will be difficult without information on the mechanism of destabilization.

PROTEIN STRUCTURE

Solution and Surface Properties

The extent of protein film formation is related to the ability to decrease surface tension, while foam stability is dependent upon the nature of the film, which in turn reflects the extent of protein-protein interactions within the film matrix. This is exemplified by β-casein, which rapidly reduces the surface tension and facilitates a large volume increase in a short amount of time; however, the foam is relatively unstable because of limited protein-protein interactions and weak films (Graham and Phillips, 1976; Kim and Kinsella, 1985). In contrast, bovine serum albumin retains considerable tertiary structure at the interface, thus maintaining an extensive intermolecular network for formation of strong films, which results in more stable foams (Damodaran, 1989).

These dichotomous structural requirements for foam formation and stability are not yet clearly distinguished. Structural features of proteins that are conducive to forming rapid and large volume foams, e.g., low molecular weight and amphipathic nature, are not necessarily ideal for promoting the protein-protein interactions that give rise to viscoelastic films and hence stable foams (Phillips, 1981; Kinsella, 1981; Kinsella and Whitehead, 1988). Upon formation of a protein-encapsulated bubble, component proteins should interact via extensive noncovalent interactions, i.e., electrostatic, hydrophobic, and hydrogen bonding, and to a certain extent via covalent disulfide bonding to form a strong viscoelastic, continuous, cohesive, "polymeric," impervious film that retains water and provides a strong structural matrix for the foam. The inherent molecular characteristics of proteins influence the formation and stability of protein-based foams, as well as the interactive forces that develop from such molecular characteristics.

Foam stability, the retention of air volume and water, is a reflection of film integrity, impermeability to gas, and viscoelastic and mechanical strength of the film (Kinsella, 1981; Kinsella amd Phillips, 1989). A thicker, denser yet elastic film retards gas diffusion, retards disproportionation, coalescence, and rupture, and slows eventual collapse. Thicker, more condensed films are formed by certain proteins depending upon the pH of the solution, e.g., BSA versus β-casein. Generally, thicker films are formed closer to the isoelectric pH when protein concentration is not limiting, although this is rare in food systems (Castle et al., 1986). Drainage of the lamellar water is a major cause of foam collapse (Halling, 1981). This may be minimized by maintaining capillary hydrostatic pressure, increasing the viscosity of the lamellar fluid, and preventing close approach and contact of the films of adjacent bubbles, which is conducive to rupture and coalescence of these bubbles. During drainage, van der Waals attractive forces between films in adjacent bubbles increase as the bubbles approaches. These attractive forces are opposed by repulsive forces, e.g., electrostatic forces, steric hindrance, volume restriction, and osmotic effects operating between protein groups. The number and disposition of charged groups, which varies among proteins, affects protein-protein interactions in films and electrostatic repulsive interactions between foam bubbles. The pH of the aqueous phase affects foaming by determining the magnitude and nature of the net charge on proteins (Kinsella, 1981; Kinsella and Whitehead, 1987; Kinsella and Phillips, 1989).

The same intrinsic forces that determine the structure and flexibility of a protein, namely, electrostatic interactions, hydrogen bonds, hydration and hydrophobic effects, and disulfide bonds, also determine the interfacial behavior and the foaming properties of a single protein and the interactions between protein mixtures at the bubble surface.

Hydrophobic Effects

Hydrophobic interactions are the main driving force responsible for the structure of proteins in solution. Despite the largely polar (hydrophilic) nature of the surface of protein molecules, a significant number of apolar residues are exposed (or accessible to solvent molecules) at the surface of proteins, such as β-lactoglobulin (Kato et al., 1983). The foaming properties, as well as other surface-active properties, of proteins correlate with surface hydrophobicity. It is attractive to consider an exposed hydrophobic portion of a protein as favoring its association into a developing film. Thus, the reduction in free energy of the system gained, when hydrophobic segments of globular proteins move into the apolar air phase in a foam, can be a critical initial event that favors adsorption. However, subsequent intermolecular interactions must then develop between contiguous polypeptides at the interface, which is important in stabilizing films (Townsend and Nakai, 1983; Kato et al., 1985). The contribution of hydrophobic effects to the functional properties of foods has received considerable attention in recent years (German and Phillips, 1989; Phillips and Kinsella, 1989; Li-Chan and Nakai, 1989).

The relationship between the surface hydrophobicity of β-Lg and the corresponding foaming properties is not entirely clear. Unfortunately, the methods used to estimate surface hydrophobicity using fluorescent binding ligands are confounded by the fact that certain natural proteins have a defined biological role of binding hydrophobic compounds. As a result, high apparent hydrophobocities may be misleading in interpreting surface functionality. Nevertheless, positive correlations between the surface hydrophobicity of β-Lg and overrun have been observed, although no correlation with foam stability was reported (Kato et al., 1983). Furthermore, while a strong correlation was observed between hydrophobicity and surface activity, there was only a weak correlation between hydrophobicity and foaming properties (Kato et al., 1981; Townsend and Nakai, 1983). The surface hydrophobicity of β-lactoglobulin, as assessed by fluorometric measurement, exhibited a significant correlation with foaming capacity when the protein in solution was unfolded by heating at 100°C in the presence of 1.5% (w/v) sodium dodecyl sulfate prior to measurement, suggesting that many globular proteins are more extensively unfolded at an air/water interface than at an oil/water interface (Townsend and Nakai, 1983; Nakai, 1983). Positive correlations between the exposed hydrophobic patches on protein surfaces and the adsorption capacity of some proteins, e.g., β-casein, have been reported (Nakai 1983).

Li-Chan and Nakai (1989) developed equations that incorporate protein hydrophobicity as a significant variable to predict functionality. This endeavor is referred to as the quantitative structure-activity relationship, QSAR (Li-Chan and Nakai, 1989). The importance of surface hydrophobicity was stressed instead of "total" hydrophobicity, which is exemplified by Bigelow's hydrophobicity parameter computed as the sum of the side chain hydrophobicities. Li-Chan and Nakai (1989) found little correlation between surface hydrophobicity and foaming. They suggested that the poor correlation was the result of the change in surface hydrophobicity upon unfolding a protein at an interface. Therefore, a measure of surface hydrophobicity in solution may be quite different from the actual surface hydrophobicity of a protein at an interface (Li-Chan and Nakai, 1989). To improve their analysis, they developed a new technique for measuring the hydrophobicity after the protein has been suitably unfolded. The resulting "exposed" hydrophobicity showed significant correlations with foam capacity (Li-Chan and Nakai, 1989). Similarly, Phillips (1992) demonstrated that the structures of β-Lg in solution prior to foaming correlated with the observed foaming properties to a greater extent than for surface hydrophobicity. Phillips (1992) observed an increased rate of proteolysis when β-Lg foams were treated with pronase as compared to the rate of proteolysis for the β-Lg solution prior to foaming. This was indicative of a more unfolded β-Lg molecule at the air/water interface with more hydrophobic groups exposed as compared to the protein in solution. Phillips (1992) showed that β-Lg unfolds at the air/water interface during foaming but in the absence of salt, refolded after the foam was collapsed.

Studies have shown that partial heat denaturation of β-Lg and BSA resulted in an increase in surface hydrophobicity, suggesting unfolding of the globular structure and enhanced foaming capacity. The effect of heat treatment is more pronounced in the case of highly structured globular proteins, e.g., lysoyme (Kato et al., 1981). The variability between proteins may arise from differences in the disposition of apolar amino acids, the proportion of apolar to polar amino acids, the differences in molecular flexibility between protein molecules, and the presence of disulfide bonds, e.g., β-casein versus β-lactoglobulin. Partial unfolding of some proteins at the interface improves their foaming properties, and limited reduction of the disulfide bonds of β-Lg improves foaming properties ostensibly by allowing greater hydrophobic contact between unfolded polypeptides at the interface (Phillips, 1988; Kella et al., 1989).

Conformational rigidity is a structural property of proteins that has an important influence on both surface adsorption and subsequent film formation. The flexibility of certain domains or segments of proteins may be a major structural factor determining surface-active properties such as film formation and foaming capacity and may help explain the disparity in correlations between foaming and hydrophobicity (Townsend and Nakai, 1983; Shimizu et al., 1983, 1986; Kato et al., 1986).

Some conformational flexibility of the protein correlates with improved surface activity since rearrangment of the protein conformation is necessary to expose hydrophobic and hydrophilic residues to the appropriate phase (Nakai et al., 1980; Kato et al., 1985). Proteins that are maintained in rigid conformations possess disulfide bonds, even when they form condensed interfacial films with good rheological properties (Nakai, 1983; Morr et al., 1985). Foaming properties of such rigid proteins generally improve after removing intramolecular disulfide bonds (Li-Chan and Nakai, 1989; Kella et al., 1989).

Electrostatic Attraction

Traditionally, it has been proposed that the net negative charge on the outer film lamella helps stabilize foams via net repulsion of adjacent films; however, electrostatic attractions within the film may also be important (Kinsella and Phillips, 1989). In fact, the frequently superior performance of naturally heterogenous protein systems, such as egg white, in which the constituent proteins have different isoelectric points and carry different net charges argues strongly that electrostatic interactions are key to their enhanced performance (Table 6.3). In egg white, electrostatic interactions contribute to excellent foaming and most importantly their relatively unique heat stability characteristics (Johnson and Zabik, 1981; Poole et al., 1984). At the natural pH of fresh egg white (pH 7–8), the basic protein lysozyme (pI 10.7) is positively charged and can interact electrostatically with negatively charged proteins (Phillips et al., 1989a). This principle has been applied to designed mixtures of net positive and negative proteins. Poole et al. (1984) demonstrated that the addition of low concentrations (0.01–0.1% w/v) of clupeine (pI 12) and lysozyme to solutions of acidic proteins (0.50%, pH 8) improved their foam volume and stability by as much as 260% and 600%, respectively. They suggested that electrostatic interactions between basic and acidic proteins in the film enhanced foam volume and stability. The addition of sucrose together with lysozyme and clupeine enhanced the foam volume and stability by as much as 420% and 700%, respectively. They suggested that sucrose decreased the hydration of the proteins thereby facilitating their movement to the interface (Poole et al., 1984).

Phillips et al. (1989b) demonstrated that lysozyme improved the foaming properties of β-Lg and whey protein isolate (WPI). The stability of β-Lg and WPI foams were increased by 124% and 114%, respectively, by adding 0.5% lysozyme. The resultant foams were in many ways superior to egg white foams. Again the very important practical property of heat stability was improved by this heterogeneous mixture. Lysozyme and clupeine improved the heat stability of β-Lg and WPI foams. Basic

TABLE 6.3   Effect of pH on Film and Foaming Properties of Bovine Serum Albumin

pH	Surface pressure	Surface yield stress (dyne/cm)	Film elasticity	Foam drainage half-life (min)
4.0	2.8	3.0	2.2	5.0
5.0	15.0	3.8	5.0	8.0
5.5	19.0	4.0	5.2	9.6
6.0	14.0	4.3	5.4	8.5
7.0	10.0	3.0	2.3	6.3
8.0	2.0	2.2	1.8	6.0

Source: Kim and Kinsella, 1985; Kinsella and Phillips, 1989.

proteins such as lysozyme may alter the structure of flexible acidic proteins at a pH between the pIs of the two proteins (Poole et al., 1987). This interaction may lead to the formation of an elastic film that is disrupted less rapidly during heating. Perhaps not surprisingly, molecular size appears to be an important variable in electrostatic interactions. Poole et al. (1987) reported that the basic amino acid arginine did not enhance foaming, and clupeine lost its capacity to enhance ovalbumin foams after 18.4% hydrolysis. These studies suggest that the oppositely charged species are not just acting as a counterion but rather as a polymeric component of this matrix and a minimum size for the basic protein may be required for enhancement of foaming properties.

Similar to the results with β-Lg and lysozyme, it was possible to provide conditions for electrostatic attraction that improved foaming by mixing succinylated (negative charge added) and native β-Lg (Phillips and Kinsella, 1991). The electrostatic interactions caused by the addition of 100% succinylated β-Lg (0.5g/100 mL) to a 2.5% solution of native β-Lg at pH 4.0 improved overrun (47%) and foam stability (61%). The use of mixtures of proteins carrying opposite charges may prove very useful for stabilizing food foams once enough is known about the electrostatic interactions involved for proper control.

Electrostatic Repulsion

Interactions such as electrostatic repulsion can have a pronounced effect on foaming. Phillips and Kinsella (1991) studied the effect of increasing the negative charge on β-Lg via succinylation. They observed 45 and 55% reduction in foam stability for 50 and 100% succinylation, respectively. This was attributed to the increase in negative charge on the surface of the β-Lg molecule following succinylation, which resulted in electrostatic repulsion. This repulsion hindered film and foam formation (Phillips and Kinsella, 1991).

Disulfide Bonds

Disulfide bonds demonstrably reduce the intrinsic flexibility of a native protein. Understanding the free thiol/disulfide interchange is of extreme importance in manipulating functional properties. In the case of soy protein, disulfide bonds not only limit molecular flexibility but also restrict foaming. German et al. (1985) established that the molecular alterations induced by reducing intersubunit disulfide bonds greatly improved film formation, foaming, and foam stability.

Recent studies indicate that disulfide bonds affect the unfolding ability and interfacial adsorption of whey proteins [90% β-lactoglobulin (w/v)], since cleavage (in a chemical denaturant environment) was accompanied by progressive increases in surface hydrophobicity, viscosity, foaming capacity, and foam stability (Kella et al., 1989). The results suggest that limited cleavage of disulfide bonds (50% bond cleavage) in proteins allows more facile unfolding at the air-water interface, which increases protein adsorption and hence foaming properties, possibly because of an increased number of points of contact with the interface (Kella et al., 1989). Since disulfide bonds affect the ability to unfold at the interface and then stabilize the surface films, the role of thiol/disulfide bond interactions are (again not surprisingly) protein dependent.

Doi et al. (1989) studied the relationship between disulfide bond formation during foaming and the foaming properties of ovalbumin. They concluded that the essential factor for formation of stable ovalbumin foams was not the formation of intermolecular disulfide bonds since disrupting these disulfide bonds did not affect foam stability. They hypothesized that the network formation by other noncovalent interactions was the main contributor to foam formation and that polymerization was just a side effect. Li-Chan and Nakai (1989) suggested that the critical roles of disulfide bonds was to stabilize protein structure, restricting the unfolding of the molecule and preventing complete exposure of buried hydrophobic regions. Phillips (1992) has shown that disulfide bond formation at an air/water interface can dramatically improve foam stability.

EXTRINSIC EFFECTS ON STRUCTURAL PROPERTIES OF PROTEINS AND FILMS

Protein concentration, film thickness, ionic strength, pH, temperature, and the presence of other components in food systems (sugars, lipids, alcohols, etc.) in addition to the physicochemical properties of the protein affect the foaming properties of proteins (Morr et al., 1985). For example, increasing the protein concentration generally produces a thicker lamellar film and results in improved overrun and foam stability, which is directly attributable to the viscosity effect.

pH Effects

The solution pH significantly affects the properties of foams by affecting the net charge of the protein and resulting film formation and film properties. The rate and extent of surface pressure development, protein structure, protein-protein interactions, film thickness, and viscoelastic properties are all affected by the net charge on the protein molecule.

Generally more rapidly formed, stronger films are obtained at pH values close to the isoelectric pH of most proteins, e.g., bovine albumin (BSA) and β-lactoglobulin (Kinsella, 1981; Halling, 1981; Waniska and Kinsella, 1985; Mitchell, 1986; Kim and Kinsella, 1985; Barbeau and Kinsella, 1986). The enhanced foam stability of many proteins is pronounced in the isoelectric pH range (Mita et al., 1977; Kim and Kinsella, 1985; German et al., 1985; Phillips et al., 1990b). Foams prepared near the IEP usually have improved foam stabilities if the particular proteins are still dispersible at that pH (Kinsella and Phillips, 1989). The highest foam stability values (50% drainage = 21.5 min) for the β-Lg solutions (0.5% conc) occurred at pH 5.0 (Phillips, 1992). These results parallel the results of Phillips et al. (1990b). They observed optimum foam stability for whey protein isolate at pH 5. The high foam stability was caused by the enhanced interfacial behavior of β-Lg at a pH close to the isoelectric point (IEP) (i.e., β-Lg IEP is 5.2). The proteins have decreased electrostatic repulsion at the interface and the more compact protein molecules can pack to a greater extent into the interface forming stronger films (Waniska and Kinsella, 1985).

Proteins typically foam best at pH levels where the molecules are flexible and less compact (Phillips, 1992). The overrun for β-Lg was highest at pH 1 and pH 9, which corresponded to pH levels where β-Lg is a flexible monomer (Kella and Kinsella, 1988; Phillips, 1992).

Foams produced under conditions where the protein was considered to be more compact, rigid, and difficult to denature had lower overrun values (Phillips et al., 1990b, Phillips, 1992). β-Lg in particular is a compact protein that is difficult to denature between pH 5 and pH 1.5 (Kella and Kinsella, 1988; Phillips et al., 1990b; Phillips, 1992). It has been shown (Fig. 6.2) that in the range pH 3–4 where the molecule is rigid and resistant to denaturation, the overrun values obtained for β-Lg are less (Phillips et al., 1990; Phillips, 1992).

β-Lg foams made at pH 5 exhibited the highest foam stability but not

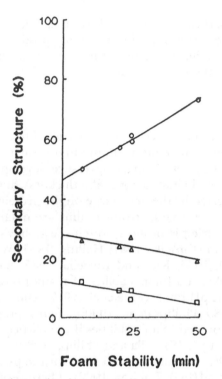

Foam Stability (min)

FIG. 6.2   Relationship between solution secondary structure and foam stability for β-lactoglobulin foamed from various salt solutions (□, α-helix; ○, β-sheet; Δ, random coil). (*From Phillips, 1992.*)

the highest overrun (Phillips, 1992). At pH 5, the pH was close enough to the IEP of β-Lg that aggregation was observed (Phillips, 1992). This manifested itself as lower observed solubility (16% reduction) and a more turbid solution (10-fold reduction in transmittance values) (Phillips, 1992). The aggregation did not adversely affect foam stability but may have caused the observed reduction in overrun. Lower overrun values have been attributed to excessive aggregation for egg white foams (Phillips et al., 1987) and also for foams formed from lysozyme, β-Lg mixtures (Phillips et al., 1989a). Aggregation probably reduces the amount of protein available for film formation but the films that are formed are considered to be thicker and more stable facilitating formation of a network structure in the protein film that results in improved rheological properties of the film, thereby increasing foam stability (Phillips, 1977; Kim and Kinsella, 1985, Phillips et al., 1989b; 1990b).

Interfacial properties such as the work of compression and the rate of adsorption were found to be positively correlated with foam stability, overrun, maximum overrun, and overrun development in the isoelectric pH range of β-lactoglobulin, while the rate of rearrangement and the average area cleared showed no significant correlation with the foaming data (Waniska and Kinsella, 1985; Phillips et al., 1990b).

β-Lactoglobulin progressively undergoes molecular expansion with increasing flexibility between pH 3 and 9, and this phenomenon has been correlated with enhanced surface activity and an increase in overrun (Shimizu et al., 1985; Waniska and Kinsella, 1985; Phillips, 1992). The rate of surface pressure development is rapid at pH 7.0, and equilibrium is achieved at 20 min in contrast to 24 hr at pH 3.0, indicating that the more flexible protein conformation enhances interfacial activity (Shimizu et al., 1985).

Surface pressure development, film elasticity, and yield stress values of BSA films were affected by changes in pH, and maximum values occurred slightly above the isoelectric pH range (5.0–6.0) (Kim and Kinsella, 1985). Maximum foam stability also coincided with this pH, reflecting the importance of minimum electrostatic repulsive forces within the film (Kim and Kinsella, 1985; Waniska and Kinsella, 1985). The collective data provide further evidence that interfacial film formation is optimized by a balance of charge on the protein molecule at pH values near the isoelectric point.

Effects of Temperature

Hydrophobic effects are reduced as the temperature is lowered (Kinsella and Phillips, 1989). The importance of temperature to hydrophobic effects and to foam formation has been demonstrated by the observed decrease in overrun of β-Lg with a decrease in temperature. Phillips (1992) also observed an eightfold reduction in the foam stability of β-Lg when the temperature was reduced from 25°C to 3°C. They attributed the decrease to reduced hydrophobic effects causing poor film formation resulting in reduced foam stability. The influence of temperature on the foaming properties of β-Lg suggested that the disrupting forces imposed on the protein during the foaming process reduced the temperature necessary for 50% unfolding (T_m) to roughly 22°C as compared to 69.2°C in solution (Phillips, 1992).

Heating β-Lg can alter its structure and its foaming properties (Phillips, 1988). Heat treatment at 80°C for 10 min had little effect on foam stability at pH 4.0 or 7.0, but at pH 5.0 the stability was improved by 65%, i.e., the stability of the foams at pH 5.0 was 100.1 and 60.8 min at

80°C and 25°C, respectively (Phillips, 1988). However, heat treatment of β-Lg (80°C) for 15 min resulted in reduced overrun at all pH levels, probably reflecting increased protein aggregation, which may have reduced the available protein for new film formation during whipping (Kato et al., 1983).

Neutral Salts

Neutral salts such as Na_2SO_4, NaCl, and NaSCN affect the physicochemical properties and interactions between proteins either by ionic strength effects, binding to the protein charged groups, or at high concentrations by altering water structure with subsequent changes in hydrophobic effects (Damodaran and Kinsella, 1982). Salts can be effective probes for studying the effects of altering hydrophobicity on foaming. Calcium is apparently more effective than equimolar concentrations of sodium in decreasing foam stability, and this has been attributed to a decrease in the thickness of the electrical double layer, which promotes coalescence of protein-coated air bubbles. Foam stability of whey protein concentrate solutions (WPC) decreased linearly with the square root of ionic strength and maximum overrun occurred at 0.05 M sodium chloride. The effects on foaming properties varied with ion species and concentration; progressive replacement of calcium with sodium reduces foaming capacity but apparently has little effect on foam stability. The effects of calcium addition in reducing foamability may indicate shielding stability by increasing the viscosity of lamellar fluid, which reduces the drainage rate (Halling, 1981). However, the addition of 10% (w/v) sucrose to β-lactoglobulin reduced the overrun and foam stability, although the specific viscosity increased fourfold (Phillips et al., 1989b). Sucrose had no apparent effect on the overrun or foam stability of whey protein isolate solutions, but sucrose improved the heat stability of β-lg and WPI foams (Phillips et al., 1989b). The stabilizing effect of sucrose on the heat stability of foams may originate from the preferential hydration of proteins in solution (Lee and Timasheff, 1981). Sucrose permits formation of a more cohesive network within protein films, instead of random aggregation that occurs upon heating (Poole et al., 1987). The effects of sucrose on protein solutions are attributed to an increase in viscosity of the solution, but it is also conceivable that sucrose, by increasing the stability of the protein, minimizes unfolding at the interface, thus reducing foaming properties. Sucrose enhances adsorption of certain proteins at air-water interfaces, and it may minimize surface denaturation thereby enhancing film strength and viscoelasticity (MacRitchie, 1978). Sucrose enhances foaming of certain proteins, as in the case of ovalbumin, a poor foaming protein at all pH values studied, which displayed markedly improved foaming properties at pH 5.25, near the isoelectric point, in the presence of sucrose and when in the presence of clupeine (Poole et al., 1987).

Lipids and other low molecular weight surfactants destabilize protein foams because of their higher surface activity, thereby displacing proteins from the interface in a competitive manner. This diminishes film thickness, interrupts film cohesiveness, and ultimately weakens the film resulting in decreased foam stability (German and Phillips, 1989; Damodaran, 1989). Many of the apparent poorly performing historical sources of functional proteins, especially with respect to foam formation and stability, are now attributable to trace contaminants that actively destabilize film formation. Phillips et al. (1989a) isolated a foam depressant fraction from milk that could completely inhibit egg white, casein, or whey protein foams. It contains protein as well as fat and was retained on a 100,000 molecular weight cutoff membrane. The foam depressant may have affected the various foams as an hydrophobic particle. The foam depressant was probably large enough to span the air bubble films thus rupturing the films by causing locally high pressures as liquid flowed away from the foam-depressant particles (Prins, 1988). Once the depressant was removed from milk protein products, the functionality of the remaining proteins was frequently excellent.

Presence of Colloidal Particles

A variety of studies have examined the role of additional colloidal particles in the continuous phase of foams. These are frequently normal consitutents in the overall aqueous system of many foods and can consist of gelled proteins, crystals of either sugars, salts, or fats, or emulsified fat globules. In each case the properties that these components provide differ depending on the particles themselves and, very importantly, on the nature of the proteins responsible for film formation.

One key aspect to the role of emulsified fat in foam formation and stability is the extent to which the emulsion interface is a denaturing surface for the proteins in question. If so, this extra surface or "environment" can actually predispose proteins to additional interactions at the bubble surface. This also favors the formation of globule networks stabilized by surface denatured protein that can in turn alter the continuous phase viscosity. This is seen practically, for example, as a difference between the foaming properties of homogenized milk in which the globule surface is primarily casein micelles versus unhomogenized milk in which the globule surface is coated by phospholipids. The stability of a foam containing emulsified fat particles may also depend on both the size and composition of the dispersed lipid particles (Anderson and Brooker, 1988; Brooker, 1990). Long-term stability of whipped cream is thought to result from formation of a three-dimensional network of fat crystals (Brooker, 1990). Since this network must develop during the foam-formation stage, the fat must be liquid initially but then be caused to crystallize during foaming. Needless to say, the physical property requirements of the fat phase are somewhat precise, and the uniquely heterogeneous nature of milkfat is thought to be important. The presence of both solid fat, to promote fat globule rupture, and liquid fat, to promote clumping, may be necessary for formation of a stable whipped cream (Buchheim and Dejmek. 1990).

NEW HORIZONS IN FOAM/FILM APPLICATIONS

With the new rapidly developing understanding of the structures and roles of discrete proteins in foaming, these principles are being applied to a host of novel applications. These include both traditional food materials and applications of biomaterials that perhaps have not even been considered before; with changing economic climates and public perceptions, opportunities are beginning to proliferate.

In addition to food foams, many materials are in fact foams. These include personal products as exotic as pharmaceutical delivery systems and as commonplace as shaving cream and packaging fillers. The inherent biosafety of proteins is an excellant selling point around which to base product development in some of these nontraditional foaming arenas.

Now that foams have been recognized as largely a viscoelastic film system, albeit a very large viscoelastic film system, the principles of foam stability and the relationships to molecular components of proteins are being applied to other film systems. The most obvious applications represent that broad area of edible polymeric films surrounding foods. The recent excitement for altering the permeability properties of food materials by including edible films around them has excited a great deal of interest in film applications.

REFERENCES

Adamson, A. 1982. *Physical Chemistry of Surfaces*, 4th ed. John Wiley and Sons, New York.
Anderson, M. and Brooker, B. E. 1988. Dairy foams. In *Advances in*

Food Emulsions and Foams, E. Dickinson and G. Stainsby (Ed.). Elsevier Applied Science, New York.

Assink, R. A., Carihan, A., and Fukushima, E. 1988. Density profiles of a draining foam by nuclear magnetic resonance imaging. *AIChE J.* 34(12): 2077.

Barbeau, W. and Kinsella, J. 1988. Ribulose bisphosphate carboxylase/oxygenase (rubisco) from green leaves: Potential as a food protein. *Food Rev. Int.* 4(1): 93.

Barfod, N. M., Krog, N., and Buchheim, N. 1989. Lipid-protein-emulsifier-water interactions in whippable emulsions. In *Food Proteins: Structure and Functional relationships*, J. E. Kinsella and W. G. Soucie (Ed.). Am. Oil Chem. Soc. Press, Champaign, IL.

Brooker, B. E. 1990. The adsorption of crystalline fat to the air-water interface of whipped cream. *Food Struct.* 9: 223.

Buchheim, W. and Dejmek, P. 1990. Milk and dairy-type emulsions. In *Food Emulsions*, K. Larsson and S. Friberg (Ed.), p. 221. Marcel Dekker, New York.

Buckingham, J. 1970. Effect of pH, concentration and temperature on the strength of cytoplasmic protein foams. *J. Sci. Food Agric.* 21: 441.

Castle, J., Dickinson, E., Murray, A., and Murray, B-S. 1986. Surface behavior of absorbed films of food products. In *Gums and Stabilisers for the Food Industry*, 3, G. O. Phillips, D. J. Wedlock, and P. A. Williams (Ed.), pp. 409–417. Elsevier Applied Science Publishers, London.

Cumper, C. 1953. The stabilization of foams by proteins. *Trans. Faraday Soc.* 49: 1360.

Damodaran, S. 1989. Interrelationship of molecular and functional properties of food proteins. In *Food Proteins: Structure and Functional Relationships*, J. E. Kinsella and W. Soucie (Ed.), pp. 21–51. Am. Oil Chem. Soc. Press, Champaign, IL.

Damodaran, S. 1990. Interfaces, protein films, and foams. In *Advances in Food and Nutrition Research*, J. E. Kinsella (Ed.). Academic Press, San Diego.

Damodaran, S. and Kinsella, J. E. 1982. Effects of ions on protein conformation and functionality. *ACS Symposium Series* 206: 327.

Dickinson, E., Murray, B. S., and Stainsby, G. 1988. Protein adsorption at air water interfaces. In *Advances in Food Emulsions and Foams*, E. Dickinson and G. Stainsby (Ed.), pp. 123–164. Elsevier Applied Science, New York.

Doi, E., Kitabatake, N., Hatta, H., and Kaseki, T. 1989. Relationship of

SH groups to functionality of ovalbumin. In *Food Proteins: Structure and Functional Relationships*, J. E. Kinsella and W. Soucie (Ed.). Am. Oil Chem. Soc., Champaign, IL.

German, J. B. and McCarthy, M. J. 1989. Stability of aqueous foams analysis using magnetic resonance imaging. *J. Agric. Food Chem.* 37(5): 1321.

German, J., O'Neill, T., and Kinsella, J. 1985. Film forming and foaming behavior of food proteins. *J. Am. Oil Chem. Soc.* 62: 1358.

German, J. B. and Phillips, L. G. 1989. Structures of proteins important in foaming. In *Food Proteins: Structure and Functional Relationships*, J. E. Kinsella and W. Soucie (Ed.). Am. Oil Chem. Soc., Champaign, IL.

Graham, D. E. and Phillips, M. C. 1976. The conformations of proteins at the air-water interface and their role in stabilizing foams. In *Foams*, R. J. Akers (Ed.), p. 233. Academic Press, New York.

Halling, P. J. 1981. Protein-stabilized foams and emulsions. *CRC Crit. Rev. Food Sci. Nutr.* 155: 13.

Johnson, T. and Zabik, M. 1981. Egg albumen protein interactions in an angel food cake system. *J. Food Sci.* 46: 1231.

Kato, A. and Nakai, S. 1980. Hydrophobicity determined by a fluorescence probe method and its correlation with surface properties of proteins. *Biochim. Biophys. Acta* 624: 13.

Kato, A., Osako, Y., Matsudomi, N., and Kobayashi, K. 1983. Changes in the emulsifying and foaming properties of proteins during heat denaturation. *Agric. Biol. Chem.* 47: 33.

Kato, A., Tsutsui, N., Matsudomi, N., Kobayashi, K., and Nakai, S. 1981. *Agric. Biol. Chem.* 45: 2755.

Kato, A., Yamaoka, H., Matsudomi, N., and Kobayashi, K. 1986. Functional properties of cross-linked lysozyme and serum albumin. *J. Agric. Food Chem.* 34: 370.

Kella, N. K. and Kinsella, J. 1988. Structural stability of β-lactoglobulin in the presence of kosmotropic salts. *Int. J. Peptide Res.* 32: 396.

Kella, N., Yang, T., and Kinsella, J. 1989. Effect of disulfide bond cleavage on structural and interfacial properties of whey proteins. *J. Agric. Food Chem.* 37: 1203.

Kim, S. H. and Kinsella, J. E. 1985. Surface activity of food proteins: Relationships between surface pressure development, viscoelasticity of interfacial films and foam stability of BSA. *J. Food Sci.* 50: 1526.

Kinsella, J. E. 1984. Milk proteins: Physicochemical and functional properties. *CRC Crit. Rev. Food Sci. Nutr.* 21: 197.

Kinsella, J. E. and Phillips, L. G. 1989. Structure function relationships in food proteins: Film and foaming behavior. In *Food Proteins: Structure and Functional Relationships*, J. E. Kinsella and W. Soucie (Ed.). Am. Oil Chem. Soc., Champaign, IL.

Kinsella, J. E. and Whitehead, D. M. 1987. Film, foaming and emulsifying properties of food proteins: Effects of modification. In *Proteins at Interfaces*. ACS Symposium Series 343, Amer. Chem. Soc., Washington, DC.

Kitabatake, N. and Doi, E. 1982. Surface tension and foaming of protein solutions. *J. Food Sci.* 47: 1218.

Lawhon, J. and Cater, C. 1971. Effect of processing method and pH of precipitation on the yields and functional properties of protein isolates from glandless cottonseed. *J. Food Sci.* 36: 372.

Lee, J. and Timasheff, S. 1981. The stabilization of proteins by sucrose. *J. Biol. Chem.* 246: 7193.

Li-Chan, E. and Nakai, S. 1989. Effects of molecular changes of food proteins on their functionality. In *Food Proteins: Structure and Functional Relationship*, J. E. Kinsella and W. Soucie (Ed.), pp. 232–251. Am. Oil Chem. Soc., Champaign, IL.

Liao, S. Y. and Mangino, M. E. 1987. Characterization of the composition, physicochemical and functional properties of acid whey protein concentrates. *J. Food Sci.* 52: 1033.

MacRitchie, F. 1978. Proteins at interfaces. *Adv. Prot. Chem.* 32: 283.

McCarthy, M. J. 1990. Interpretation of the magnetic resonance imaging signal from a foam. *AIChE J.* 36(2): 287.

McCarthy, M. J., Charoenrein, S., German, J. B., McCarthy, K. L., Reid, D. S. 1991. Phase volume measurements using magnetic resonance imaging. *Water Relationships in Foods: Advances in the 1980s and Trends for the 1990s*. American Chemical Society Symposium on Water Relationships in Foods, Dallas, TX, April 10–14, 1989, H. Levine and L. Slade (Ed.). Plenum Press, pp. 615–626. New York.

McCarthy, M. J. and Heil, J. R. 1990. Mobility of water in foams determined by magnetic resonance spectroscopy. AIChE Symposium Series.

McGee, J., Long, S., and Briggs, W. 1984. Why whip egg whites in copper bowls? *Nature* 308: 607.

Mita, T. E., Nikai, K., Hiraoka, T., Matsao, S., and Matsumoto, H. 1977. Physiochemical studies on wheat protein foams. *J. Coll. Interface Sci.* 59: 172.

Morr, C., German, J. B., Kinsella, J. E., Regenstein, J., Van Buren, J., Kilara, A., Lewis, B., and Mangino, M. 1985. A collaborative study to develop a standardized food protein solubility procedure. *J. Food Sci.* 50: 1715.

Nakai, S. 1983. Structure-function relationships of food proteins with an emphasis on the importance of protein hydrophobicity. *J. Agric. Food Chem.* 31: 676.

Phillips, L. 1988. A study of the foaming properties of proteins. M. S. thesis, Cornell University, Ithaca, NY.

Phillips, L. G. 1992. Relationship between structural, interfacial and foaming properties of β-lactoglobulin. Ph.D. dissertation, Cornell University, Ithaca, NY.

Phillips, L. G., Davis, M. J., and Kinsella, J. E. 1989a. The effects of various milk proteins on the foaming properties of egg white. *Food Hydrocolloids* 3(3): 163.

Phillips, L. G., German, J. B., O'Snail, T. E., Foegeding, E. A., Harwalkar, V. R., Kilara, A., Lewis, B. A., Mangino, M. E., Morr, C. V., Regenstein, J. M., Smith, D. M., and Kinsella, J. E. 1990a. Development of a standardized method for the measurement of the foaming properties of food proteins: A collaborative study. *J. Food Sci.* 55(5): 1441.

Phillips, L. G., Haque, Z., and Kinsella, J. E. 1987. A method for the measurement of foam formation and stability. *J. Food Sci.* 52: 1047.

Phillips, L. G., Hawks, S. E., and Kinsella, J. E. 1991a. The effects of shear and temperature on the foaming properties of β-lactoglobulin. Presented at Am. Chem. Soc. meeting, New York City, Aug. 25–30.

Phillips, L. G. and Kinsella, J. E. 1991. Effects of succinylation on β-lactoglobulin foaming properties. *J. Food Sci.* 55(6): 1735.

Phillips, L. G., Schulman, W., and Kinsella, J. E. 1990b. pH and heat treatment effects on foaming of whey protein isolate. *J. Food Sci.* 55(4): 1116.

Phillips, L. G., Yang, S. T., and Kinsella, J. E. 1991b. Effects of neutral salts on the stability of whey protein isolate foams. *J. Food Sci.* 55(2): 588.

Phillips, L. G., Yang, S. T., Schulman, W., and Kinsella, J. E. 1989b. Effects of lysozyme, clupeine and sucrose on the foaming properties of whey protein isolate and β-lactoglobulin. *J. Food Sci.* 54(3): 743.

Phillips, M. C. 1981. Protein conformation at liquid interfaces and its role in stabilizing emulsions and foams. *Food Technol.* 35: 50.

Pilhofer, G. 1991. M. S. thesis, University of California-Davis.

Poole, S., West, S., and Fry, J. 1987. Effects of basic proteins on the denaturation and heat gelation of acidic proteins. *Food Hydrocolloids* 1(4): 301.

Poole, S., West, S., and Walters, C. 1984. Protein-protein interactions: Their importance in the foaming of heterogeneous protein systems. *J. Sci. Food Agric.* 35: 701.

Precht, D. 1986. Fat crystal structure in cream and butter. In *Thin Liquid Films*, P. Ivanov (Ed.), p. 331. Marcel Dekker, New York.

Prins, A. 1987. Theory and practice of formation and stability of food foams. In *Food Emulsions and Foams*, E. Dickinson (Ed.). Royal Society of Chemistry, Burlington House, London.

Prins, A. 1988. Principles of foam stability. In *Advances in Food Emulsions and Foams*, E. Dickinson and G. Stainsby (Ed.). Elsevier Applied Science, New York.

Shimizu, M., Ametani, A., Kaminogawa, S., and Yamauchi, K. 1986. The topography of α_{s1}-casein adsorbed to an oil-water interface: An analytical approach using proteolysis. *Biochim. Biophys. Acta* 869: 259.

Shimizu, M., Saito, M., and Yamauchi, K. 1985. Emulsifying and structural properties of β-lactoglobulin of different pHs. *Agric. Biol. Chem.* 49: 189.

Shimizu, M., Saito, M., and Yamauchi, K. 1986. Hydrophobicity and emulsifying activity of milk proteins. *Agric. Biol. Chem.* 50: 791.

Shimizu, M., Takahashi, T., Kaminogawa, S., and Yamauchi, K. 1983. Adsorption onto an oil surface and emulsifying properties of bovine α_{s1}-casein in relation to its molecular structure. *J. Agric. Food Chem.* 31(6): 1214.

Song, K. B. and Damodaran, S. 1987. Structure-function relationship of proteins: Adsorption of structural intermediates of BSA at the air/water interface. *J. Agric. Food Chem.* 35: 236.

To, B., Helbig, N. B., Nakai, S., and Ma, C. Y. 1985. Modification of whey protein concentrate to simulate whippability and gelation of egg white. *Can. Inst. Food Sci. Technol. J.* 18: 150.

Townsend, A. A. and Nakai, S. 1983. Relationships between hydrophobicity and foaming characteristics of food proteins. *J. Food Sci.* 48: 588.

Wang, J. and Kinsella, J. E. 1976. Functional properties of novel proteins: Alfalfa leaf proteins. *J. Food Sci.* 41: 286.

Waniska, R. D. and Kinsella, J. E. 1985. Surface properties of β-lactoglobulin: Adsorption and rearrangement during film formation. *J. Agric Food Chem.* 33: 1143.

Waniska, R. D. and Kinsella, J. E. 1979. Foaming properties of proteins: Evaluation of a column aeration apparatus using ovalbumin. *J. Food Sci.* 44: 1398.

Yasumatsu, K., Sawanda, K., Moritaka, S., Misaki, M., Toda, J., Wada, T., and Ishii, K. 1972. Whipping and emulsifying properties of soybean products. *Agric. Biochem. Chem.* 36: 719.

7
Protein Interactions in Gels: Protein–Protein Interactions

Denise M. Smith

Michigan State University
East Lansing, Michigan

INTRODUCTION

The typical characteristics and quality of many foods are determined by the properties of protein gels formed during heating. The textural attributes, yield, quality, and sensory characteristics of processed products such as frankfurters, cheeses, yogurt, and custards are the direct result of the formation of protein gels during processing. The properties a gel may impart to a food system vary with the type of protein, interactions with other ingredients, and processing procedures used. The type and quality of a gel formed is a direct result of specific responses of the proteins to applied forces encountered during preparation, processing, and storage. Protein responses are further modified by interactions with food ingredients, such as other proteins, salts, fats, hydrocolloids, and starches. Effects of food ingredients and processing variables on protein gelation in food systems are only poorly understood. Most research has been in single component systems in which variables can be easily controlled; however, protein interactions in a food are very complex and not fully exploited during product development. When the biochemical basis of protein gelation is understood, food scientists can more easily and scientifically engineer the desired textures and functional properties into food systems.

Gels exhibit very diverse microstructural and mechanical properties and, as such, are very difficult to define. Several review articles have attempted to define gels and characterize gel networks (Ziegler and Foegeding, 1990; Clark, 1992). Clark (1992) defined a gel as a material containing a continuous and well-defined solid network that is assembled from particles or polymers embedded in an aqueous solvent. Two types of gels are commonly found in foods (Clark, 1992). Polymeric networks are formed from linear polymers cross-linked by physical aggregation of protein through noncovalent interactions, while particulate networks are formed when partially unfolded globular proteins aggregate randomly into grapelike clusters or linearly to form a "string of beads" (Doi, 1993). A protein gel network is generally formed via noncovalent cross-linkages such as hydrophobic interactions, hydrogen bonds, or electrostatic interactions and less frequently by covalent interactions such as disulfide bonds (Clark, 1992). The relative contribution of each type of bond to a gel network varies with the properties of the protein and environmental conditions during heating.

The objective of this chapter is to describe protein gelation as a series of biochemical events, focusing on some of the biochemical mechanisms involved in the heat-induced gelation of thermally irreversible gels. Most food proteins form thermally irreversible gels (Ziegler and Foegeding, 1990); however, some proteins, such as conalbumin, β-lactoglobulin, and α-lactalbumin may form either thermally irreversible or thermally reversible gels depending on the specific conditions. A thermally irreversible gel is a viscoelastic solid formed during heating, which does not revert to a viscous liquid on reheating; thus the chemical bonds or the nonspecific chain entanglements formed during gelation are not reversed by cooling or reheating.

Heat-induced gelation has been simplified into a two-step process that involves unfolding of the protein molecule followed by aggregation into a cross-linked gel network during heating (Ferry, 1948), such that

$$x\ P_N \rightarrow x\ P_D \rightarrow (P_D)_x \tag{1}$$

where x is the number of protein molecules, P_N is the native protein, and P_D is the denatured protein. Many mathematical models describing gelation from a mechanical or rheological perspective have been reviewed recently (Clark, 1992; Clark and Lee-Tuffnell, 1986), including the Flory-Stockmayer theory (Flory, 1941; Stockmayer, 1943, 1944), mean field theory (Clark, 1992), and percolation theory (Stauffer et al., 1982).

The Flory-Stockmayer model describes gelation as a sudden event that occurs when the degree of cross-linking between polymers reaches a critical value, called the gel point, at which viscosity diverges to infinity. The percolation theory assumes that monomers form small aggregates and at a critical threshold of bonding the gel point is reached, after which the aggregates cross-link throughout the percolation lattice. Thus, the two-step process of Ferry (1948) can be expanded to more completely include aspects of some theoretical models by dividing aggregation into a series of steps leading to the final gel structure (Fig. 7.1). In this model suggested by Foegeding and Hamann (1992), proteins unfold and aggregate to form a progressively more viscous solution. Once the gel point is reached, a continuous gel network is formed which exhibits properties of a viscoelastic solid. The viscoelastic properties of the material may change with time to form an equilibrium matrix or the final gel structure.

Different biochemical techniques have been used to study both protein unfolding and aggregation. Differential scanning calorimetry (DSC) is most commonly used to monitor protein unfolding during heating. Specific changes in protein structure during heating have been studied using optical techniques, such as Fourier transform infrared spectroscopy, Raman spectroscopy, and circular dichroism. Recently, monoclonal antibodies have been used to detect specific changes in protein structure. These techniques are used to monitor changes in protein secondary structure due to heat or environmental conditions. Aggregation has been followed using turbidity measurements, light scattering, or other form of spectroscopy. Clark (1992) described dynamic testing as the method of choice for monitoring structure formation of gels. Small strain dynamic testing is a nondestructive technique often used to follow the transition

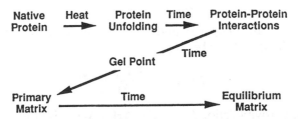

FIG. 7.1 Schematic representation of heat-induced gelation of thermally irreversible proteins. (*Adapted from Foegeding and Hamann, 1992.*)

from viscous liquid to viscoelastic solid during gelation. The use of dynamic testing to monitor gel formation has been extensively reviewed (Hamann, 1987; Ziegler and Foegeding, 1990; Hamann, 1991). Final gel structure is often observed by scanning or transmission electron microscopy.

DIFFERENTIAL SCANNING CALORIMETRY

Differential scanning calorimetry is used to measure the heat capacity of a protein as a function of temperature. It can be used to detect the thermal transition or unfolding temperatures(s) of proteins, to quantify the heat taken up or enthalpy of the conformational transition (Donovan, 1984), and to detect the existence of multiple domains within a protein. Thermal transition temperatures are determined from peaks in the heat capacity curve or endotherm caused by unfolding of the molecule. The enthalpy change associated with thermal unfolding, calculated from the area under each peak in the endotherm, is positively associated with the exposure of buried hydrophobic amino acids to solvent and negatively associated with the rupture of secondary and tertiary structure hydrogen bonds (Privalov and Gill, 1988). Polypeptides containing more than about 200 amino acid residues usually fold into two or more independent cooperative units called domains. The domains contain about 100–200 amino acid residues and may be connected by one or two polypeptide chains. In a multidomain protein, the domains can be considered independent of each other; that is, one domain can unfold independently of another domain in the same molecule.

The ratio of the calculated van't Hoff enthalpy to the calorimetric enthalpy is used to determine the existence of multiple domains within a protein (Privalov, 1979, 1982). The calorimetric enthalpy (ΔH_{cal}) is calculated by integrating the area under a DSC endothermic curve and is described by

$$\Delta H_{cal} = \int \Delta C_p \, dT \qquad (2)$$

where C_p is the heat capacity and T is the temperature. The van't Hoff enthalpy (ΔH_{vH}) is described by

$$\Delta H_{vH} = \frac{4RT_m^2 C_{p(max)}}{\Delta H_{cal}} \qquad (3)$$

If the cooperativity ratio, defined as $\Delta H_{vH}/\Delta H_{cal}$, is close to 1, then the protein contains a single domain. Multiple domains are present if the ratio is below 1. A ratio greater than 1 generally suggests aggregation or other form of molecular interaction.

The number of domains within a protein can be calculated by deconvolution of the endotherm into several transitions using a least squares fitting procedure (Freire and Biltonen, 1978a,b; Ramsey and Freire, 1990), such that $\Delta H_{vH} = \Delta H_{cal}$ and assuming a two-state unfolding process. Each domain is defined by a melting temperature, ΔH_{cal} and ΔH_{vH}.

Ideally, protein denaturation, which is primarily an endothermic process, should be studied in dilute solutions to minimize intermolecular interactions. Intermolecular interactions, such as aggregation, are exothermic and may partially mask the unfolding endotherm (Privalov, 1979). Due to a lack of instrumental sensitivity, many food protein studies using DSC were performed at high concentrations in which exothermic aggregation may alter the actual pattern of the endotherm and result in inaccurate calculation of enthalpy.

Although its use has not been extensively reported in the food science literature, differential scanning microcalorimetry is an excellent technique for studying protein denaturation in dilute solutions. Protein concentrations of 1 mg/mL in a 1-mL sample cell are used to measure changes in heat capacity of 10^{-5} calories per degree (Krishnan and Brandts, 1978). This technique is commonly used by biochemists when studying protein transitions and protein domain structure. Unfolding of myosin (Bertazzon and Tsong, 1989), actin (Bertazzon et al., 1990), and apo-α-lactalbumin (Xie et al., 1991) have been studied using differential scanning microcalorimetry. Unfortunately for food scientists, many of these experiments were performed under conditions not commonly found in food systems.

SINGLE DOMAIN PROTEINS

When heated, many small, compact globular proteins denature or unfold as a single cooperative system in a two-state process (Privalov, 1982). A single transition temperature is detected by differential scanning calorimetry and $\Delta H_{vH}/\Delta H_{cal}$ is 0.95–1.0 (Privalov, 1982). It is assumed that single-domain proteins unfold cooperatively, such that any partial unfolding of structure destabilizes the remaining structure, which must simultaneously unfold to a random coil. In this theoretical model, a protein is either in a native or denatured state and the population of intermediate states does not usually exceed 5% (Privalov, 1989). Thus, a single domain protein is stable up to a certain critical temperature. Once this temperature is reached, the protein unfolds and may then aggregate to form a gel network. Single domain proteins include ovalbumin, α-lactalbumin, lysozyme, and β-lactoglobulin (Privalov, 1979).

The relationship between heat-induced denaturation as determined by differential scanning calorimetry and structure development monitored by dynamic testing of a single-domain protein can be illustrated using the egg albumen protein ovalbumin. Ovalbumin is a globular protein with a molecular weight of about 45,000, which comprises about 54% of the total egg white protein. This protein is responsible for many of the functional properties associated with egg albumen and thus plays an important role in the formation of scrambled eggs, omelets, custards, and other egg-containing food products.

Ovalbumin contains a single domain as evidenced by a cooperativity ratio of 0.95 (Privalov, 1982). Ovalbumin in 0.15 M NaCl, pH 7.0, exhibited a single DSC transition peak at 83.9°C and the gel point occurred at 95.4°C as indicated by a rapid, linear increase in storage moduli as detected by dynamic testing (Arntfield et al., 1989). The DSC endotherm peak representing protein unfolding had returned to the baseline prior to structure development, suggesting that unfolding of the single ovalbumin domain occurred prior to gel network formation (Fig. 7.2).

By changing solution conditions, a single-domain protein can form a variety of gel structures that range from highly ordered to randomly aggregated (Doi, 1993; Hermansson, 1988). For example, 5.0% ovalbumin solutions at pH 7.5 form soft, transparent gels when heated in 10–20 mM NaCl, stronger, translucent gels in 30–40 mM NaCl, and turbid, soft gels at higher NaCl concentrations (Doi, 1993). Gel properties of α-lactalbumin and β-lactoglobulin also change with solution conditions (Hermansson, 1988). Differences in the extent of protein unfolding

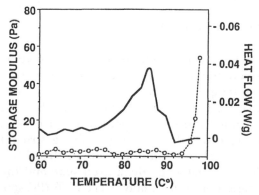

FIG. 7.2 Protein unfolding, as monitored by differential scanning calorimetry, and structure development, as monitored by storage modulus, of 10% ovalbumin during heating at pH 7.0. (*Adapted from Arntfield et al., 1989.*)

and aggregation caused by changes in environmental conditions result in different final gel structures.

STRUCTURE OF DENATURED PROTEINS

In reality, most "denatured" proteins are not completely unfolded and some ordered structure may exist after heating (Tanford, 1968; Clark and Lee-Tuffnell, 1986). The transition temperature of the muscle protein G-actin is 57.2°C at pH 8.0 as measured by DSC; however, actin is known to retain about 60% of its native helical structure when heated to 90°C. Actin is completely unfolded in 5 M quanidine hydrochloride or 8 M urea (Bertazzon et al., 1990). Byler and Purcell (1989) observed changes in protein secondary structure during heating of β-lactoglobulin and bovine serum albumin by Fourier-transform infrared spectroscopy. They observed that none of the proteins was totally unfolded into a disordered state when heated above its transition temperature as determined by DSC.

Privalov (1979, 1989) found that the secondary structure of denatured proteins may differ due to environmental conditions, such as pH. Herald and Smith (1992) reported a DSC transition temperature of 90.4°C for ovalbumin at pH 7.0 and 9.0; however, the secondary structure of ovalbumin at the two pHs was different when heated to 90°C. At pH 9.0 ovalbumin contained significant quantities of α-helix and β-sheet, whereas at pH 7.0 little native structure was observed at the transition temperature. Storage moduli and microstructure of the final gels were different (Herald and Smith, 1992; Herald, 1991) which can be attributed to pH-induced differences in unfolded structure and chemical bonding during cross-linking.

The formation of nonnative, ordered structures of hydrogen-bonded β-sheets has been observed when gelling proteins were heated above their transition temperatures. Infrared spectroscopy has been used to identify the formation of intramolecular hydrogen bonds during heat-induced gelation. Clark et al. (1981) suggested that the formation of a band at 1620 cm^{-1} correlated with protein aggregation and the formation of hydrogen-bonded β-sheet. These authors observed that formation of a second, less intense band at 1680 cm^{-1} together with the band at 1620 cm^{-1} indicated formation of antiparallel β-sheet. The formation of ordered structure during gelation has been reported for bovine serum albumin (Clark et al., 1981; Byler and Purcell, 1989), ovalbumin (Chirgadze and Nevskaya, 1976; Herald and Smith, 1992), β-lactoglobulin (Byler and Purcell, 1989), chicken myosin (Wang, 1993), and surimi

(Niwa, 1992). Hydrogen-bonded β-structure was not observed when α-lactalbumin or casein were heated under nongelling conditions. Herald and Smith (1992) showed that the intensity of the 1620 and 1680 cm^{-1} peaks increased as heating temperature increased and correlated with the development of elastic character of ovalbumin gels, as measured by an increase in storage moduli (Fig. 7.3).

Small changes in protein conformation due to heating can be detected using monoclonal antibodies (mAbs). Monoclonal antibodies specific for the native structure or the denatured molecule would help identify epitopes in the molecule that change during heat-induced denaturation. Monoclonal antibodies that react with both native and denatured forms of a protein would help identify regions of the protein which do not change on heating. Thus, heat-induced changes in very specific portions of a protein can be detected.

Ikura et al. (1992) prepared five mAbs which recognized epitopes specific to soluble heat-denatured ovalbumin. The authors identified five

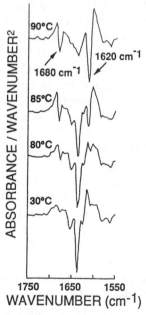

FIG. 7.3 Second derivative infrared spectra of 1.75% S-ovalbumin in 25mM KD$_2$PO$_4$, pD 7.0, heated for 15 min at the indicated temperature. Note increase in bands at 1620 and 1680 cm^{-1} with temperature. (*Adapted from Herald and Smith, 1992.*)

specific regions of the molecule that changed structurally during heating by isolating and sequencing peptides from ovalbumin that reacted specifically with the mAbs (Fig. 7.4). These regions were attributed to conformational changes on the surface and to exposure of interior portions of the molecule.

Kaminogawa et al. (1989) used similar techniques to monitor the denaturation process of β-lactoglobulin, a major whey protein that imparts gelling properties to several food systems. The binding affinity of mAbs 21B3 and 31A4 increased when β-lactoglobulin was heated above 67°C, whereas the binding affinity of mAbs 61B4 and 62A6 decreased as temperature was increased above 80°C. The transition temperature of β-lactoglobulin determined by DSC was about 73–74°C under similar buffer conditions (Foegeding et al., 1992). The epitope for mAb 31A4

Ac - G S I G A A S M E F C F D V F K E L K V H H A N E N I F Y C P I A I M S

37   A L A M V Y L G A K D S T R T Q I N K V V R F D K L P G F G D S I E A Q

73   C G T S V N V H S S L R D I L N Q I T K P N D V Y S F S L A S R L Y A E

109 E R Y P I L P E Y L Q C V K E L Y R G G L E P I N F Q T A A D Q A R E L

145 I N S W V E S Q T N G I I R <u>N V L Q P S S V D S Q T A M V L V N A I V F</u>
                                      7F1

181 <u>K</u> G L W E K A <u>F K D E D T Q A M P F R V T E Q E S K P V Q M M</u> Y Q I G L
                       RC9

217 F R V A S M A S E K M K I L E L P F A S G T M S M L <u>V L L P D E V S G L</u>
                                                          6C9

253 <u>E Q L E S I I N F E K L T E W</u> T S S N V M E E R K I K V Y L P R M K M E
                          6FA

289 E K Y N L T S V L M A M G I T D V F S S S A N L S G I S S A E S L K <u>I S</u>
                                                                    D11

325 <u>Q A V H A A H A E I N E A G R E V V G S A E A G V D A A S V S E E F R A</u>

361 <u>D H P F L F C I K</u> H I A T N A V L F F G R C V S P - O H

FIG. 7.4 Amino acid sequence of ovalbumin. Underlined sequences indicate epitopes recognized by monoclonal antibodies (7F1, RC9, 6C9, 6FA, and D11) specific to heat-denatured ovalbumin. (*Adapted from Ikura et al., 1992.*)

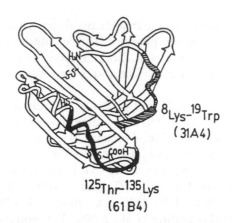

FIG. 7.5 Schematic representation of the three-dimensional structure of β-lactoglobulin. Region with hatch marks represents that portion of the molecule recognized by monoclonal antibody 31A4. Region in black represents the epitope recognized by monoclonal antibody 61B4. (*Adapted from Kaminogawa et al., 1989.*)

was in the region of Lys^8-Trp^{19}, which is primarily random coil, whereas that for 61B4 was in the region of Thr^{125}-Lys^{135}, which is a helix region (Fig. 7.5). The authors suggested that β-lactoglobulin unfolded during heating such that a conformational change occurred in the random coil region of Lys^8-Trp^{19} followed by a structural change in the helical region Thr^{124}-Lys^{135}. Monoclonal antibody techniques present the protein chemist with very powerful probes for monitoring the denaturation process of proteins.

MULTIPLE DOMAIN PROTEINS

Domains may exhibit distinct thermal stabilities and may alter the stability of a neighboring domain. The native structure of each domain is stable up to a certain critical temperature, after which the domain unfolds as shown by a large increase in the DSC endotherm. The transition temperature of each domain and the order in which domains unfold within a protein may change with environmental conditions (Potekin and Privalov, 1982). Bertazzon and Tsong (1990) also observed that domains may change size with pH when a portion of one domain may migrate to another. Changes in both domain transition temperature and the order in which domains unfold may alter gel properties.

The muscle protein myosin can be used to illustrate the effect of

unfolding of multiple domains on final gel properties. Myosin is responsible for the many of the functional properties observed in muscle foods
and has been extensively studied. Myosin has a molecular mass of about
5.21×10^5 and is composed of six subunits: two heavy and four light
chains (Fig. 7.6). Each heavy chain has a molecular mass of about 200
kilodaltons (Squire et al., 1990). The amino-terminal residues of the
heavy chain fold into a globular head region. The remainder of the
heavy chain is called the rod or tail region. The four light chains, each
with a molecular mass of about 17–23 kilodaltons, are noncovalently
associated with the head region. The rod regions of the two heavy chains
form a dimeric, extended coiled-coiled helical structure. The rod contains hinge regions that are susceptible to proteolysis. Hydrolysis of myosin heavy chain with trypsin yields light meromyosin (LMM) and heavy
meromyosin (HMM). Hydrolysis with papain yields two identical globular heads (subfragment 1) and myosin rod (LMM and HMM S-2).

Wang (1993), using differential scanning microcalorimetry, reported
10 two-state transitions or domains in the temperature range of
44.2–70.8°C for chicken breast muscle myosin in 0.6 M NaCl, 0.05 M
Na phosphate buffer, pH 6.5 (Fig. 7.7). The domains were assigned
to different portions of the myosin molecule. We also used dynamic
rheological testing to monitor structure formation when myosin was
heated under identical conditions. The gel point of myosin was 53.5°C,
however, the storage modulus underwent a series of transitions in the
temperature range between 53.5 and 75°C. Structure formation was not
complete until storage modulus reached a maximum at 75°C.

By comparing the unfolding of myosin domains to corresponding
changes in storage moduli during heating, it was possible to look at the
influence of myosin domains on gel structure. Unfolding of the first
three myosin domains (as indicated by the first three deconvoluted peaks)

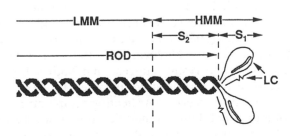

FIG. 7.6 Schematic representation of the myosin molecule (LMM, light
meromyosin; LC, light chain; HMM, heavy meromyosin; S_1, subfragment 1:
S_2, subfragment 2).

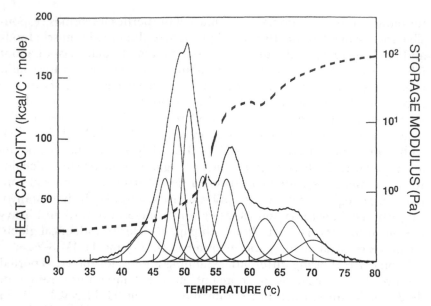

FIG. 7.7 Differential scanning calorimeter endotherm and dynamic rheogram (storage modulus) of 1% chicken breast muscle myosin in 0.6 M NaCl, pH 6.5, heated from 30 to 80°C at 1°C/min. (*Adapted from Wang, 1993.*)

contributed very little to gel formation, as storage modulus did not increase until these domains were completely unfolded. These domains were assigned to the rod region of the myosin molecule. Bertazzon and Tsong (1989) observed that myosin rod was about 95% reversible on heating, thus suggesting that unfolded domains from the rod do not aggregate to form a gel network. The initial large increase in storage modulus corresponded to the unfolding of subfragment 1 at 50.7°C and subsequent interactions between subfragment 1 and the unfolded rod. Bertazzon and Tsong (1989) observed that the rod was thermally irreversible when heated with subfragment 1 due to cross-linking between these fragments. These results suggest that both subfragment 1 and certain domains of the rod must be unfolded for protein aggregation and network formation to occur. As heating temperature was increased, unfolding of additional domains modified the network structure as evidenced by changes in storage modulus. Final network structure was not attained until all myosin domains were unfolded.

A gel network is dependent on the extent of protein unfolding, thus heating a multidomain protein to different endpoint temperatures should result in protein gels of different physical properties. Different gel microstructures and rheological properties were observed when chicken breast salt-soluble protein (Wang and Smith, 1992) and myosin (Wang, 1993) were heated isothermally between 45 and 75°C in 0.6 M NaCl, pH 6.5, for 30 min. Wu et al. (1991) reported that isothermal heating of myosin from 44 to 70°C produced gels with different rate constants, equilibrium shear moduli, and mechanical energy loss. Myosin in 0.5 M NaCl, 10 mM Na phosphate buffer, pH 7.0, formed highly elastic gels with maximum shear moduli when heated isothermally at 48–50°C. On either side of these temperatures, shear moduli decreased such that below 44 or above 70°C, shear moduli were not detected. Both Wu et al. (1991) and Wang (1993) attributed differences in physical properties of myosin gels to differences in the extent of myosin domain unfolding with temperature. Thus, by careful selection of processing temperatures to selectively unfold certain protein domains, food scientists have another technique to modify the textural properties of gel-based foods.

REFERENCES

Arntfield, S. D., Murray, E. D., Ismond, M. A. H., and Bernatsky, A. M. 1989. Role of the thermal denaturation-aggregation relationship in determining the rheological properties of heat induced networks for ovalbumin and vicilin. *J. Food Sci.* 54: 1624–1631.

Bertazzon, A. and Tsong, T. Y. 1989. High-resolution differential scanning calorimetric study of myosin, functional domains, and supramolecular structures. *Biochemistry* 28: 9784–9790.

Bertazzon, A. and Tsong, T. Y. 1990. Study of effects of pH on the stability of domains in myosin rod by high-resolution differential scanning calorimetry. *Biochemistry* 29: 6453–6459.

Bertazzon, A., Tian, G. H., Lamblin, A., and Tsong, T. T. 1990. Enthalpic and entropic contributions to actin stability: Calorimetry, circular dichroism, and fluorescence study and effects of calcium. *Biochemistry* 29: 291–298.

Byler, D. M. and Purcell, J. M. 1989. FTIR examination of thermal denaturation and gel formation in whey proteins. *SPIE* 1145: 415–417.

Chirgadze, Y. N. and Nevskaya, N. A. 1976. Infrared spectra and resonance interactions of amide I vibration of the anti-parallel chain pleated sheet. *Biopolymers* 15: 607–625.

Clark, A. H. 1992. Gels and gelling. Ch. 5. In *Physical Chemistry of Foods,*

H. G. Schwartzberg and R. W. Hartel (Ed.), pp. 263–305. Marcel Dekker, New York.

Clark, A. H. and Lee-Tuffnell, C. D. 1986. Gelation of globular proteins. Ch. 5. In *Functional Properties of Food Macromolecules,* J. R. Mitchell and D. A. Ledward, pp. 203–272. Elsevier, New York.

Clark, A. H., Saunderson, D. H. P., and Suggett, A. 1981. Infrared and laser-Raman spectroscopic studies of thermally-induced globular protein gels. *Int. J. Peptide Protein Res.* 17: 353–364.

Doi, E. 1993. Gels and gelling of globular proteins. *Trends Food Sci. Tech.* 4: 1–5.

Donovan, J. W. 1984. Scanning calorimetry of complex biological structures. *Trends Biochem. Sci.* 8: 340–344.

Flory, P. J. 1941. Molecular size distribution in three dimensional polymers. I. Gelation *J. Am. Chem. Soc.* 63: 3083–3091.

Ferry, J. D. 1948. Protein gels. *Adv. Protein Chem.* 4: 1–79.

Foegeding, E. A. and Hamann, D. D. 1992. Physicochemical aspects of muscle tissue behavior. Ch. 8. In *Physical Chemistry of Foods*, H. G. Schwartsberg and R. W. Hartel (Ed.), pp. 423–441. Marcel Dekker, New York.

Freire, E. and Biltonen, R. L. 1978a. I. Theory and application to homogeneous systems. *Biopolymers* 17: 463–479.

Freire, E. and Biltonen, R. L. 1978b. Statistical mechanical deconvolution of thermal transitions in macromolecules. II. General treatment of cooperative phenomena. *Biopolymers* 17: 481–496.

Hamann, D. D. 1987. Methods for measurement of rheological changes during thermally induced gelation of proteins. *Food Technol.* 41(1): 100–108.

Hamann, D. D. 1991. Rheology. A tool for understanding thermally induced protein gelation. Ch. 15. In *Interactions of Food Proteins*, N. Parris and R. Barford (Ed.), pp. 212–227. ACS Symposium Series 454, American Chemical Society, Washington, DC.

Herald, T. J. 1991. Effect of pH, salt type and denaturants on the denaturation properties, storage modulus, secondary structure and microstructure of hen egg S-ovalbumin heat-induced gels. Ph.D. dissertation, Michigan State University, East Lansing, MI.

Herald, T. J. and Smith, D. M. 1992. Heat-induced changes in the secondary structure of hen egg S-ovalbumin. *J. Agric. Food Chem.* 40: 1737–1740.

Hermansson, A. M. 1988. Gel structure of food biopolymers. Ch. 3. In

Food Structure-its Creation and Evaluation J. M. V. Blanschard and J. R. Mitchell (Ed.), pp. 25–40. Butterworths, London.

Ikura, K., Higashiguchi, F., Kitabatake, N., Doi, E., Narita, H., and Sasaki, R. 1992. Thermally induced epitope regions of ovalbumin identified with monoclonal antibodies. *J. Food Sci.* 57: 635–639.

Kaminogawa, S., Shimizu, M., Ametani, A., Hattori, M., Ando, O., Hachimura, S., Nakamura, Y., Totsuka, M., and Yamauchi, K. 1989. Monoclonal antibodies as probes for monitoring the denaturation process of bovine β-lactoglobulin. *Biochim. Biophys. Acta* 998: 50.

Krishnan, K. S. and Brandts, J. F. 1978. Scanning calorimetry. *Meth. Enzymol.* 49: 3–14.

Niwa, E. 1992. Chemistry of surimi gelation. Ch. 16. In *Surimi Technology*, T. C. Lanier and C. M. Lee (Ed.), pp. 389–427. Marcel Dekker, New York.

Potekhin, S. A. and Privalov, P. L. 1982. Co-operative blocks in tropomyosin. *J. Mol. Biol.* 159: 519–535.

Privalov, P. L. 1979. Stability of proteins. Small globular proteins. *Adv. Protein Chem.* 33: 167–241.

Privalov, P. L. 1982. Stability of proteins. Proteins which do not present a single cooperative system. *Adv. Protein Chem.* 35: 1–104.

Privalov, P. L. 1989. Thermodynamic problems of protein structure. *Ann. Rev. Biophys. Biophys. Chem.* 18: 47–69.

Privalov, P. L. and Gill, S. J. 1988. Stability of protein structure and hydrophobic interaction. *Adv. Protein Chem.* 39: 191–239.

Ramsay, G. and Freire, E. 1990. Linked thermal and solution perturbation analysis of cooperative domain interactions in protein. Structural stability of diphtheria toxin. *Biochemistry* 29: 8677–8683.

Squire, J. M., Luther, P. K., and Morris, E. P. 1990. Organization and properties of the striated muscle sarcomere. Ch.1. In *Molecular Mechanisms of Muscular Contraction*, J. M. Squire (Ed.), pp. 1–48. CRC Press, Boca Raton, FL.

Stauffer, D., Coniglio, A., and Adam, M. 1982. Gelation and critical phenomena. *Adv. Polm. Sci.* 44: 103–158.

Stockmayer, W. H. 1943. Theory of molecular size distribution and gel formation in branched-chain polymers. *J. Chem. Phys.* 11: 45–55.

Stockmayer, W. H. 1944. Theory of molecular size distribution and gel formation in branched-chain polymers. II. General cross linking. *J. Chem. Phys.* 12: 125–131.

Tanford, C. 1968. Protein denaturation. *Adv. Protein Chem.* 23: 121–275.

Wang, S. F. 1993. Gelation properties of chicken breast muscle myosin. Ph.D. dissertation, Michigan State University, East Lansing.

Wang, S. F. and Smith, D. M. 1992. Functional properties and microstructure of chicken breast salt soluble protein gels as influenced by pH and temperature. *Food Structure* 11: 273–285.

Wu, J. Q., Hamann, D. D., and Foegeding, E. A. 1991. Myosin gelation kinetic study based on rheological measurements. *J. Agric. Food Chem.* 39: 229–236.

Xie, D., Bhakuni, V., and Freire, E. 1991. Calorimetric determination of the energetics of the molten globule intermediate in protein unfolding: Apo-α-lactalbumin. *Biochemistry* 30: 10673–10678.

Ziegler, G. R. and Foegeding, E. A. 1990. The gelation of proteins. *Adv. Food Res.* 34: 203–298.

8

Protein–Polysaccharide Interactions

David A. Ledward

University of Reading
Whiteknights, Reading, England

INTRODUCTION

Though a great deal of research has been devoted to the study of proteins and polysaccharides in both solution and, where appropriate, gelled systems, much of this work has been concerned with the behavior of a single type of macromolecule. However, food products often contain a mixture of two or more hydrocolloids, such mixtures having improved eating and/or processing qualities. The desired effects may be achieved by mixing two (or more) polysaccharides or proteins. Whether the mixture consists of two different polysaccharides, two different proteins, or a protein and a polysaccharide, the general effects and their interpretation are similar. Protein-polysaccharide interactions are thus not specific classes of interactions but are typical of the types of interactions that may occur between different types of polymer in solution. It is probably true to say that, with the possible exception of mixed polysaccharide gels (Morris, 1990,), it is the potential applications of protein-polysaccharide mixtures in food technology that are arousing most interest at the present time (Ledward, 1979; Tolstoguzov, 1986, 1988a,b, 1991; Aguilera, 1992).

It is now about 15 years since I last reviewed this topic (Ledward, 1979), and it was with some trepidation that I reread the original review. The field has certainly developed tremendously in the intervening years, and although I was pleased to see that some of the ideas and views expounded then had survived the test of time, I was also very well aware that other views needed modification and a more lateral view of the interactions was needed than previously shown.

It is generally accepted that the behavior of mixed polymer systems that do not interact chemically is usually dominated by the enthalpy of segment-segment interactions (Morris, 1990). In many cases the potential interactions are endothermic in nature and thus are not favored. These repulsive forces can lead to the mutual exclusion of each polymer from the domain of the other, i.e., the effective concentration of both polymers may increase. Tolstoguzov and coworkers refer to this phenomenon as thermodynamic incompatibility, and at sufficiently high concentration it may lead to phase separation (Tolstoguzov, 1986, 1988a,b, 1991).

This thermodynamic incompatibility is normally most apparent between different classes of polymer such as proteins and polysaccharides. These systems have been widely studied, because if one or both of the polymers possess gelling ability, they have the potential to create multi-textured products (Tolstoguzov, 1986). Ziegler and Foegeding (1990) have summarized the types of gel networks that can form with two different gelling agents.

If the mixing process is exothermic, then the interactions will be attractive, and this can lead to the formation of soluble and/or insoluble complexes. This normally occurs only for polymers of opposite net charge and has been most extensively studied for proteins and acidic polysaccharides such as alginate, pectate, and carboxymethylcellulose (Ledward, 1979), or sulfated ones such as carrageenan (Wills et al., 1988). Pavlovskaya et al. (1992) have recently shown that ultrasound may be a useful technique to aid our understanding of both compatible and incompatible systems.

A further possible reaction between a protein and a polysaccharide involves the direct formation of covalent bonds between the two polymers. Though not common in food systems such linkages may be of paramount importance in dictating the textural characteristics of, for example, extruded soy products (Ledward and Mitchell, 1988) or in producing new functional ingredients (Kobayashi et al., 1990; Kato et al., 1990; Kato and Kobayashi, 1991).

Before discussing these different types of interaction in more detail, it is worth reiterating the warning given by Morris (1990) when discussing mixed polymer gels. He rightly points out that in some cases the appearance of polymer synergism may be spurious, the true interactions being between one of the macromolecules and small ions (or molecules) introduced into the system as a contaminant or counterion with the other polymer. Such effects will not be considered in detail in this review.

PHYSICOCHEMICAL INTERACTIONS

Thermodynamic Incompatibility

As stated previously, the behavior of mixed polymer solutions is largely dictated by the energies of interaction between chains and since the number of energetically unfavorable contacts that a given chain can make increases with increasing flexibility (as well as increasing size), conformationally mobile polymers are more effective in excluding other polymers than are more constrained ones. Thus, most polysaccharides are relatively inefficient in excluding even flexible protein chains such as gelatin. The exception is dextran, in which the sugar residues are separated by three covalent linkages compared to the two in most other polysaccharides. It is not surprising therefore that most work has utilized this polysaccharide to exclude or fractionate proteins from solution (Tolstoguzov, 1986; Albertsson, 1971). Even when there is no evident phase separation, the apparent protein concentration can be increased so that, for example, gelatin solutions will gel more rapidly (i.e., form the collagen like triple helix) in the presence of low concentrations of dextran. At least at short aging times stronger gels are also formed in the presence of only 0.2% dextran of molecular weight 65,000 (Fig. 8.1). On increasing the dextran concentration further, the strength starts to decrease (Fig. 8.2). The reversal, with regard to the effect of dextran concentration on rigidity, occurs at the concentration at which phase separation takes place. As would be expected, the effect is very dependent on molecular weight (Fig. 8.2). Other examples of the protein (and polysaccharide) exclusion involve starch (Christianson et al., 1981) and could, considering the wide use made of starch in the food industry, be very important industrially.

The above discussion is concerned with the concentration range in which both polymers, i.e., the protein and the polysaccharide, coexist in a single phase, the onset of phase separation only occurring at higher concentrations. As a very broad and not strictly correct generalization, the total polymer concentration needs to be at least 4% before phase

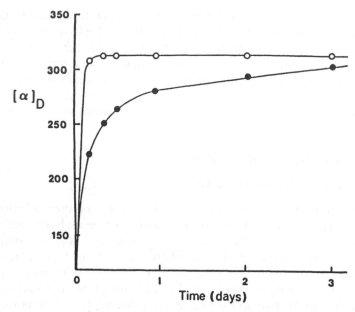

FIG. 8.1   Rate of collagen-fold formation in gelatin (4%; 6°C) in the presence (o) and absence (●) of dextran (molecular weight 65,000; 0.2%) as monitored by the time coarse of optical rotation change (*From Tolstoguzov et al., 1974.*)

separation occurs (Tolstoguzov, 1986; Morris, 1990). This phenomenon of phase separation has been widely studied by Tolstoguzov and coworkers in Russia (Tolstoguzov, 1986, 1988a,b, 1991; Suchkov et al., 1981; Semenova et al., 1991; Burova et al., 1991), and Fig. 8.3 shows a typical phase diagram for two thermodynamically incompatible polymers. The curved line is often referred to as the binodal or "cloud-point" curve, and the concentrations of polymer, i.e., the polysaccharide and protein, that fall below this line will coexist is a single phase, although each polymer domain will exclude the other polymer. At any mix of concentration above the binodal line, the system will spontaneously separate into two phases—one phase polysaccharide enriched the other protein enriched. The composition of the phases will be given by the point of intersections of the tie line going through the initial concentration and the binodal curve. All mixtures whose initial composition fall on the same tie line will also separate into two phases with the same final concentrations, although the relative *volume* of the two phases will reflect the initial concentration of the two polymers (Tolstoguzov, 1986; Morris, 1990).

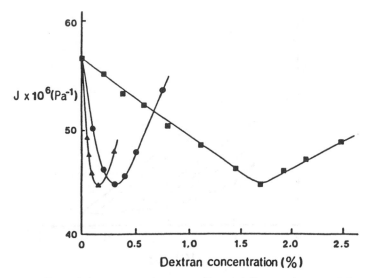

FIG. 8.2   Effect of dextran on the compliance J (inversely proportion to the gel strength) of gelatin gels (10%) after gelation for 30 min. Results are shown for varying concentrations of dextrans of molecular weight 500,000 (■), 65,000 (●) and 15–20000 (▲). (*From Tolstoguzov et al., 1974.*)

That is to say, the length of the tie line between the initial concentration and the composition of one of the phases is inversely proportional to the volume of that phase, i.e., directly proportional to the volume of the other phase. With our present knowledge of these systems, the family of tie lines for a given mixture must be empirically derived. Since the expanded polysaccharide coils occupy far larger volumes than the more compact protein molecules, it is not surprising that with similar concentrations of each polymer in the initial mix the final concentration of protein in the protein-enriched phase is usually far higher than the final concentration of polysaccharide in the protein-depleted phase (Tolstoguzov, 1986, 1988a,b, 1991). The volume fraction of the polysaccharide phase will be much higher, however. This is shown in Fig. 8.3 by the concentration ranges shown for the polysaccharide and protein.

The effectiveness of a polysaccharide in leading to phase separation of a protein will depend on:

1.   The relative charges on the polymers (reflecting attractive/repulsive forces) and thus the pH of the system.

2.   The flexibility of the chain: with both casein and soya the order of effectiveness for anionic polysaccharides is dextran sulfate >

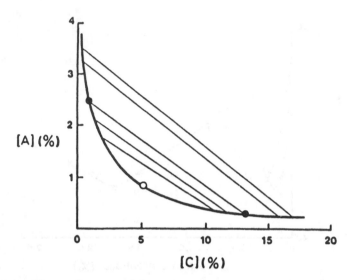

FIG. 8.3 Phase diagram of the system water-casein (C)–sodium alginate (A) at Ph 7.2 and 25°C. The bold line is the binodal or "cloud-point" curve and the faint lines are "tie lines" joining points (●) representing the composition of coexisting casein-rich and alginate rich phase. With decreasing concentration the tie lines converge to the critical point (o), below which the system remains as a single homogenous phase. (*From Suchov et al., 1981.*)

gum arabic > alginate > carboxymethyl-cellulose (CMC) > pectin (Tolstoguzov, 1986).

3.   Molecular size: higher molecular weight polymers are more effective.

4.   Salt concentration: in general with charged polysaccharides, incompatibility is increased with increasing salt concentration (Morris, 1990).

Energetically Favored Interactions

When two polymers of opposite net charge are mixed, as will be the case with anionic polysaccharides at or above their pK value and proteins below their isoelectric point, the enthalpy of mixing is exothermic and the charge-charge interactions will lead to complex formation (Ledward, 1979). This phenomenon has been studied by several research groups (Ledward, 1979; Tolstoguzov, 1986; Hill and Zadow, 1974), and it is well established that the interactions are primarily electrostatic in nature. For example, in studies carried out several years ago we found that at pH 6.0 and low ionic strength (0.05 M), pectate, alginate, and CMC were capable of modifying the structure of both myoglobin (pI 6.9) and bovine serum albumin (pI 4.9) (Imeson et al., 1977) leading to spectral changes in myoglobin and decreased thermal stability in both proteins. The decrease in thermal stability was greater for myoglobin ($4°C$) than for BSA ($2°C$). Burova et al. (1991) also found that charged polysaccharides (methylcellulose, dextran T-500, alginate, pectin, CMC, and carrageenan) at pH 4.2–6.0 and low ionic strength decreased the thermal stability of the 11S globulin from broad beans. Under conditions of incompatibility no effect was seen. Imeson et al. (1977) found that with BSA at pH 6.0, Sephadex chromatography separated the protein and polysaccharide. However, following heat denaturation at this pH, much stronger interactions were formed, giving rise to stable high molecular weight complexes which inhibited protein-protein aggregation and hence precipitation. Although the protein did not precipitate following heat denaturation, on subsequent cooling no reversal to the native form was observed; the spectra of the myoglobin systems were typical of the heat-denatured form of the protein (Fig. 8.4).

Thus the interactions are sufficiently strong in the heated system to prevent reformation of the energetically favored native structure on cooling. These interactions were undoubtedly electrostatic in nature since they increased in strength with decreasing pH (from 7 to 5), i.e., with increasing net charge on the protein and with decreasing ionic strength.

Wills ct al. (1988) further extended this work by studying K-carrageenan/myoglobin systems. They claimed from analysis of the spectra that only two protein components were present in the system, namely, native and denatured myoglobin (Fig. 8.5). This was deduced since the Soret peak did not undergo a gradual shift in wavelength with decreasing pH, rather the pattern was consistent with the existence of two absorbing species in solution. The existence of the isobestic points at 383 and 432 nm supported this contention. This suggests a one-step denaturation of myoglobin, which proceeds without stable intermediates. If so the apparent equilibrium constant, K will be proportional to C^n where C is the denaturant concentration (Schechter and Epstein, 1968), i.e.,

$$logK = constant + nlogC$$

Apparent K values could be obtained from the absorption spectra (Wills et al., 1988) and a linear relationship was found between logK and the log of the hydrogen ion concentration ($-pH$) (Fig. 8.6). This implies

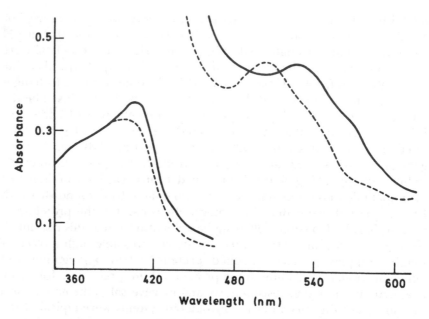

FIG. 8.4 Absorption spectra of myoglobin (0.1%) and alginate (0.1%) solutions at pH 6.0, 0.001 μM at 20°C (---) and 83°C (—). To record spectra from 350–450 nm the solutions were diluted 1:10. The spectra are typical of metmyoglobin (20°C) and heat-denatured myoglobin (83°C). (*From Imeson et al., 1977.*)

that the hydrogen ions act as the denaturant (which at these pH values is obviously incorrect), but the relationship is rational since the hydrogen ion concentration should relate to the charge and its distribution on the macromolecules (presumably the myoglobin since the carrageenan will be fully ionized at these pH values). In addition, the polysaccharide must complex with the denatured protein (and free hematin) to prevent their aggregation and subsequent precipitation. This work does *not* show, however, whether stable complexes are formed with the native protein or whether in fact the negatively charged polysaccharide primarily causes destabilization in much the same way as any other denaturant, and the unfolded globin then binds to the polysaccharide. Attempts to fractionate myoglobin/polysaccharides systems at pH 6.0 and low ionic strength on Sephadex G-100 were unsuccessful since the hematin bound to the Sephadex (Imeson et al., 1977). This would suggest that free "native" myoglobin (which should elute normally) was not present in the solution, although the spectra indicated that both native and denatured forms

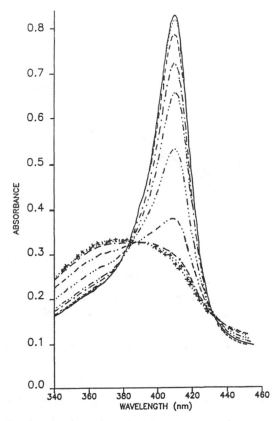

FIG. 8.5 Absorption spectra of myoglobin (0.01%) in the presence of K-carrageenan (0.01%) at pH range 7.2 (−) to 5.0 (++) and ionic strength less than 0.001 M. (*From Willis et al., 1988.*)

were present (Fig. 8.5). However, if the denatured polysaccharide-bound protein did bind to the Sephadex, then any native protein in the system would subsequently denature (and thus bind to the column) to restore the equilibrium as elution proceeded. It must be remembered that the hematin group in myoglobin possesses a high positive charge, which will have a tendency to bind to the anionic polysaccharide. Other globular proteins will not possess this partially buried positive charge, and thus the results with myoglobin may not be generally applicable. It is often assumed that *native* globular proteins will bind to anionic polysaccharides at low pH (see, e.g., Tolstoguzov, 1986) but convincing evidence has not, as far as I know, been found. Tolstoguzov (1991) has argued that since

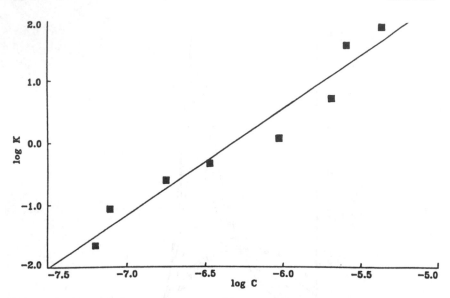

FIG. 8.6 Log of the apparent equilibrium constant (for metmyoglobin-denatured myoglobin), logK, against the log of the hydrogen ion concentration, logC, for solutions of 0.01% myoglobin in the presence of 0.01% K-carrageenan at an ionic strength < 0.001 M. (*From Wills et al., 1988.*)

the thermal denaturation of trypsin inhibitor (a globular protein) in the presence of dextran sulfate and pectate is irreversible, binding of the native protein takes place on the assumption presumably that an indirect effect would be reversible. However, if, following denaturation, the protein binds to the polysaccharide, then renaturation would not be expected, as is found with myoglobin (Wills et al., 1988). Recent work from Kelly et al. (1993) has confirmed, using ultracentrifugation, light scattering, and intrinsic viscosity measurements, that denatured BSA binds strongly to alginate even at pH 7.0 when the protein will carry a small *net negative* charge. The size of the soluble complexes increased with increasing severity of heating. The results with native, globular BSA were less clear-cut, since at about 20°C, there was an apparent interaction leading to increased molecular size as judged by analytical ultracentrifugation and viscosity measurements. However, on heating to 40°C the apparent size diminished, and on recooling to 20°C no increase in size was noted. At 20°C the polymers will attempt to exclude one another, and this may be responsible for the results obtained since the actual concentrations of the polymers will be higher than measured; on raising the temperature to 40°C the endothermic repulsive forces will diminish and the effect will be lost—why it is not thermoreversible is unclear. Perhaps we should keep an open mind as to whether unfolding (at least partially) of the protein is a prerequisite for complex formation involving globular proteins.

Whatever the mechanism, the carboxylate or sulfate groups of the polysaccharides must be involved together with some, or all, of the positively charged protein residues, i.e., ϵ-amino, α-amino, guanidinium, and imidazole, and the actual strength of the interactions will be related to the number and distribution of these sites as well as the overall charge on the protein. Thus, following denaturation, the available number of such sites increases as "buried" basic groups are liberated, and also the flexibility of the "random coil" denatured state permits configurational adjustments to maximize the interactions, yielding stable complexes.

The structure of the polysaccharide is of course also important to the effectiveness of the interactions, and at pH 3.5–4.0 Imeson et al. (1978) found alginate to be more effective than pectate and CMC. The charge per unit residue is obviously of importance (Soshinsky et al., 1992), but the increased flexibility of alginate compared to CMC may account for its superior effectiveness while the kinks introduced into the pectate molecule by the neutral sugars may reduce the number of freely accessible carboxy groups in the pectate compared to the alginate molecule (Imeson et al., 1978; Ledward, 1979). Kelly et al. (1993) found that at pH 7.0 there was no evidence of interaction between denatured BSA and pectate (although alginate bound strongly) further supporting the above hypothesis. Such steric considerations are discussed in more detail when considering the chemical reactivity of these polysaccharides (see page 239).

Soluble-Insoluble Complexes

When solutions of a globular protein and anionic polysaccharide are mixed at a pH value under which the polymers carry opposite net charges, an insoluble complex is usually formed (Ledward, 1979; Tolstoguzov, 1986). However, if these solutions are mixed at a higher pH, when both carry a net negative charge, then no complex formation is expected or indeed usually seen. However, if the high pH mixture is slowly acidified to a pH value below the protein's isoelectric point, the "complex" formed may be soluble (Fig. 8.7). Similar results can be obtained by varying the ionic strength of the system. These have been termed M (insoluble or mixed) and T (soluble or titration) complexes (see, e.g., Tolstoguzov, 1986). Both M and T complexes are very stable and can be stored for weeks without any tendency to dissolve (M com-

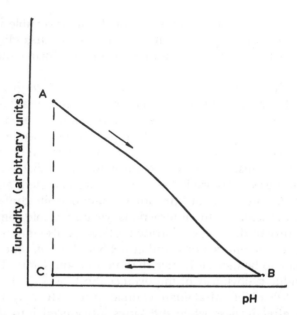

FIG. 8.7   Schematic representation of the effect of mixing a protein and a polysaccharide solutions at low pH indicating the formation of an insoluble complex (A), the solubilization of this complex by raising the pH (B), and the formation of a soluble complex, by the slow reduction of the pH to the original low value (C).

plexes) or precipitate (T complexes). The nonequilibriun nature of these complexes is claimed to be due to the fact that on mixing the two polymers at low pH, the protein molecules attach themselves noncooperatively and more or less uniformly to the anionic polysaccharide by electrostatic (ionic) bonds so that the net charge on the complex decreases. Thus the complexes are usually insoluble, unless the amount of protein bound is very low (so the complexes still carry a high negative charge) as may happen at high polysaccharide-to-protein ratios (e.g., 10:1). These complexes will only become insoluble if the pH is lowered below the pK value of the anionic groups on the polysaccharide. In the formation of T complexes, however, the protein binding is thought to be cooperative (Tolstoguzov, 1986) and thus nonuniform. This is because an additional protein molecule, given time, will bind adjacent (or as near as possible) to an already attached protein molecule since this is energetically favored over binding to a free site some distance from an already bound protein. This hypothesis to explain the relative properties of "M and T complexes" has been widely advocated by Tolstoguzov and coworkers and may be valid in systems containing high molecular weight polysaccharides and relatively high concentrations of low molecular weight proteins. As discussed earlier, this hypothesis assumes binding of the native protein. With higher molecular weight, more flexible proteins (e.g., gelatin), insoluble complexes usually form at low pH (Tolstoguzov, 1990), irrespective of the method of formation.

The soluble complexes formed at pH 5–7 with anionic polysaccharides and denatured myoglobin or BSA (see previous section) or gelatin (Tolstoguzov, 1990) are probably soluble because, though the protein is tightly bound, the complexes still carry large negative charges and thus have little tendency to aggregate. As the pH is lowered further and the total positive charge increases, the complexes will become electrically neutral and precipitation should occur. It must be borne in mind that exposing a mixture of a charged polysaccharide and protein to conditions under which the protein unfolds generates a flexible random coil that will interact with the polysaccharide as potential positively charged binding sites are exposed. Such sites will not be available on the native protein or previously denatured proteins that have self-associated.

When the numerous binding sites on the proteins (and to some extent the polysaccharides) are considered along with the potential conformation of the protein under different solvent conditions (pH, polysaccharide concentration, etc.) then one must reserve judgment on the hypothesis proposed by Tolstoguzov (1986) involving binding of the *native* globular protein to acidic polysaccharides at low pH.

From the above discussion it is readily apparent that proteins and polysaccharides can physically interact to give:

1. A single phase system containing the different polymers in domains in which the two polymers mutually exclude one another
2. A single phase system containing soluble complexes
3. A liquid two-phase system in which the two components are primarily in the different phases
4. An insoluble complex and soluble phase depleted in both components

CHEMICAL INTERACTIONS

Although there are numerous potentially reactive groups on both proteins and polysaccharides, conditions are not usually, in food processing systems, conducive to their reacting to produce stable covalent linkages.

However, when such reactions can occur they are usually of paramount importance in establishing the textural and other qualities of the product.

One of the most researched areas has been the formation of covalent amide bonds between propylene glycol alginate (PGA) and gelatin under mildly alkaline conditions to produce gels that are stable to temperatures of at least 100°C (McKay et al., 1985). The amide bonds are apparently formed between the carboxy groups of the PGA and uncharged amino ones (presumably the ϵ-amino groups of lysine or hydroxylysine since at the pH used, around 9.6, the arginine groups are fully ionized and there are few histidine residues in gelatin).

$$
\begin{array}{ccc}
\overset{\displaystyle O}{\underset{\displaystyle \diagdown}{\overset{\displaystyle \parallel}{R.C}}} & + & R^1 - (CH_2)_4 - NH_2 \qquad pH\ 9.6 \\
\quad OC_3H_6OH & & \qquad\qquad \longrightarrow
\end{array}
$$

PGA ϵ-amino group of lysine on gelatin

$$
\begin{array}{cc}
\overset{\displaystyle O}{\overset{\displaystyle \parallel}{R.C}} & + CH_3.CHOH.CH_2OH \\
\quad \diagdown \\
\quad NH - (CH_2)_4 - R^1
\end{array}
$$
Amide linkage

It would appear that it is the ester group that is involved in the reaction since the unesterified alginate will not undergo this reaction (Morris, 1990). It is imperative that once the gels have been formed, which occurs within 4–5 sec at pH 9.6 and faster at higher pH, by diffusion or direct addition of alkali to the warm (50°C) solution that the pH is reduced as loss of structure on storage can take place due to both protein deamination and depolymerization of polyurethane esters following β-elimination at the higher pH.

PGA at alkaline pH will also react with other polymers that contain hydroxyl groups (McDowell, 1970), presumably by the formation of ester linkages.

$$
\begin{array}{ccc}
\overset{\displaystyle O}{\overset{\displaystyle \parallel}{R.C}} & + & R^{11} - OH \longrightarrow \\
\quad \diagdown \\
\quad OC_3H_6OH
\end{array}
$$

$$R.C\overset{\displaystyle\nearrow O}{\underset{\displaystyle\searrow}{}}\qquad\qquad + \qquad\qquad CH_3.CHOH.\ CH_2OH$$

$$OR^1$$

Transesterification of PGA and hydroxyl group

These reactions are primarily of importance in reactions of non–amino-containing polymers, such as starch and polyvinyl alcohol, not proteins.

A further phenomenon that appears to be due to specific interaction between a protein and polysaccharide is the observation that 1% alginate markedly affects the extrusion behavior of soy grits and flour. The addition of the alginate markedly reduces the extruder torque and product temperature in both single and twin screw extrusion (Smith et al., 1982; Berrington et al., 1984; Imeson et al., 1985). This effect was shown to be due to a significant reduction in viscosity (Berrington et al., 1984). It was especially noticeable that of numerous polysaccharides studied (including guar gum, locust-beam gum, CMC, pectin, and carrageenan), only alginate had a measurable impact on extrusion behavior at the 1% level.

Oates et al. (1987a,b) studied the phenomenon in more detail and convincingly demonstrated that heating soy in the presence of alginate, at temperatures similar to those experienced in extrusion processing, led to the formation of water (Fig. 8.8), which would account for the decrease in viscosity observed during extrusion processing in the presence of this polysaccharide. Alginate consists of both polymannuronic and polyguluronic residues, and it is seen from Fig. 8.9 that the flat ribbonlike polymannuronate sequences allow much easier access to the carboxy groups than the highly buckled polyguluronate regions. Pectate has a backbone geometry very similar to that of polyguluronate and thus also allows only restricted access to the carboxy groups. It is therefore not too surprising that, if these carboxy groups (or their esters) are involved in the reactions, "high M" alginates are far more reactive than "high G" samples (or pectates). This is the likely explanation for the observation that, in the PGA-gelatin system, high M alginates produce stronger gels than those containing high G alginates (McKay et al., 1985) The highly significant increase in water content seen on heating soy isolate in the presence of high M alginate (Table 8.1) suggests that the carboxy groups of the alginate may be involved in the reactions giving rise to water formation. However, it may not be the carboxy groups per se that are the reactants in the heated soy alginate mixes since Oates and Ledward (1990) have shown that high M alginates are far less stable to heat than high G ones. The degraded products, unlike the native materials, are able to undergo browning reactions with both glycine and lysine, and

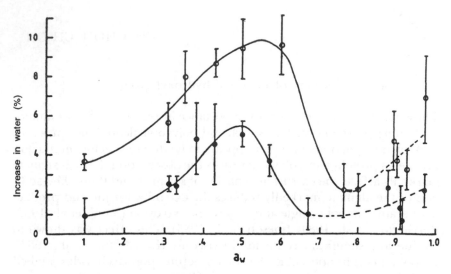

FIG. 8.8   Increase in apparent water content (determined by drying in vacuo at 70°C) as a function of initial water activity, a_w for soy isolate + 2% high mannuronate alginate (o) and soy isolate alone (●). All values are means and the error bars are the standard errors of the differences. (*From Oates et al., 1987a.*)

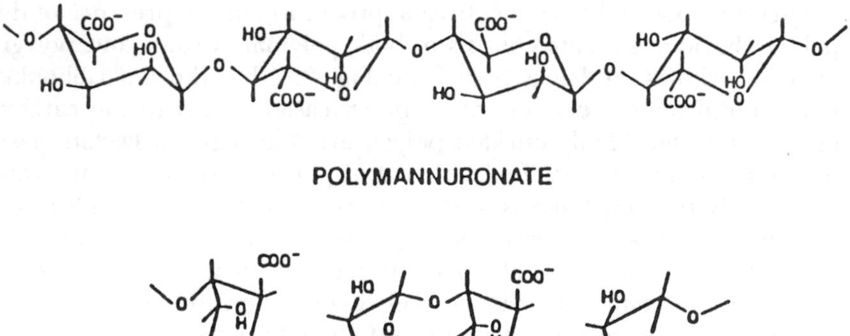

POLYMANNURONATE

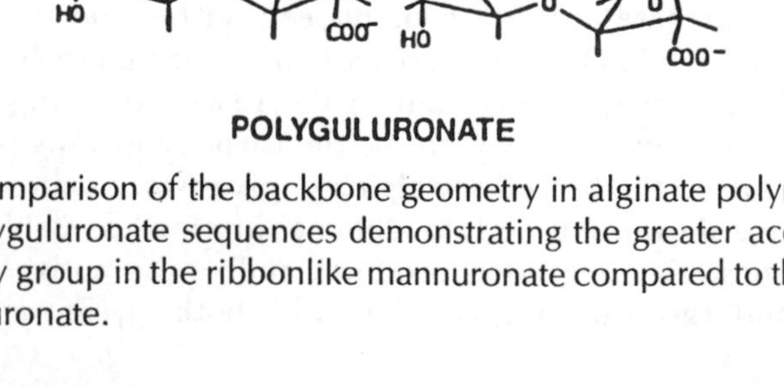

POLYGULURONATE

FIG. 8.9   Comparison of the backbone geometry in alginate polymannuronate and polyguluronate sequences demonstrating the greater accessibility of the carboxy group in the ribbonlike mannuronate compared to the "buckled" polyguluronate.

TABLE 8.1 Percentage Moisture Content of Soy Isolate and Soy Isolate Plus 2% Hydrocolloid After Heating at 185°C for 35 Min

Sample	Moisture[a] (g/100 g solids)
Isolate	12.2 ± 1.2
Isolate + high M-aginate	20.4 ± 6.6
Isolate + high G-alginate	12.0 ± 0.7
Isolate + CMC	12.2 ± 6.3
Isolate + pectate	15.8 ± 3.5

[a]Values are the means ± standard deviation of 20 determinations.
Source: Adapted from Oates et al., 1987a.

thus it is possible that the reactant group is an anhydro end group formed by cleavage of the glycosidic bond, such groups are known to be highly reactive and able to undergo browning reactions with amino groups.

If these groups on the polysaccharide are able to participate in nonenzymic browning reactions, then the likely reactants on the protein would be the amino groups. Lysine is an obvious candidate, but since this grouping is known to partake in such reactions irrespective of the presence of alginate (Cheftel, 1986), other groups are also presumably involved. Oates et al. (1987b) have presented convincing evidence that glutamic acid (or amide) groups are involved since of a whole range of proteins tested (pancreatin, ovalbumin, hemoglobin, whole blood albumen, casein, and gelatin) only soy and gluten (proteins very rich in glutamic acid residues) showed significant formation of additional water in the presence of high M alginate. It is interesting to note that in soy and gluten most of the glutamic acid residues exist in nature in the amide form, and it is tempting to suggest that these are the reactive groups. In the preparation of soy isolate from soy flour there is a variable amount of conversion of the amide to the acid, and it has been observed that soy isolates vary quite markedly in their ability to produce water when heated with high M alginate (Oates et al., 1987b). This lends indirect support to the contention that the reactive groups on the protein are glutamate residues. This contention about the potential role of glutamate residues in reacting with accessible carboxy groups on the polysaccharide may be of more than academic interest, since it is well established that the glutamate-rich proteins (soy and gluten) are more amenable to texturization by extrusion than most other proteins (Ledward and Mitchell, 1988). Though glutamate may be involved, Oates et al. (1987b) also observed that mixtures of glutamic acid and a high M alginate produced a significant increase in water content on heating at 170°C for 35 min. No increase was seen, however, in the presence of a high G alginate or in several other systems containing other amino acids (aspartic acid, lysine, or glycine) and high M alginate. This suggests that the acid grouping itself possesses reactivity and the amide form is not essential. Although there may be debate about whether the amide and/or acid form of glutamic acid is involved in the reactions, that this is the reactive site seems certain. For example, heating mixtures of soy isolate and 2% high M or high G alginate at 185°C for 35 min showed decreases in glutamic acid content from about 20 to 10 g/100 g in the high M systems; no change was seen in the high G systems (Oates et al., 1987b). There was also a significant *increase* in the aspartic acid content of the high M system, suggesting that one mechanism involves decarboxylation of the glutamic acid. The only other amino acid residue to show a significant decrease on heating was lysine; the decrease in this case was independent of the presence of alginate. Though the detailed chemistry of these reactions has yet to be elucidated, the covalent linkages established during the reactions, which produce water as a by-product, may be of key importance in dictating the texture of extruded proteins (Ledward and Mitchell, 1988).

Although the results found by Oates et al. mainly involved heating mixtures of soy isolate and polysaccharide, in extrusion processing soy grits are usually used. These will already contain simple and complex polysaccharides, and Ledward et al. (1990) extended the earlier studies to look at soy flour and soy samples devoid of soluble (dialyzed against water) and both soluble and insoluble (isolated protein) sugars. The results are summarized in Table 8.2. All samples browned quite markedly when heated at 185°C for 35 min, with the untreated flour being the most reactive. However, consideration of the results quoted in Table 8.2 led the authors to conclude that:

1.   Browning in a low amide isolate (some loss of ammonia occurred during preparation) is, at least in part, due to reactions involving lysine.
2.   The higher molecular weight carbohydrates present in the flour induce further browning reactions, some involving lysine and some glutamate residues.
3.   The low molecular weight, dialyzable, components in the flour induce still further browning, involving, to some extent, both lysine and glutamic acid (not necessarily amide) residues.

Though the detailed chemistry of these reactions is not understood,

TABLE 8.2 Chemical Changes Induced in Soy Protein by Heat Treatment at 185°C for 35 Min

Parameter	Isolate		Dialyzed flour		Untreated flour	
	Before	After	Before	After	Before	After
Water (g/100 g protein)	12.36 ± 0.86	13.30 ± 0.86	15.26 ± 0.70	18.70 ± 0.49	18.28 ± 0.88	25.63 ± 0.63
NH_3 (g/100 g N)	9.34 ± 1.51	8.57 ± 2.60	13.08 ± 0.70	7.54 ± 0.20	13.66 ± 0.10	7.56 ± 0.20
Lysine (g/100 g protein)	5.99 ± 0.48	4.19 ± 0.58	4.60 ± 0.58	2.48 ± 0.17	5.32 ± 0.70	0.85 ± 0.21
Glutamic acid (g/100 g protein)	20.45 ± 1.44	21.46 ± 1.37	20.96 ± 0.84	18.42 ± 0.34	18.42 ± 1.62	14.20 ± 0.85
Reflectance + 420 nm	32.5 ± 6.6	14.0 ± 4.9	41.8 ± 0.6	23.5 ± 3.2	61.3 ± 1.8	11.5 ± 0.2

All values are the means standard deviation of 6 determinators. Samples were heated at 160°C for 30 min.
Source: Adapted from Ledward et al., 1990.

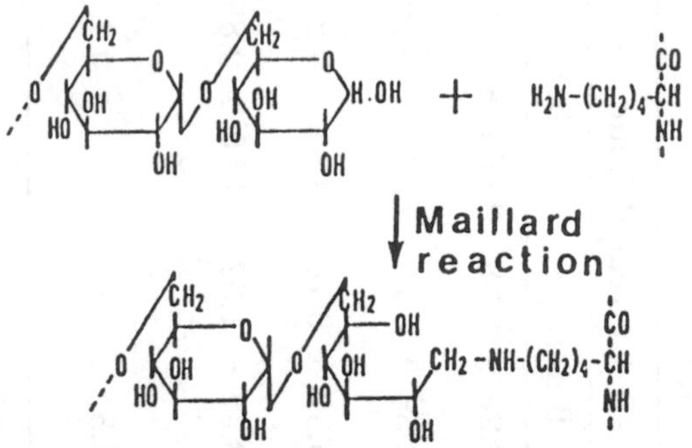

FIG. 8.10 Proposed mechanism for the formation of dextran-protein complexes by controlled dry heating at 60–65°C for 3 weeks. (*From Kato and Kobayashi, 1991.*)

their possible role in texturization of extruded products (especially those between the protein and high molecular weight nondialyzable material) may be very significant (Ledward and Mitchell, 1988). A full understanding of their nature should allow more control of the extrusion process.

Since 1990 several workers (Kobayashi et al., 1990; Kato et al., 1990; Kato and Kobayashi, 1991; Dickinson and Galazka, 1991) have described the preparation of covalent protein-polysaccharide complexes prepared by controlled "dry" heating of certain proteins, β-lactalbumin (Dickenson and Galazka, 1991), soy (Kobayashi et al., 1990), and ovalbumin (Kato et al., 1990) and a polysaccharide (dextran and PGA). These soluble complexes are formed by a Maillard reaction between the two biopolymers and possess excellent functional, especially emulsifying properties. The actual nature of the covalent linkage(s) has not yet been determined, but since the molecular weight of the ovalbumin-dextran complexes are in the range 130,000–230,000, Kato and Kobayashi (1991) suggest that the bond forms between the reducing end group of the dextran molecule and the ε-amino group of lysine (Fig. 8.10) so that only one or two moles of dextran can bind per mole of protein. The low water content of the system presumably ensures that the amino group is in the reactive undissociated form.

USE IN FOOD SYSTEMS

Protein Recovery

As early as 1949, Gortner reported that when negatively charged polymers are added to protein solutions of low pH, an interaction occurs, causing precipitation of the reactants. Since that time a number of workers have shown that it is possible to recover proteins from waste streams, and from other dilute protein solutions, by reaction with acidic polysaccharides. For example, Smith et al. (1962) precipitated the protein of soy bean whey effluent with a number of edible polysaccharides, including alginic acid, and Gillberg and Tornell (1976) have described a method of preparing rapeseed protein isolates using acidic polymers to precipitate the proteins following an initial alkali extraction stage. They suggest a number of possible precipitants, including sodium alginate, carboxymethylcellulose, and carrageenans, which are capable of yielding protein recoveries of about 90%.

Recovery of protein from cheese whey has been extensively studied, and it has been shown that any one of a number of materials, including sodium alginate, CMC, and polyacrylic acid (Shank and Cunningham, 1968; Hansen et al., 1971), may be used to precipitate the protein. However, by far the most widely studied system uses CMC. It has been shown that the amount of protein precipitated is very dependent on the pH, ionic strength, and degree of substitution of the cellulose derivative. Under favorable conditions more than 90% of the protein can be removed from solution (Hansen et al., 1971; Hill and Zadow, 1974) and the protein extracted by this method has excellent functional properties (Morr et al., 1973).

Hansen and Balachandran (1983) have reviewed how the insoluble complex produced by treatment of whey with CMC can be solubilized by subsequent treatment of the dried complex with ammonia. This, it is claimed, would be an energetically less costly method of preparing the functional soluble product than treatment of the wet precipitate with alkali and subsequent spray drying.

Our investigation into the use of anionic polysaccharides for protein precipitation showed that sodium alginate is, under most conditions, both the most efficient and the least affected by increasing salt concentration (Fig. 8.11), and, as discussed earlier, this may be related to both charge and steric effects. However, at low ratios of protein to polysaccharide, pectate is more effective than alginate in precipitating protein. This phenomenon is probably attributable to the low stability of concentrated pectate solutions, especially under conditions of high ionic strength. Consequently, the effect of resolubilization of the protein-alginate complex in the presence of excess alginate is not observed with pectate solutions.

In an investigation of the ability of alginate and pectate to precipitate blood plasma proteins, it was found that, as expected, the efficiency of precipitation was very dependent on pH, ionic strength, and the ratio of

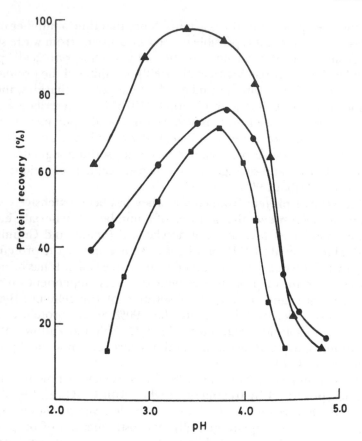

FIG. 8.11   Effect of pH on the recovery of blood plasma by sodium alginate
(▲), sodium pectate (●), and CMC (■) at an ionic strength of 0.16. Protein
concentration, 1.2%; polysaccharide concentrate 0.4%. (*From Imeson et al.,*
1978.)

protein to polysaccharide (Imeson et al., 1978). Even at high ionic
strength good recoveries could be obtained with both polysaccharides at
pH 3.8.

Protein Fractionation and Purification

Precipitation of whey proteins by CMC is selective and may be used
as a method of separating the protein components. Thus, Hidalgo and
Hansen (1971) found that at pH 4.0 the precipitated complex was composed entirely of β-lactoglobin and CMC, the α-lactalbumin remaining
in solution. On decreasing the pH to 3.2, further addition of CMC

precipitated the α-lactalbumin (Hidalgo and Hansen, 1971). This same principle has been used by Tolstoguzov et al. (1974) to fractionate other mixed protein solutions, e.g., the yeast proteins from *Candida guilliermondii* (Bogracheva et al., 1982, 1983).

It is also readily apparent that the "thermodynamic incompatibility" of many protein polysaccharide solutions will serve to simply and cheaply concentrate the protein component (Fig. 8.3) and has been used to concentrate the proteins in skim milk (Antonov et al., 1982; Tolstoguzov, 1986). In this process the protein content could be concentrated to about 30% by mixing with a 3% solution of pectin, the pectin solution being diluted to 1–2%.

Inhibition of Protein Precipitation

Acidic polysaccharides inhibit the precipitation of proteins at pH values in the normal isoelectric range (Ganz, 1974), and thus it is possible to use polysaccharides to maintain a protein in solution under conditions that would normally lead to precipitation.

The ability of acidic polysaccharides to inhibit protein precipitation at the isoelectric point of the proteins is used in the preparation of fruit-flavored milk drinks (Ledward, 1979). This is necessary because the acidity of fruit is antagonistic to milk proteins. As the pH decreases these proteins are destabilized, and at about pH 4.6, the isoelectric point of casein, they are completely precipitated. However, fruits give a satisfactory flavor only between pH 4.0 and 5.0. Thus, in order to prepare an acceptable product it is essential to dissolve the milk proteins in this pH range. This can readily be achieved by formation of the appropriate protein-polysaccharide complex, of decreased isoelectric point, which is consequently soluble at pH 4.5. This complex may, of course, precipitate at lower pH values when the overall net charge approaches zero. CMC is particularly effective in keeping milk proteins in solution.

One area with potential for the use of protein-polysaccharide interactions in product development has been opened up by the observation that, under appropriate conditions, polysaccharides will inhibit the precipitation of some water-soluble proteins following thermal denaturation (Imeson et al., 1977). Thus even after heating, a soluble system is obtained in which both components in the complex will have useful functional properties, properties that may well differ from those of the individual components (Ledward, 1979).

Food Preservation

Since proteins will interact to form complexes with polysaccharides of opposite *net* charge, it is not surprising that enzymic activity can be affected quite markedly by the presence of an anionic polysaccharide at low pH (Ledward, 1979). Studies have shown that a number of charged polysaccharides (pectin, carrageenan, and κ-carrageenan) modify the structure of a number of enzymes (trypsin, horseradish peroxidase) and thus may be considered to act as competitive inhibitors (Gatfield and Shute, 1976). Thus it has been suggested that anionic polysaccharides could be used as a method of enzyme inactivation in foods, either as an individual treatment or in conjunction with more conventional techniques such as blanching (Ledward, 1979).

Texturization

Perhaps the area that has aroused most interest in recent years is the potential to utilize protein/polysaccharide interactions to generate textured products (Lefebvre and Thebaudin, 1992). Much earlier work made use of the polysaccharide's gelling ability, the other ingredients, including the protein, merely being dispersed throughout the gel. For example, Ishler et al. (1963) produced meatlike fibers of alginate by extruding a protein-alginate solution, containing flavorings and colors, through fine spinnerets into a calcium salt–acetic acid bath, followed by further coagulation in hot water. On drying, the fibers possessed good textural properties and could be heated (cooked) with little loss of texture. Many variations and modifications to this general technology have been published, and, not unexpectedly, the physical characteristics of the fibers are very dependent on the composition of the coagulating bath, as well as the composition of the spinning dope (Lawrie and Ledward, 1984).

Imeson et al. (1979) found that, when a solution of blood plasma (6%) and sodium alginate (2%) was extruded into unbuffered coagulating baths of calcium chloride (pH ∼ 8), there was a rapid increase in the shear strength of the fiber bundles with increasing salt concentration up to 3% calcium chloride. Above this level the strength of the fibers was independent of the calcium concentration. When dopes were extruded into 5% calcium chloride, fiber strength was independent of pH in the range 4–8. Below pH 4, however, the fiber bundles rapidly decreased in strength, exhibiting a minimum value at pH 3.5. Decreasing the pH still further increased fiber strength once more. The protein content of the fibers increased significantly, from about 6% at pH 3.5 to about 15% at pH 2. The rheological properties of the spun alginate–blood plasma fibers were found to vary in a complex manner with the guluronic acid block content and the molecular weight (viscosity) of the alginate (Lawrie and Ledward, 1984).

It is likely that, at neutral pH, the protein is merely trapped within the calcium alginate filaments since it can be easily washed out (Imeson et al., 1979). At lower pH, however, the carboxy groups of the alginate will tend to exist in the undissociated form (since the pK values of these groups are ~ 4). Consequently, extrusion into a bath of about this pH would be expected to involve the formation of fibers containing a high proportion of weaker alginic acid filaments. As the pH is reduced still further, however, the precipitating conditions become similar to those employed in conventional protein fiber production and the acid-denatured proteins precipitate irrespective of the presence of the alginate. The protein fibers could then coexist with the calcium alginate and alginic acid filaments to give bundles of increased strength. Thus, at pH 2, there is almost complete recovery of both alginate and protein (Imeson et al., 1979). In addition, at low pH, there is a distinct possibility of electrostatic protein-polysaccharide interactions occurring (Imeson et al., 1977, 1978), and, although these may be insignificant at high ionic strengths, they may help to explain the increased strength (and protein content) of fibers produced by spinning plasma-alginate mixtures into baths containing 1.5% calcium chloride at pH 3.5 compared to spinning into baths containing 3 or 4% calcium chloride at this pH (Table 8.3).

Although the role of protein-polysaccharide complex formation in the above system is a little tenuous, such complex formation is undoubtedly a key factor in other processes that may have commercial potential. For example, gelatin and alginate or pectate will, at low pH, form a mixed gel (Tolstoguzov, 1986; Muchlin et al., 1976; Tschmak et al., 1976). This is illustrated in Fig. 8.12, where it can be seen that the gel formed 2 hr

TABLE 8.3   Effect of Coagulating Bath $CaCl_2$ Concentration at pH 3.5 on the Properties of Fibers Prepared from Plasma (6.0%)-Alginate (2.0%) Mixtures

$CaCl_2$ (%)	Warner-Bratzler[a] shear strength	Moisture (%) (N/cm^2)	Protein (%)	Alginate (%)	Ash (%)
5	42.2 ± 7.4	89.15	4.43	5.14	1.28
4	7.1 ± 1.4	93.92	2.61	2.69	0.78
3	9.0 ± 1.3	95.41	1.75	2.18	0.66
1.5	9.2 ± 1.0	95.19	2.99	1.16	0.66

[a]Mean ± standard deviation of eight measurements from two tows.
Source: Lawrie and Ledward, 1984.

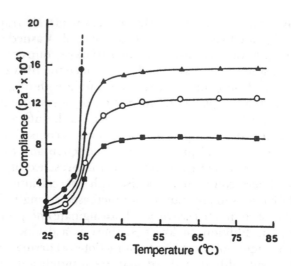

FIG. 8.12 Thermal stability of complex gels of gelatin (5%) and sodium alginate (2.5%) at pH 3.9 after aging for 2 hr (●), 1 day (▲), 3 days (○), and 5–8 days (■). (*From Tolstoguzov, 1986.*)

after mixing the solutions melts over the expected temperature range (30–40°C), but on aging the gel for a day or more, a thermostable network is formed that resists heating to at least 80°C, although some loss in rigidity (increase in compliance) is seen at 30–40°C. The strength of the residual network increases with aging time, and since it can be destroyed at high ionic strength or by raising the pH above the isoelectric point of the gelatin, it is presumably stabilised by electrostatic protein-polysaccharide linkages (alginate alone will not gel under these conditions). Under mildly acidic conditions the interaction of gellan gum and gelatin results in a synergistic enhancement of gel strength (Shim, 1985). The optimum ratio of the two polymers is 1:1, and a total polymer concentration of 0.2 to 0.4% yields acceptably textured gels. Thus the total hydrocolloid requirement for specific applications may well be reduced with concomitant cost savings.

Although the gelatin-polysaccharide interactions discussed above will yield a gel network, charge-charge interactions normally result in the formation of an insoluble complex (precipitate). This behavior is used commercially in microencapsulation. For example, both gum arabic and gellan gum will form a coacervate with gelatin that will deposit around dispersed oil droplets to produce microencapsulated spheres (Glicksman, 1969; Shim, 1985; Chilvers and Morris, 1987).

Perhaps the area that has received the most attention regarding the practical application of protein-polysaccharide complex formation has been that involving milk proteins to generate textured milk products. The most widely studied milk protein-polysaccharide interaction is probably that between κ-casein and κ-carrageenan, a reaction that has been used as the basis of some dessert milk products. This interaction is of interest as it will persist even at pH values above the isoelectric point of casein, i.e., in solutions in which both the protein and polysaccharide carry net negative charges. However, all the available evidence suggests that the driving force for the interaction is electrostatic (Snoeren et al., 1975). Grindrod and Nickerson (1968) and Payens (1972) have demonstrated that κ-casein, which constitutes only 12% of the casein in milk, is the only protein to interact with carrageenan.

Even though all caseins have similar amino acid compositions and thus isoelectric points, it is believed that only the κ-casein reacts under these conditions since it possesses an extensive region of 95 amino acid residues, in which at neutral pH, 13 positively charged and only one negatively charged residue are situated (Mercier et al., 1971, 1973). This region can thus complex with the negatively charged sulfate groups on the sulfated guar, and locust bean gum will stabilize some or all of these proteins (Ledward, 1979). Though not fully understood, it is unlikely that these phenomena involve the formation of a protein-polysaccharide complex, but more likely that the calcium ions neutralize the sulfate groups on the polysaccharide, permitting the chain to form a "gel"-like structure, which physically serves to inhibit protein aggregation (Ledward, 1979).

A practical and commercial application of protein-polysaccharide complex formation is the use of propyleneglycol alginate (PGA), in which 70–85% of the carboxylate groups are esterified, to stabilize salad dressings (McDowell, 1975). Since PGA is less susceptible than underivitized alginate to precipitate under acid conditions, the resulting complexes, formed at low pH with the proteins in the dressing, will tend to produce a hydrated network rather than a precipitate. The strength of the gel network decreases with increasing esterification, i.e., decreasing number of carboxylate groups, allowing the flow properties of the dressing to be controlled by use of a suitable PGA sample.

Although the formation of complexes between proteins and anionic polysaccharides is useful in the generation of textured products, the effects may not always be desirable. For example, it is established that the efficiency of some polysaccharide gelling agents may be reduced in the presence of proteins. This effect has been observed when alginates are used as thickeners in meat and fish products (Edlin and Rocks, 1969).

For example, at pH 6.3 in the presence of 0.006 M calcium, the viscosity of a 1% alginate solution was more than twice that of similar solution containing 1% myoglobin (Ledward, 1979; Hughes et al., 1980). However, whether this is due to the myoglobin competing with the polysaccharide and thus inhibiting gelation or the myoglobin chelating the calcium ions, making them unavailable for polysaccharide gel formation, is not known.

Most of the above discussion on texturization has been concerned with energetically favored interactions, but thermodynamically incompatible polymers have tremendous potential as a means of generating nonhomogeneous product texture. These biphasic systems can be gelled to give a range of products, depending on the states of the continuous and dispersed phases, which will depend on many factors including mixing conditions, the density difference between the phases, interfacial tension, the viscosity of the two phases, and the time prior to gelation (small droplets will coalesce prior to gelation). Because of the complex interplay of these factors, individual systems are largely managed on an empirical basis. Recently gelatin-based fat replacers have been marketed from mixtures of gelatin and locust bean gum or guar gum in which the gelatin gel exists as particles (average size 10–30 μm) in a continuous galactomannan matrix. These systems are claimed to possess fatlike characteristics as they melt in the mouth, possess a smooth, short plastic texture, and retain a high viscosity after melting (Anonymous, 1993).

A further interesting and potentially useful development is the use of flow to stretch and orientate the dispersed phase prior to gelation and thus generate gelled products with, it is claimed, the anisotopic, fibrillar character of such materials as meat (Tolstoguzov, 1988). The general principles involved are outlined in Fig. 8.13. The product obtained, after rapid gelation by whatever technique is appropriate, e.g., heating, cooling, or the use of a setting bath of appropriate composition, may be fibers (if only the dispersed phase sets), a gel filled with liquid channel or capillaries (if only the continuous phase gels), or a homogeneous matrix filled with orientated fibers (if both phases gel). As well as affecting the perceived texture, the formation of a gel filled with liquid channels has attractions for the production of flavored products (since liquids have far better flavor release characteristics than gels) and products with increased juiciness (Morris, 1990).

Texture can also be imparted to foods by the formation of covalent linkages between PGA and gelatin. Applications for this interaction include binding of fish protein concentrate to produce a product resembling fish muscle (Unilever Ltd., 1976) and the formation of sausage skins using PGA and dispersions, slurries, or pastes of collagen (Collagen Products Ltd., 1976).

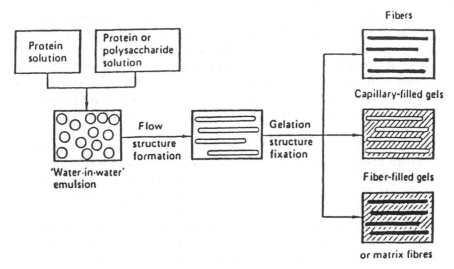

FIG. 8.13 Principle of "spinneretteless spinning" for the production of orientated, multitextured products. (*From Tolstoguzov, 1986.*)

Stabilization of Foams and Emulsion

In 1984 it was reported by the Leatherhead Food Research Association in England that certain basic (high isoelectric point) proteins, namely, lysozyme (pI = 10.7) and clupeine (pI = 12) and derivatives of β-lactoglobulin (pI > 9), dramatically improved the foaming ability of some acidic proteins (pI 4.6–6.6). This work has been reviewed by Poole and Fry (1987). The effect was only observed under conditions where the two species possessed different net charges, i.e., at pH values between their isoelectric points and at low ionic strength. At pH values approaching the isoelectric point of either species, the enhancing effects diminished. Such foams could tolerate quite high lipid contents (Poole and Fry, 1987) and would have obvious potential in the development of dessert products. Such basic proteins also enhanced the gelling ability of certain acidic proteins (Poole and Fry, 1987).

The problems of obtaining either basic proteins of good functional (foaming, emulsifying, and gelling) ability or basic polysaccharides generally has limited the potential of extending these findings to protein-polysaccharide systems, but the effects would be manifest if such compounds could be found. For example, Burova et al. (1991) found that at pH 4.2–6.0 and low ionic strength, several acidic polysaccharides improve both the emulsifying ability and the stability of emulsions of the llS globulin from broad beans. Unfortunately, the complexes responsible for these effects were poorly soluble even at high polysaccharide concentrations. At pH 7.6, where the protein and the polysaccharides are incompatible, no such effects were seen. Dickinson and Galazka (1991) found that both propylene glycol and dextran sulfate, but not dextran, were capable of forming weak complexes at pH 7 with β-lactoglobulin, but these were not particularly useful in stabilizing emulsions prepared with the protein, presumably due to the presence of unabsorbed polysaccharide.

However, these workers were able to prepare complexes of the protein and both PGA and dextran (but not dextran sulfate) by heating at 60°C for 3 weeks. These soluble complexes, which were believed to contain a covalent Maillard linkage between the two polymers, were found to give excellent emulsion stability. As discussed earlier, workers in Japan have also developed a range of soluble polysaccharide-protein complexes (mainly dextran based) by controlled dry heating at 60°C for several weeks (Kobayashi et al., 1990; Kato et al., 1990; Kato and Kobayashi, 1991). Such covalently bound complexes, of molecular weight about 200,000 (ovalbumin-dextran) or 450,000 (soy-dextran), were found to have excellent functional properties and may be very useful in the development of new food products.

The potential uses of protein-polysaccharide are as many and varied as the types of possible interactions, and continuing research should allow for fuller exploitation in, among other areas, the food industry.

REFERENCES

Aquilera, J. M. 1992. Generation of engineered structures in gels. In *Physical Chemistry of Foods*, H. G. Schwartzberg and R. W. Hartel (Ed.), pp. 387–421. Marcel Dekker, New York.

Albertsson, P. A. 1971. *Partition of Cell Particles and Macromolecules*, 2nd ed. Wiley-Interscience, New York.

Anon. 1993. Slimgel: A new gelatin based fat replacer. *Eur. Food & Drink Rev.* (Spring): 9–10.

Antonov, J. A., Grinberg, V. J., Zhuraoskaya, N. A., and Tolstoguzov, V. B. 1982. Concentration of protein skimmed milk by the method of membraneless isobaric osmosis. *Carbohydr. Polymers* 2: 81–91.

Berrington, D., Imeson, A. P., Ledward, D. A., Mitchell, J. R., and Smith, J. 1984. The effect of alginate inclusion on the extrusion behaviour of soya. *Carbohydr. Polymers* 4: 443–460.

Bogracheva, T. J., Grinberg, V. J., and Tolstoguzov, V. B. 1982. Use of polysaccharides to remove lipids from the protein globulin fraction of bakers yeast. *Carbohydr. Polymers* 2: 163–178.

Bogracheva, T. J., Grinberg, V. J., and Tolstoguzov, V. B. 1983. Compatibility of gum acacia with macromolecular components of the globulin fraction of bakers yeast. *Nahrung* 27: 735–740.

Burova, T. V., Grinberg, N. V., Grinberg, V. Y., Leontiev, A. L., and Tolstoguzov, V. B. 1991. Effects of polysaccharides upon the functional properties of llS globulin of broad beans. *Carbohydr. Polymers* 18: 101–108.

Chilvers, G. R. and Morris, V. J. 1987. Coacervation of gelatin-gellan gum mixtures and their use in microencapsulation. *Carbohydr. Polymers* 17: 111–120.

Cheftel, J. C. 1986. Nutritional effects of extrusion cooking. *Food Chem.* 20: 263–283.

Christianson, D. D., Hodge, J. E., Osbourne, D., and Detroy, R. W. 1981. Gelatinization of wheat starch as modified by xanthan gum, guar gum and cellulose gum. *Cereal Chem.* 58: 513–517.

Collagen Products Ltd. 1976. British Patent, 1450-687.

Dickinson, E. and Galazka, V. B. 1991. Emulsion stabilization by ionic and covalent complexes of β-lactoglobulin with polysaccharides. *Food Hydrocolloids* 5: 281–296.

Edlin, R. L. and Rocks, J. K. 1969. U.S. Patent No. 3,480,450.

Ganz, A. J. 1974. How cellulose gum reacts with proteins. *Food Eng.* 46(6): 67–69.

Gatfield, I. L. and Shute, R. 1976. Circular dichroism studies on the interaction between horseradish peroxidase and polygacturonic acid in aqueous solution. *Lebensm.-Wiss. Technol.* 9: 363–365.

Gillberg, L. and Tornell, B. 1976. Preparation of rapeseed protein isolates. *J. Food Sci.* 41: 1070–1075.

Glicksman, M. 1969. *Gum Technology in the Food Industry*. Academic Press, New York.

Gortner, R. A. 1949. *Outlines of Biochemistry*. John Wiley, New York.

Grindrod, J. and Nickerson, T. A. 1968. Effect of various gums on skim milk and purified milk proteins. *J. Dairy Sci.* 51: 834–841.

Hansen, P. M. T., Hidago, J. E., and Gould, I. A. 1971. Reclamation of whey protein with carboxymethylcellulose. *J. Dairy Sci.* 54: 830–834.

Hansen, P. M. T. and Balachandran, R. 1983. Precipitation and recovery of whey protein with carboxymethylcellulose and preparation of a soluble complex by ammonia adsorption. In *Upgrading waste for feeds and foods*, D. A. Ledward, A. J. Taylor, and R. A. Lawrie (Ed.), pp. 85–91. Butterworths, London.

Hidalgo, J. E. and Hansen, P. M. T. 1971. Selective precipitation of whey proteins with carboxymethylcllulose. *J. Dairy Sci.* 54: 1270–1274.

Hill, R. D. and Zadow, J. G. 1974. The precipitation of whey proteins by carboxymethylcellulose of differing degrees of substitution. *J. Dairy Res.* 41: 373–380.

Hughes, L., Ledward, D. A., Mitchell, J. R., and Summerlin, C. 1980. The effect of some meat proteins on the rheological properties of pectate and alginate gels. *J. Texture Stud.* 11: 247–256.

Imeson A. P., Ledward, D. A., and Mitchell, J. R. 1977. On the nature of the interactions between some anionic polysaccharides and proteins. *J. Sci. Food Agric.* 28: 661–667.

Imeson A. P., Ledward, D. A., and Mitchell, J. R. 1979. The effects of calcium and pH on spun fibres produced from plasma-alginate mixtures. *Meat Sci.* 3: 287–294.

Imeson, A. P., Richmond, P., and Smith, A. C. 1985. The extrusion of soya with alginate using a twin screw extruder. *Carbohydr. Polymers* 5: 329–340.

Imeson, A. P., Watson, P. R., Mitchell, J. R., and Ledward, D. A. 1978. Protein recovery from blood plasma by precipitation with polyuronates. *J. Food Technol.* 13: 331–338.

Ishler, N. H., MaCallister, R. V., Szczesniak, A. S., and Engel, E. 1963. U.S. Patent No. 3,093,483.

Kato, A. and Kobayashi, K. 1991. Excellent emulsifying properties of protein-dextran conjugates. In *Microemulsions and Emulsions in Food.*, pp. 213–229. ACS Symposium Series 448.

Kato, A., Sasaki, Y., Furata, R., and Kobayashi, K. 1990. Functional protein-polysaccharide conjugate prepared by controlled dry heating of ovalbumin-dextran mixtures. *Agric. Biol. Chem.* 54: 107–112.

Kelly, R., Gudo, E., Mitchell, J., and Harding, S. 1993. Investigations into intereactions between bovine serum albumin with a sodium alginate and a pectin using dynamic light scattering, sedimentation ultracentrifugation and viscometry. *Carbohydr. Polymers* (in press).

Kobayashi, K., Kato, A., and Matsudomi, N. 1990. Developments in new functional food materials by hybridization of soy protein to polysaccharide. *Nutr. Sci. Soy Protein* 11: 23–28.

Lawrie, R. A. and Ledward, D. A. 1984. Texturisation of recovered proteins. In *Upgrading Waste for Feed and Foods.* D. A. Ledward, A. J. Taylor, and R. A. Lawrie (Ed.), pp. 163–182. Butterworths, London.

Ledward, D. A. 1979. Protein-polysaccharide interactions. In *Polysaccharides in Food*. J. M. V. Blanshard and J. R. Mitchell (Ed.), pp. 205–217. Butterworths, London.

Ledward, D. A. and Mitchell, J. R. 1988. Protein extrusion-more questions than answers. In *Food Structure—Its Creation and Evaluation*. J. R. Mitchell and J. M. V. Blanshard (Ed.), pp. 219–230. Butterworths, London.

Ledward, D. A., Rasul, S., and Mitchell, J. R. 1990. Chemical changes during extrusion processing. In *Processing and Quality of Foods*, Vol. 1, P. Zeuthen, J. C. Cheftel, C. Ericson, T. R. Gormley, P. Linko, and K. Paulus (Ed.), pp. 276–281. Elsevier, London.

Lefebvre, A. C. and Thebaudin, J. Y. 1992. Texture formation by association of proteins and polysaccharides, for application in meat product manufacture. *Viandes et Produits Carnes* 13: 41–48.

McDowell, R. H. 1970. New reactions of propylene glycol alginate. *J. Soc. Cosmetic Chem.* 21: 441–447.

McDowell, R. H. 1975. New developments in the chemistry of alginates and their use in food. *Chem Ind.*: 395–391.

McKay, J. E., Stainsby, G., and Wilson, E. L. 1985. A comparison of the reactivity of alginate and pectate esters with gelatin. *Carbohydr. Polymers* 5: 223–236.

Mercier, J. C., Grosclaude, F., and Ribadeau-Duman, B. 1971. Structure primaire de la caseine a_s-bovine. *Eur. J. Biochem.* 23: 41–51.

Mercier, J. C., Brignon, G., and Ribadeau-Duman, B. 1973. Structure primaire de la caseine K_B-bovine. *Eur. J. Biochem.* 35: 222–235.

Morr, C. V., Swensen, P. E., and Richter, R. L. 1973. Functional characteristics of whey protein concentrates. *J. Food Sci.* 38: 324–330.

Morris, E. R. 1990. Mixed Polymer gels. In *Food Gels*, P. Harris (Ed.), pp. 291–359. Elsevier, London.

Muchlin, M. A., Wajnermann, E. S., and Tolstoguzov, V. B. 1976. Complex gels of proteins and acid polysaccharides, *Nahrung* 20: 313–320.

Oates, C. G., Ledward, D. A., Mitchell, J. R., and Hodgson, I. 1987a. Physical and chemical changes resulting from heat treatment of soya and soya alginate mixtures. *Carbohydr. Polymers* 7: 17–33.

Oates, C. G., Ledward, D. A., Mitchell, J. R., and Hodgson, I. 1987b. Glutamic acid reactivity in heated protein and protein-alginate mixtures. *Int. J. Food Sci. Technol.* 22: 477–484.

Oates, C. G. and Ledward, D. A. 1990. Studies on the effect of heat on alginate. *Food Hydrocolloids* 4: 215–220.

Pavlovskaya, G. E., McClements, D. J., and Povey, M. J. 1992. Preliminary study of the influence of dextran on the precipitation of legumin from agueous salt solutions. *Int. J. Food Sci. Technol.* 27: 629–636.

Payens, T. A. J. 1972. Light scattering of protein reactivity of polysaccharides especially of carrageenans. *J. Dairy Sci.* 55: 141–150.

Poole, S. and Fry, J. C. 1987. High performance protein foaming and gelation systems. In *Development in Food Proteins—5*, B. J. F. Hudson (Ed.), pp. 257–298. Elsevier, London.

Schechter, A. N. and Epstein, C. J. 1968. Spectral studies on the denaturation of myoglobin. *J. Mol. Biol.* 35: 567–589.

Semenova, M. G., Pavlovskaya, G. E., and Tolstoguzov, V. B. 1991. Light scattering and thermodynamic phase behavior of the system lls globulin-κ-carrageenan-water. *Food Hydrocolloids* 4: 469–479.

Shank, J. L. and Cunningham, W. H. 1968. U.S. Patent No. 3,404,142.

Shim, J. L. 1985. Gellan gum gelatin blends. U.S. Patent No. 4,517,216.

Smith, A. K., Nash, A. M., Eldridge, A. C., and Wolf, W. J. 1962. Recovery of soybean whey protein with edible gums and detergents. *J. Agric. Food Chem.* 10: 302–304.

Smith, J., Mitchell, J. R., and Ledward, D. A. 1982. Effect of the inclusion of polysaccharides on soya extrusion. *Prog. Food Nutr. Sci.* 6: 139–147.

Snoeren, T. H. M., Payens, T. A. J., Jevnink, J., and Both, P. 1975. Electrostatic interaction between κ-carrageenan and κ-casein. *Milchwissenschaft* 30: 393–396.

Snoeren, T. H. M., Both, P., and Schmidt, D. G. 1976. An electronmicroscopic study of carrageenan and its interaction with κ-casein. *Neth. Dairy Milk J.* 30: 132–141.

Soshinsky, A. A., Antonov, Y. A., Tolstoguzov, V. B., and Malovikova, A. A. 1992. Water-insoluble complexes of ribulose-1,5-bisphosphate carboxylase of alfalfa with pectin. *Nahrung* 36: 150–156.

Suchkov, V. V., Grinberg, V. Ya., and Tolstoguzov, V. B. 1981. Steady-state viscosity of the liquid two phase disperse system water-casein-sodium alginate. *Carbohydr. Polymers* 1: 39–53.

Tolstoguzov, V. B. 1986. Functional properties of protein-polysaccharide mixtures. In *Functional Properties Food Macromolecules*, J. R. Mitchell and D. A. Ledward (Ed.), pp. 385–415. Elsevier, London.

Tolstoguzov, V. B. 1988a. Some physico-chemical aspects of protein processing into foodstuffs. *Food Hydrocolloids* 2: 339–370.

Tolstoguzov, V. B. 1988b. Creation of fibrous structures by spinneretteless spinning. In *Food Structure—Its Creation and Evaluation*,

J. M. V. Blanshard and J. R. Mitchell (Ed.), pp. 181–196. Butterworths, London.

Tolstoguzov, V. B. 1990. Interactions of gelatin with polysaccharides. In *Gums and Stabilisers for the Food Industry—5*, G. O. Phillips, D. J. Wedlock, and P. A. Williams (Ed.), pp. 157–175. IRL Press, Oxford.

Tolstoguzov, V. B. 1991. Functional properties of food proteins and the role of protein-polysaccharide interaction. *Food Hydrocolloids* 4: 429–468.

Tolstoguzov, V. B., Belkina, V. P., Gulov, V. Ja, Titova, E. F., and Belavzeva, E. M. 1974. State of phase, structure and mechanical properties of the gelatinous system water-gelatin-dextran. *Die Stärke* 26: 130–138.

Tschumak, G. Ja., Wajnermann, E. S., and Tolstoguzov, V. B. 1976. The structure and properties of complex gels of gelatin and pectin. *Nahrung* 20: 321–328.

Unilever Ltd. 1976. British Patent 1443 513.

Wills, G. D., Ledward, D. A., and Mitchell, J. R. 1988. Interactions between κ-carrageenan and myoglobin. *Proc. 34th Congress of Meat Science and Technology, Brisbane, Part B*, pp. 322–324.

Ziegler, G. R. and Foegeding, E. A. 1990. The gelation of proteins. *Adv. Food Nutr. Res.* 34: 203.

9

Chemical and Enzymatic Modification of Proteins for Improved Functionality

Fakrieh Vojdani and John R. Whitaker
University of California
Davis, California

INTRODUCTION

In addition to providing the eight essential amino acids (plus histidine for infants) required by adults, as well as nitrogen, proteins also provide flavor and functionality in foods. Functionality of proteins is defined as any physicochemical property that affects the processing and behavior of proteins in a food system, as judged by the quality attributes of the final product. The particular functionality required varies with the food system and application. No single protein can meet all the functional properties required in various foods (Kinsella, 1982). In fluid milk, the desired functionality is met ideally by the micellar state of the casein–calcium phosphate complex, while in cheese it is the precipitated caseins, and their subsequent derivatives, that are responsible for the difference among cottage cheese, Cheddar cheese, and blue cheese. Egg, muscle, and wheat proteins provide completely different functional properties in foods. Some of the desired functional properties of proteins are shown in Table 9.1.

261

TABLE 9.1 Desired Functional Properties of Proteins in Food Systems

Functional property	Mode of action	Food system example
Solubility	Protein solvation	Beverages
Water absorption and binding	Hydrogen bonding of water, entrapment of water	Meats, sausages, breads, cakes
Viscosity	Thickening, water binding	Soups, gravies
Gelation	Protein matrix formation and setting	Meats, curds, cheese
Cohesion-adhesion	Protein acts as adhesive material	Meats, sausages, baked goods, pasta products
Elasticity	Hydrophobic bonding in gluten, disulfide links in gels	Meats, bakery
Emulsification	Formation and stabilization of fat emulsions	Sausages, bologna, soup, cakes
Fat absorption	Binding of free fat	Meats, sausages, donuts
Flavor binding	Adsorption, entrapment, release	Simulated meats, bakery, etc.
Foaming	Form stable films to entrap gas	Whipped toppings, chiffon desserts, angel cakes

Source: Kinsella, 1982.

The diverse functional properties of proteins in foods are a result of the physicochemical properties of the proteins and their changes during the processing of foods. Numerous attempts have been made to correlate the physicochemical properties of proteins with their functional properties in food systems (Table 9.2), with only partial success. These correlations are based on studies of the changes in the physicochemical properties of selected proteins before and after processing. Properties measured include solubility, surface tension, surface hydrophobicity and hydrophilicity, size, rates of denaturation, charge distribution, amino acid sequences, rates of diffusion and orientation at surfaces, primary,

Table 9.2 Molecular Characteristics Desirable in Proteins for Films and Foams

Soluble: facilities rapid diffusion to interface
Large: allows more interactions in the interface, stronger films
Amphipathic: provides unbalanced distribution of charged and apolar residues for improved interfacial interactions
Flexible domains: facilitate phase behavior and unfolding at interface
Interactive regions: disposition of different functional segments facilitate secondary interactions in the air, the interface and aqueous phases
Disposition of charged groups: affects protein:protein interactions in the film and charge repulsion between neighboring bubbles
Retention of structure: enhances overlap and segmental interactions in films
Polar residues: provides hydratable (glycosyl) or charged residues to keep bubbles apart, binding and retain water

Source: Kinsella and Phillips, 1989.

secondary, tertiary, and quaternary structures, and foaming and emulsifying properties.

Experimental approaches include the comparative study of proteins with diverse physicochemical properties, and the chemical modification of proteins by the removal of glyco- (Kato et al., 1987a) or phosphate groups (Kato et al., 1987b), the oxidation of sulfhydryl groups of cysteine to disulfide bonds to form cross-links (Swaisgood and Horton, 1989), the cross-linking of polypeptide chains by other means [by transglutaminase, peroxidase, or polyphenol oxidase (Matheis and Whitaker, 1984a, b)], by covalent attachment of alkyl, aryl, and other groups (primarily to the ε-amino groups of lysyl residues; see Table 9.3), and by selective hydrolysis of peptide and other amide bonds with proteases and/or acids. Some of these methods are summarized in Table 9.3. The newest, and most specific, techniques involve single amino acid mutations of proteins by recombinant DNA technology.

Although scientists have been resourceful in their use of chemical and enzymatic methods to modify proteins, they have a long way to go in order to match the more than 135 ways in which nature modifies proteins during and after protein translation (Table 9.4).

In this review, we shall confine our discussion to four methods only for chemically and enzymatically modifying proteins, with the intent of providing some detail, rather than a broad coverage of all possible methods found in the literature. The four methods covered are: (1) the use

TABLE 9.3   Side Chain Modification of Proteins

Side chain or group	Procedure (reagent)	Ref.
Amino (ε- and α)	Amidination (ethyl acetimidate)	Hunter and Ludwig (1962); Wallace and Harris (1984)
	Reductive alkylation (aldehyde + NaBH$_4$)	Means and Feeney (1968); Jentoft and Dearborn (1979)
	Acylation (acetic anhydride; succinic anhydride)	Grossberg and Pressman (1963); Buttkus et al. (1965)
	Cross-linking (glutaldehyde + NaBH$_4$)	Peters and Richards (1977)
Amide (Asn and Gln)	Hydrolysis (deamidases; sulfonic acids)	Hamada et al. (1988); Shih (1990a)
Carboxyl (Asp and Glu)	Water-soluble carbodiimide + nucleophile (gly ethyl ester)	Hoare and Koshland (1967)
	Transglutaminase (amine; amino acid)	Motoki et al. (1986)
Guanidino (Arg)	Dicarbonyls (2,3-butanedione; p-hydroxyphenyglyoxal)	Riordan (1973); Yamasaki et al. (1980)
Imidazole (His)	Diethyl pyrocarbonate (ethoxyformic anhydride)	Melchior and Fahrney (1970)
Indole (Trp)	N-Bromosuccinimide	Spande and Witkop (1967)
Phenol (Tyr)	Crosslinking (peroxidase + H$_2$O$_2$; polyphenol oxidase + O$_2$ + phenol)	Matheis and Whitaker (1984a,b)
Thiol (Cys-SH)	Crosslinking (sulfhydryl oxidase; lipoxygenase + O$_2$ + linoleic acid; O$_2$; disulfides)	Feeney and Whitaker (1988); Swaisgood and Horton (1989)
	Carboxymethylation (iodoacetate; iodoacetamide)	Brake and Wold (1962)

TABLE 9.3   (Continued)

Side chain or group	Procedure (reagent)	Ref.
Thioether (Met)	Oxidation (H_2O_2)	Stauffer and Etson (1969)
Glycosyl (Ser, Thr, Asn)	Deglycosylation (endo-β-N-acetyl-glucosaminidase H for Ser and Thr; β-N-acetylglucosaminidase F for Asn)	Trimble et al. (1987); Elder and Alexander (1989)
	Glycosylation	Ginsburg (1989)
Phosphorylation (Ser, Thr)	Dephosphorylation (alkaline phosphatase; protein phosphatase)	Krebs (1986)
	Phosphorylation (protein kinase; $POCl_3$)	Bingham and Farrell (1974); Matheis et al. (1983); Hunter and Cooper (1985)

TABLE 9.4   In Vivo Nonhydrolytic Enzyme-Catalyzed Posttranslational Modifications of Amino Acid Residues of Proteins

Amino acid (group)	Typical modifications	Total modifications known
Arginine (Guanidino)	methyl- (mono- and di-) ADP-ribosyl- citrulline ornithine	6
Lysine (ε-NH_2)	glucosyl- phospho- pyridoxyl- biotinyl- lipoyl- acetyl- methyl- (mono-, di-, tri-) δ-hydroxyl- δ-glycosyl- cross-links	33

TABLE 9.4 (Continued)

Amino acid (group)	Typical modifications	Total modifications known
Histidine (imidazole)	methyl- (1- and 3-) phospho- (1- and 3-) iodo- flavin-	6
Proline	4-hydroxy- 3-hydroxy- 3,4-dihydroxy- 4-glycosyloxy-	5
Phenylalanine	β-hydroxy- β-glycosyloxy-	2
Tyrosine (benzene ring/hydroxyl)	β-hydroxy- β-glycosyloxy- sulfono- iodo- (mono-, di-) bromo- (mono-, di-) chloro- (mono-, di-) bis-ether adenylyl uridylyl- RNA	18
Serine (hydroxyl)	phospho- glycosyl- methyl- phosphopantetheine- ADP-ribosyl	8
Threonine (hydroxyl)	phospho- glycosyl- methyl-	6
Cysteine (sulfhydryl)	cystine glycosyl- dehydroalanyl heme flavin seleno-	7
Aspartic acid/ Glutamic acid (carboxyl)	γ-carboxyl- β-phospho- methyl	3

Table 9.4   *(Continued)*

Amino acid (group)	Typical modifications	Total modifications known
Asparagine/ Glutamine (amide)	glycosyl R-NH- pyrrolidone	4
Carboxyl-terminal residue (carboxyl)	-amide -amino acid	11
Amino-terminal residue (amino group)	acetyl- formyl- glucosyl- amino acyl- pyruvyl- α-ketobutyryl- methyl- glycuronyl- murein	19
Total modifications		135[a]

[a]Including 3 each for aspartic acid and glutamic acid and 4 each for asparagine and glutamine.
Source: Whitaker, 1977; Uy and Wold, 1980.

of highly specific proteases to produce controlled size peptides, (2) the deamidation of protein-bound asparagine and glutamine residues, (3) the effect of charge on proteins before and after deglycosylation and dephosphorylation of proteins, and (4) recombinant DNA–derived proteins. Emphasis will be on changes in the functional properties of these proteins. No attempt is made to cover all the literature in these areas. Method 1 is based specifically on research done in the authors' laboratories.

ENZYMATIC HYDROLYSIS OF PROTEINS TO PRODUCE CONTROLLED-SIZE PEPTIDES

There are numerous publications on the enzymatic hydrolysis of proteins to improve their solubility, foaming, and emulsifying properties. Early research was primarily on the caseins to provide extensively hydrolyzed preparations for intravenous and other clinical feeding and for use in soups, bouillon, and sauces. Early on, investigators were faced with the problem of bitter peptides. Substantial research indicated that bitterness results from lower molecular weight hydrophobic peptides with a hydrophobic index > 1200. Location of the more hydrophobic amino acid residues near the C-terminal end of the peptide also enhances the bitterness. Some enzymes, such as trypsin, chymotrypsin, and subtilisin, produce more bitter peptides than does pronase (Adler-Nissen, 1986b). Another major problem with the earlier work resulted from the use of crude protease preparations containing several types of both endo- and exo-splitting enzymes, as well as nonproteolytic enzymes that affected the flavor.

Limited Proteolysis

The extensively hydrolyzed products described above, with increased solubility, have few if any functional properties. More recently, it has become clear that the extent of proteolysis must be limited if foaming and emulsifying properties are to be maintained or enhanced. Three methods are available for producing limited hydrolysis products. The method used most frequently is to limit the degree of hydrolysis (DH; percentage of the total peptide bonds hydrolyzed) by stopping the reaction at a desired time. In this method a variety of products are produced, ranging in size from the initial protein to the smallest peptide possible, based on the specificity of the enzyme used and the relative rates of peptide bond hydrolysis.

Let us take as an example β-lactoglobulin, one of the two major whey proteins in milk (Fig. 9.1). There are three arginine residues in the protein of 18,300 daltons: Arg-40, Arg-124, and Arg-148. Assuming equal opportunity for attack by endoprotease Arg-C at any one of the peptide bonds on the carbonyl side of the arginine residues, six peptides would be formed in the first step (Fig. 9.2), seven peptides would appear in the second step (with two peptides the same as in step one), and four peptides in the third (and last step). The final four peptides are the same as four of the seven peptides present after step two (see diagrams at bottom of Fig. 9.2).

There are 15 lysine residues in β-lactoglobulin, so that 30 peptides could be formed in a first step by endoprotease Lys-C. This number would increase in subsequent steps and then decrease to 16 peptides when all of the peptide bonds on the carbonyl side of the lysine residues are hydrolyzed. Table 9.5 shows the nature of these peptides. Therefore, only 11 peptides ranging in size from 5 to 34 amino acids are found. Four of these peptides (#3, 4, 13, and 16) have 13, 33, 34, and 21 amino

```
                          10                                              20
H.Leu-Ile-Val-Thr-Gln-Thr-Met-Lys-Gly-Leu-Asp-Ile-Gln-Lys-Val-Ala-Gly-Thr-Trp-Tyr-
                               2               4           2

                           30                                             40
    Ser-Leu-Ala-Met-Ala-Ala-Ser-Asp-Ile-Ser-Leu-Leu-Asp-Ala-Gln-Ser-Ala-Pro-Leu-Arg-
                           4               4                                  1

                             50                                           60
    Val-Tyr-Val-Glu-Glu-Leu-Lys-Pro-Thr-Pro-Glu-Gly-Asp-Leu-Glu-Ile-Leu-Leu-Gln^a-Lys-
             3   3       2           3       4       3                    2

                          70                                              80
Trp-Glu-Asn-Asp^b-Glu-Cys-Ala-Gln-Lys-Lys-Ile-Ile-Ala-Glu-Lys-Thr-Lys-Ile-Pro-Ala-
    3       4   3           2   2               3   2       2

                             90                                           100
    Val-Phe-Lys-Leu-Asp-Ala-Ile-Asn-Glu-Asn-Lys-Val-Leu-Val-Leu-Asp-Thr-Asp-Tyr-Lys-
        2   4               3       2               4       4       2

                             110                                  [SH] 120
Lys-Tyr-Leu-Leu-Phe-Cys-Met-Glu-Asn-Ser-Ala-Glu-Pro-Glu-Gln-Ser-Leu-Val^c-Cys-Gln-
2                       3               3       3

[SH]                       130                                            140
Cys-Leu-Val-Arg-Thr-Pro-Glu-Val-Asp-Asp-Glu-Ala-Leu-Gly-Lys-Phe-Asp-Lys-Ala-Leu-
            1       3           4   4   3               2       4   2

                             150                                          160
    Lys-Ala-Leu-Pro-Met-His-Ile-Arg-Leu-Ser-Phe-Asn-Pro-Thr-Leu-Gln-Glu-Glu-Gln-Cys-
    2                        1                                      3   3

     162
    His-Ile.OH
```

^a Variant C, His.
^b Variant B and C, Gly.
^c Variant B and C, Ala.

FIG. 9.1 Primary structure of β-lactoglobulin A. Superscripts a, b, and c indicate changes in amino acid residues in variants B and C. The numbers under the amino acid residues indicate specificity for the carbonyl group of the peptide bond for the following enzymes: 1, endoprotease Arg-C; 2, endoprotease Lys-C; 3 and 4, endoprotease Glu-C [specific for glutamic acid residues in 50 mM ammonium bicarbonate buffer; specific for both glutamic acid and aspartic acid residues in 50 mM phosphate buffer (Drapeau, 1976)]; 1 and 2, trypsin (specific for both arginine and lysine residues). (*Adapted from Braunitzer et al., 1973, and Whitney et al., 1976.*)

acid residues, respectively, could have some secondary structure, and would be expected to provide more functionality than the remaining 10 lower molecular weight peptides. Dialysis with a cut-off–size membrane of 1000 could give these four peptides free of the remaining lower molecular weight peptides and amino acids. Note that two lysines (#6 and 12), one dipeptide (#8), and two tripeptides (#14 and 15) are formed. We cannot expect all the susceptible peptide bonds to be hydrolyzed at the same rate, so the intermediate peptide composition can

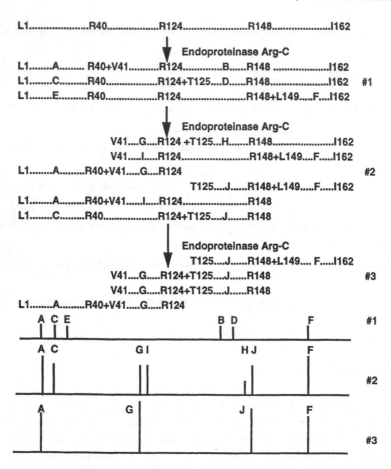

FIG. 9.2 Predicted sequence of hydrolysis of the three susceptible peptide bonds of β-lactoglobulin by endoprotease Arg-C. Endoprotease Arg-C has specificity for residues Arg_{40} (R_{40}), R_{124}, and R_{148}. Assuming equal susceptibility at the three arginine residues, six peptides would be formed in the first step, seven in the second, and four peptides would be present at complete hydrolysis (three peptide bonds hydrolyzed). A through J are the peptides formed. The bottom diagrams give a schematic indication of the peptides formed at each step.

be more or less complex than the two cases presented here. If a protease has specificity for more than one amino acid residue (for example, trypsin with specificity for both arginine and lysine residues), then more peptides would be formed, or if there were two or more different proteases in the enzyme preparation used more peptides would be formed.

Two methods can be used to give reproducible, controlled-size peptides from protein. The first method is use of a semipermeable membrane reactor of selected cut-off size limit (Adler-Nissen, 1986a). When peptides are produced that can diffuse out of the reactor, they will not be degraded further.

The second method of making reproducible, controlled-size peptides is the use of highly specific proteases so that only a few bonds are hydrolyzed and further reaction stops. We shall now discuss unpublished data from our laboratory using highly specific proteases to produce reproducible, controlled-size peptides from β-lactoglobulin (Vojdani, 1992).

TABLE 9.5 Peptides Formed from β-Lactoglobulin A Hydrolysis by Endoprotease Lys-C[a]

Peptide	No. amino acids	Hydrophobicity
1. H·Leu-Ile-Val-Thr-Gln-Thr-Met-Lys-COOH	8	5620
2. H·Gly-Leu-Asp-Ile-Gln-Lys-COOH	6	5100
3. H·Val-Ala-Gly-Thr-Trp-Tyr-Ser-Leu-Ala-Met-Ala-Ala-Ser-Asp-Ile-Ser-Leu-Leu-Asp-Ala-Gln-Ser-Ala-Pro-Leu-Arg-Val-Tyr-Val-Glu-Glu-Leu-Lys-COOH	33	5410
4. H·Pro-Thr-Pro-Glu-Gly-Asp-Leu-Glu-Ile-Leu-Leu-Gln-Lys-COOH	13	6070
5.[b] H·Trp-Glu-Asn-Asp-Glu-Cys-Ala-Gln-Lys-COOH	9	3610
6. H·Lys-COOH	1	—
7. H·Ile-Ile-Ala-Glu-Lys-COOH	5	7290
8. H·Thr-Lys-COOH	2	4050
9. H·Ile-Pro-Ala-Val-Phe-Lys-COOH	6	8460
10. H·Leu-Asp-Ala-Ile-Asn-Glu-Asn-Lys-COOH	8	4540
11. H·Val-Leu-Val-Leu-Asp-Thr-Asp-Tyr-Lys-COOH	9	6540

TABLE 9.5   (Continued)

Peptide	No. amino acids	Hydrophobicity
12. H·Lys-COOH	1	—
13.[c] H·Tyr-Leu-Leu-Phe-Cys-Met-Glu-Asn-Ser-Ala-Glu-Pro-Glu-Gln-Ser-Leu-Val-Cys-Gln-Cys-Leu-Val-Arg-Thr-Pro-Glu-Val-Asp-Asp-Glu-Ala-Leu-Gly-Lys-COOH	34	4930
14. H·Phe-Asp-Lys-COOH	3	6533
15. H·Ala-Leu-Lys-COOH	3	6480
16.[b] H·Ala-Leu-Pro-Met-His-Ile-Arg-Leu-Ser-Phe-Asn-Pro-Thr-Leu-Gln-Glu-Glu-Gln-Cys-His-Ile-COOH	21	5420

[a]Complete hydrolysis, based on specificity of endoprotease Lys-C, assumed.
[b]Peptides 5 and 16 would be covalently linked via a disulfide bond between Cys_{66} and Cys_{160} (Fig. 9.1) unless β-lactoglobulin (or the peptides) is treated with β-mercaptoethanol or dithiothreitol.
[c]Peptide 13 would have an internal disulfide bond between Cys_{106} and Cys_{119} or between Cys_{106} and Cys_{121} unless reduced. See footnote b.

β-Lactoglobulin A was isolated from the milk of a single Holstein cow in the University of California, Davis, herd that did not contain β-lactoglobulin gene B; the caseins were removed by isoelectric precipitation (pH 4.6) followed by further separation of the whey on a Sephadex G-100 column at pH 7.5. β-Lactoglobulin separated as the dimer (36,600 MW) from α-lactalbumin (14,200 MW) and the other minor whey proteins. Traces of other proteins were removed on a DEAE-cellulose column and a hydroxyapatite column. β-Lactoglobulins A and B can be separated on a Mono-Q column on an FPLC instrument.

The endoproteases used were: (1) endoprotease Arg-C (*Staphylococcus aureus*) with specificity for peptide bonds in which glutamic and aspartic acid residues provide the carbonyl group (in 50 mM ammonium carbonate and ammonium acetate buffer only glutamic acid residues; peptide bonds of both glutamic acid and aspartic acid residues are hydrolyzed in 50 mM phosphate buffer), (2) endoprotease Lys-C (*Lactobacter enzymogens*), (3) endoprotease Arg-C (mouse submaxillary gland), and (4) trypsin (bovine). All four endoproteases were from Sigma Chemical Co.

Polypeptides of β-Lactoglobulin by Specific Proteases.   A typical curve for the hydrolysis of a protein by an endoprotease is shown in Fig. 9.3 for the action of endoprotease Lys-C on α-lactalbumin. Aliquots were removed from the reaction at various times, and the number of peptide bonds hydrolyzed were determined by the 2,4,6-trinitrobenzene method (TNBS) (Fields, 1972). The results are expressed as degree of hydrolysis. In addition, the molecular weights and number of different-size peptides formed by the proteolysis were determined using size-exclusion chromatography on a Superose-12 column (1.5 × 30 cm) run on a Pharmacia FPLC (fast protein liquid chromatography) instrument (Fig. 9.4).

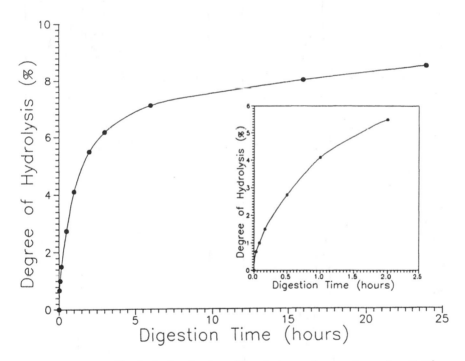

FIG. 9.3   Rate of hydrolysis of α-lactalbumin by endoproteinase Lys-C. The insert shows the rate of hydrolysis during the first 2.5 hr. The reaction was performed in 0.1M phosphate buffer, pH 8.1, containing 2% (1.1 mM) α-lactalbumin, 11.1 μM enzyme, and 0.02% sodium azide at 37°C. Aliquots were removed at the indicated times, the reaction stopped by heating at 98°C for 10 min. in presence of 1% SDS and the number of peptide bonds hydrolyzed determined by the 2, 4, 6-trinitrobenzene sulfonate method (Fields, 1972).

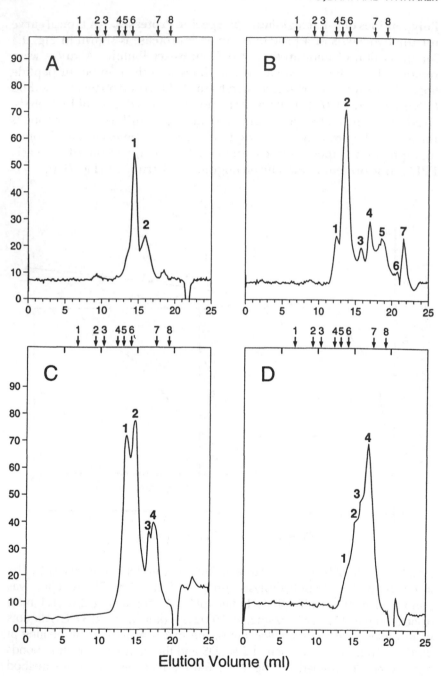

Elution Volume (ml)

As shown in Fig. 9.4A, two peaks (12,600 and 2500; Table 9.6) were found when β-lactoglobulin was hydrolyzed with endoprotease Arg-C to a DH of 2. β-Lactoglobulin (MW = 18,300) eluted at the point marked 5 at top of the graph and α-lactalbumin (MW 14,200) eluted at point 6. Therefore, β-lactoglobulin had been hydrolyzed since the largest polypeptide (12,600) eluted after α-lactalbumin. The large peak is undoubtedly polypeptide $Val_{41} \rightarrow Arg_{124}$ (Fig. 9.1) since the disulfide bonds were not reduced, while the smaller peak is probably a mixture of polypeptides $Leu_1 \rightarrow Arg_{40}$, $Thr_{125} \rightarrow Arg_{148}$ and $Leu_{149} \rightarrow Leu_{162}$ (Table 9.6).

Trypsin hydrolysis of β-lactoglobulin to DH 3 gave at least seven polypeptides (Fig. 9.4B) with evidence for some other minor polypeptides. Peak 1 is probably a dimer of β-lactoglobulin (MW 36,600), while peak 2 appears to be monomeric β-lactoglobulin (MW 18,300) that did not undergo proteolysis because of the two disulfide bonds (Fig. 9.1). Three other polypeptides of 5300, 2400, and 500 were produced (Table 9.6). Figure 4C is for a DH 5 β-lactoglobulin hydrolysate by endoprotease Lys-C. There are four distinct polypeptides. Polypeptide A has a MW of ~11,500. Peptide 2 (Fig. 4C) is about 10,000 MW, while polypeptides 3 and 4 are about 2100 and 1300, respectively (Table 9.6). The chromatogram for a DH 5 β-lactoglobulin hydrolysate by endoprotease Glu-C (Fig. 9.4D) indicates a minimum of four poorly resolved polypeptides ranging in weight from ~13,500 (1st shoulder) to ~2000 for the major peak (Table 9.6).

FIG. 9.4 Peptides formed by specific enzymatic hydrolysis of β-lactoglobulin as determined on a size exclusion anion exchange chromatography column (Mono-Q HR 5/5) on an FPLC system (Pharmacia Co.). The column was equilibrated in 0.02 M Tris-HC1 buffer, pH 8.0. Following loading of the sample, the column was washed with 5 mL of the equilibration buffer, then with a linear NaCl gradient buffer (0.0–70% 1 M NaCl in starting buffer for 35 min (1 mL/min) and then with 1 M NaCl. The absorbance was recorded at 214 nm. The standards used were (numbered at top of graphs): (1) Blue Dextran 2000, MW 2 million; (2) human transferrin, MW 80,000; (3) bovine serum albumin, MW 66,000; (4) ovalbumin, MW 43,000; (5)β-lactoglobulin, MW 18,300; (6) α-lactalbumin, MW 14,200; (7) bradykinin, MW 1,060; (8) prolylglycine, MW 172. (A) Endoproteinase Arg-C (reaction conditions; as given in Fig. 9.3 except at pH 8.0 and 1/213 (M/M) enzyme to substrate ratio; DH = 2). (B) Trypsin (reaction as in Fig. 9.3; DH 3). (C) Endoproteinase Lys-C (reaction as in Fig. 9.3 except at pH 8.65 and 0.3 unit activity/mL; DH 5). (D) Endoproteinase Glu-C (reaction as in Fig. 9.3, except at pH 7.8 and 1/35 to 1/150 (M/M) E:S; DH 5).

TABLE 9.6 Polypeptide Molecular Weights from Endoprotease Hydrolysis of β-Lactoglobulin[a]

Protease	Peak no.	Molecular weight (daltons)	Amount (%)
Endo Arg-C	1	12,600	71.7
	2	2,500	28.3
Trypsin[b]	1	36,600[c]	10.0
	2	18,000	45.7
	3	5,300	6.11
	4	2,400	12.3
	5	500	17.1
Endo Lys-C	1	11,500	34.3
	2	10,500	39.8
	3	2,100	7.92
	4	1,300	18.1
Endo Glu-C (Shoulder)	1	13,500	8.77
(Shoulder)	2	7,200	28.9
(Shoulder)	3	3,500	11.9
	4	2,000	50.5

[a]See Fig. 9.4 for Superose gel filtration chromatograms.
[b]Two fractions of 8.77% total eluted from the column well after the total liquid volume.
[c]Probably β-lactoglobulin dimer joined by disulfide bonds, since done at pH 8, which remained intact on gel filtration column.

Secondary Structures.   Figure 9.5 shows circular dichroism data for β-lactoglobulin (Fig. 9.5E) and β-lactoglobulin hydrolysates by endoproteases Arg-C (Fig. 9.5A; DH 2), endoprotease Glu-C (Fig. 9.5B; DH 5), trypsin (Fig. 9.5C; DH 3), and endoprotease Lys-C (Fig. 9.5D; DH 5) ranging from pH 3 to 10. The differences in molecular ellipticity results shown in Fig. 9.5 indicate that the hydrolysates differed in secondary structure from β-lactoglobulin.

Table 9.7 shows the calculation of the secondary structures of β-lactoglobulin and the β-lactoglobulin hydrolysates at pH 7 (from Fig. 9.5). The secondary structures of the DH 2 hydrolysate by endoprotease Arg-C were almost identical to that of β-lactoglobulin. There appears to be a small decrease in the β-sheet content. The most noticeable effect of the other three endoproteases was in the marked decrease in the α-helix content, while the β-sheet and β-turn contents of the hydrolysates were

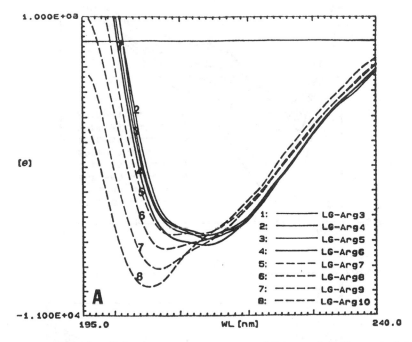

FIG. 9.5   Circular dichroism measurements of β-lactoglobulin and of the enzymatic hydrolysates (Fig. 9.4). The spectra were recorded on a Jasco J-600 spectropolarimeter (Japan Spectroscopic Co., CTD) at 190–350 nm in 0.1-cm quartz cells with 1.0-nm band width, sensitivity of 20 mdeg/FS, time constant of 2.0 sec, step resolution of 0.2 nm, and scan speed of 20 nm/min. The protein samples (0.01–0.07 mM; based on MW of β-lactoglobulin) were prepared by diluting stock solutions (Fig. 9.3) in 10 mM citrate (pH 3–5), 10 mM phosphate (pH 6–8), or 10 mM borate (pH 9–10) buffer. The readings were taken following equilibration in the buffer for 30 min. (A) Endoproteinase Arg-C, DH 2. (B) Endoproteinase Glu-C, DH 5. (C) Trypsin, DH 3. (D) Endoproteinase Lys-C, DH 5. (E) β-lactoglobulin. The numbers at end of insert 1-8 are the pH values used.

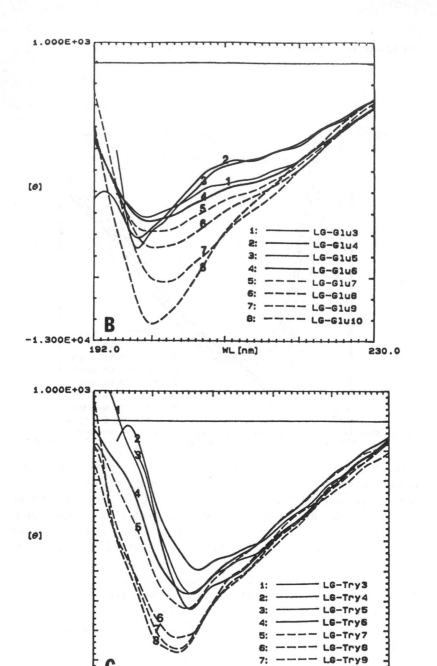

FIG. 9.5 (*Continued*)

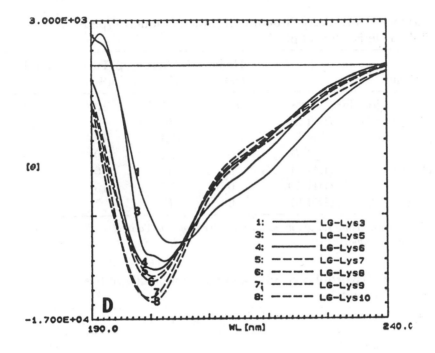

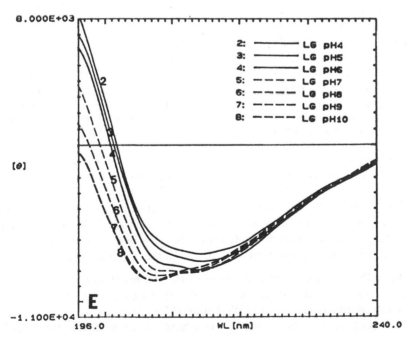

TABLE 9.7   Secondary Structure of β-Lactoglobulin Type A
and Its Derivatives at pH 7

Protein		α-Helix (%)	β-Sheet (%)	β-Turn (%)	Random (%)
β-Lactoglobulin		15	45	10	30
β-Lg Arg-C	(DH 2)	15	40	10	35
β-Lg Glu-C	(DH 5)	5	55	10	30
	(DH 7)	0	52	10	46
β-Lg Try	(DH 3)	5	50	10	35
	(DH 10)	0	27	15	58
β-Lg Lys-C	(DH 5)	0	40	7	53

We thank Dr. Yang for giving us a copy of the program, Yang et al., 1986, used to calculate the data.

TABLE 9.8   Hydrophobic Sites and Hydrophobicity of the Surface
of Several Proteins[a]

Protein		Number of sites	Hydrophobicity
β-Lactoglobulin (β-Lg)		0.75	1627
β-Lg Try	(DH 3)	0.78	924
β-Lg Try	(DH 10)	2.50	162
β-Lg Arg-C	(DH 2)	0.59	1522
β-Lg Lys-C	(DH 5)	1.77	327
β-Lg Glu-C	(DH 5)	2.00	219
β-Lg Glu-C	(DH 7)	3.20	185
Bovine serum albumin		5.00	1180

[a]Measured with *cis*-parinaric acid.

about the same as for β-lactoglobulin. β-Lg Glu-C (DH 5) and β-Lg Try-C (DH 3) retained about 5% α-helix content (in contrast to 15% for β-lactoglobulin). β-Lg Glu-C (DH 7), β-Lg Try (DH 10), and β-Lg Lys-C (DH 5) hydrolysates had no α-helix content. In summary, it is remarkable that the β-sheet and β-turn structures of β-lactoglobulin hydrolysates were largely intact, indicating that the large polypeptides can refold since the protein was denatured in order to achieve proteolysis.

Surface Properties.   Table 9.8 demonstrates the effect of proteolysis on the number of hydrophobic sites on the surface of β-lactoglobulin and β-lactoglobulin hydrolysates. We found 0.75 hydrophobic sites per monomer on β-lactoglobulin at pH 7.0, where the protein is a dimer. β-Lg trypsin, DH 3 had an identical value (0.78), while β-Lg trypsin, DH 10 had 2.50 hydrophobic sites per average molecule. This is consistent with the more extensive proteolysis. β-Lg Arg-C hydrolysate (DH 2) had a lower hydrophobic site number (0.59) than β-lactoglobulin, consistent with some of its functional activities described below. β-Lg Lys-C (DH 5) and β-Lg Glu-C (DH 5 and 7) hydrolysates had a significantly higher number of hydrophobic sites than β-lactoglobulin, consistent with the circular dichroism data (and functionality behavior). The value of 5.0 hydrophobic sites for bovine serum albumin under our conditions agrees with that of other researchers (Sklar et al., 1977).

Table 9.9 lists the surface and exposed (0.5% SDS, pH 7.0, 90°C for 10 min) hydrophobicity values for β-lactoglobulin (types A and AB), β-

TABLE 9.9 Surface and Exposed Hydrophobicity of Proteins by Fluorometric Method Using cis-PnA and ANS[a]

Protein	cis-PnA		ANS	
	Surface	Exposed[b]	Surface	Exposed[b]
β-Lg (native) type A	1627	848	54	37
β-Lg type AB	1560	1200	58	98
β-Lg Arg-C (DH 2)	1522	302	82	142
β-Lg Glu-C (DH 5)	219	434	207	80
β-Lg Glu-C (DH 7)	185	573	120	48
β-Lg Trypsin (DH 3)	924	606	47	60
β-Lg Trypsin (DH 10)	162	280	56	29
β-Lg Lys-C (DH 5)	327	514	41	38
BSA	1180	1123	219	—

[a]Measured by the method of Kato and Nakai (1980) using cis-parinaric acid (cis-PnA) and by the method of Hayakawa and Nakai (1985) using 8-anilinonaphthalene sulfonate (ANS).
[b]Protein denatured by heating 0.5% protein in 25 mM phosphate buffer, pH 7.0, containing 0.5% sodium dodecyl sulfate (SDS) at 90°C for 10 min.

lactoglobulin hydrolysates, and bovine serum albumin, as determined using *cis*-parinaric acid and 8-anilinonaphthalene sulfonate. The results are consistent with the data in Table 9.8. *cis*-Parinaric acid is a more sensitive probe than is anilinonaphthalene sulfonate. The surface values and exposed values are generally about the same except for β-lactoglobulin (type A) and β-Lg Arg-C and for β-Lg Glu C (DH 5 and 7) when using anilinonaphthalene sulfonate. We note further that the exposed values are higher for β-Lg Glu-C (DH 5 and 7) for *cis*-parinaric acid, but the reverse is true when anilinonaphthalene sulfonate is the probe.

The surface properties of β-lactoglobulin and its enzymatic hydrolysates were determined by measuring the surface tension (Wilhelmy plate method) (Graham and Phillips, 1979a–c) and the rates of diffusion and rearrangement of the materials at the water-air interface (Tables 9.10 and 9.11). Except for β-Lg Glu-C (DH 7), the surface tension values are not greatly different. All preparations reduced the surface tension at the water-air interface (72 mN/m for the buffer only). Except for the values for β-Lg Glu-C (DH 5) and β-Lg Try (DH 10) for average area cleared per molecule (Å) and apparent number of amino acid residues penetrating the surface, the other preparations are significantly different from each other and from β-lactoglobulin by these measurements. All of the values were less than those for β-lactoglobulin, as expected, except for β-Lg Lys-C (DH 5), where the values are much higher. We do not have an explanation for this observation.

Solubility and Functionality Properties of Polypeptides Formed by Endoprotease Arg-C.   In Figs. 9.6–9.9 we compare the solubility, foam-

TABLE **9.10**   Surface Properties of β-Lactoglobulin and Its Derivatives at pH 5

Protein		Average area cleared per molecule (Å^2)	App. no. amino acid residues penetrating surface	Surface pressure (mN/m)
β-Lactoglobulin (β-Lg)		318	21.2	24.0
β-Lg Arg-C	(DH 2)	267	17.8	23.8
β-Lg Glu-C	(DH 5)	202	13.5	22.5
	(DH 7)	236	15.7	19.8
β-Lg Lys-C	(DH 5)	482	32.1	23.0
β-Lg Try	(DH 10)	204	13.6	22.8

ing, and emulsion properties of β-lactoglobulin and its enzymatic hydrolysates. Figure 9.6 shows these properties for a β-lactoglobulin hydrolysate (DH 2) by endoprotease Arg-C. There is essentially no change in solubility between β-lactoglobulin and the DH 2 hydrolysate over the pH range 2–10 (Fig. 9.6A). Both are nearly 100% soluble under the conditions used. The hydrolysate has a little lower emulsion activity index (E.A.I.) than does β-lactoglobulin at pH 4–5, but it is better at pH 3 than β-lactoglobulin and significantly better at pH 6 and above. For example, at pH 8, E.A.I. is $\sim$680 m^2/g (better than bovine serum albumin of $\sim$600) for the hydrolysate and significantly better than β-lactoglobulin at 460 m^2/g. The emulsion produced by the hydrolysate is stable at all pHs except 4 (Fig. 9.6C), while β-lactoglobulin emulsions are markedly unstable in the pH range 5–6 (near the pI, where the repulsive forces are at a minimum). Except at pH 4, 5, and 6 (near the pI), the hydrolysate has a significantly lower foam capacity than does β-lactoglobulin (Fig. 9.6D). The β-lactoglobulin–produced foam is significantly more stable than the hydrolysate-produced foam at all pHs, except at pH 5 where the values are slightly higher (Fig. 9.6E). In summary, the DH 2 hydrolysate of β-lactoglobulin produced by endoprotease Arg-C has essentially the same solubility as β-lactoglobulin from pH 2 to 10, and is less effective as a foaming agent under the conditions used. However, it is a significantly better emulsifying agent than is β-lactoglobulin, particularly in the region of pH 5–6. Apparently, the modification has changed the pI of the hydrolysate sufficiently at pH 5–6 to enhance the charge to satisfy the DLVO requirements for a good balance between charge repulsion and attractive van der Waal forces (hydrophobicity).

Solubility and Functional Properties of Polypeptides Formed by Endoprotease Glu-C.   Figure 9.7 shows the effect of selectively hydrolyzing β-lactoglobulin with endoprotease Glu-C on its solubility, emulsifying, and foaming properties. Surprisingly, the solubilities of the DH 5 and DH 7 hydrolysates decreased substantially from those of β-lactoglobulin in the pH range 4–7 (to pH 8 for DH 5). For example, at pH 5 the hydrolysates are only about 50% as soluble as β-lactoglobulin (Fig. 9.7A). The E.A.I. values for the DH 5 hydrolysate are better at pH 5–6 and pH 9–10 than for β-lactoglobulin (Fig. 9.7B), while the DH 7 hydrolysate is not as good as β-lactoglobulin except at pH 6. The DH 5 hydrolysate has very good emulsion stability at pH 5–6 where β-lactoglobulin has very poor stability (Fig. 9.7C). The DH 5 hydrolysate and β-lactoglobulin have about the same foam capacity at pH 2–7, while the DH 5 hydrolysate is a little better at pH 8–10 (Fig. 9.7D). However, the DH 7 hydrolysate has significantly better foaming capacity than β-lactoglobulin at all

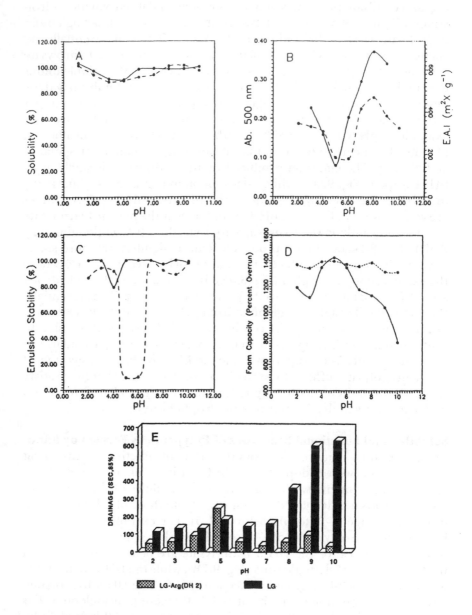

pHs. At pH 2–3 and 6–10 the foaming capacity is very high. The DH 7 hydrolysate produces a more stable foam at pH 3, 4, and 6 than does β-lactoglobulin (Fig. 9.7E). Therefore, in spite of a decrease in solubility on modification of β-lactoglobulin by endoprotease Glu C, the DH 5 hydrolysate has better emulsifying properties (especially at pH 5–6) and foaming properties (pH 3–4 and pH 6) than β-lactoglobulin. It appears that the additional negative charges at the C-terminal end by glutamic acid are responsible for these improved properties. The solubility data indicate that there may be more hydrophobic interaction in the pH range of 4–7, which is confirmed by the increase in hydrophobic binding sites (Table 9.8).

Solubility and Functional Properties of Polypeptides Formed by Trypsin. Figure 9.8 illustrates the effect of selective hydrolysis of β-lactoglobulin by trypsin on the solubility and functional properties. Very limited hydrolysis (DH 1.5 and 3.1; Fig. 9.8A) of β-lactoglobulin by trypsin reduced the solubility at pH 4 (near the pI). The DH 6.4 hydrolysate had the same solubility at pH 4 as β-lactoglobulin, while the DH 10 hydrolysate had better solubility. Above pH 5, all hydrolysates had better solubility than β-lactoglobulin. At pH 3, the DH 10 hydrolysate had a higher solubility than β-lactoglobulin. However, it should be noted that all preparations, including β-lactoglobulin, had 85% or better solubility (Fig. 9.8A). There was an increase in E.A.I. for the DH 3 and 10 hydrolysates above pH 5. At pH 6, 9, and 10 the increases were significant (Fig. 9.8B). As shown in Fig. 9.8C, the pH of the minimum emulsion stabilities differ among β-lactoglobulin and the hydrolysates. β-Lactoglobulin has minimum emulsion stability at pH 5–6. For the DH 3 hydrolysate the minimum is at pH 5; therefore, at pH 6, the DH 3 hydrolysate gives stable emulsions, while β-lactoglobulin does not. The DH 10 hydrolysate

FIG. 9.6 Functional properties of β-lactoglobulin (•---•) and of endoproteinase Arg-C hydrolysate (•—•), DH 2, of β-lactoglobulin (see Fig. 9.3). (A) Solubilities were determined for 0.1% solutions, after equilibration at desired pH for 1 hr at 23°C. Samples were centrifuged using microfilterfuge tubes containing 0.2μm nylon 66 membrane filters. The protein content of the filtrates was determined by the Lowry method (Lowry et al., 1951) with bovine serum albumin as reference standard. (B) Emulsion activity index (E.A.I.) determined by the method of Pearce and Kinsella (1978). (C) Emulsion stability determined by the method of Pearce and Kinsella (1978). (D) Foam capacity as described by Vojdani (1992). (See symbols used at bottom of Fig. 9.6E.)

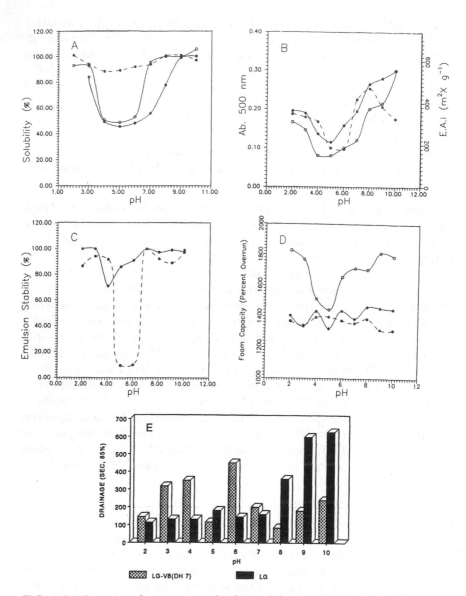

FIG. 9.7 Functional properties of β-lactoglobulin (•---•) and of endoproteinase Glu-C hydrolysates (•—•), DH 5; ○—○, DH 7) of β-lactoglobulin. The
experimental procedures were as described in Fig. 9.6. (A) Solubility. (B)
Emulsion activity index (E.A.I.). (C) Emulsion stability. (D) Foam capacity.
(E) Foam stability. (See symbols used at bottom of Fig. 9.7E.)

has a minimum emulsion stability at pH 4, while the emulsions are relatively stable at pH 5 and 6. The shift in pH optimum for emulsion stability, caused by positively charged arginine or lysine residues at the C-terminal end of the polypeptides, is marked. The DH 3 hydrolysate and β-lactoglobulin have similar foam capacities at all pHs (Fig. 9.8D). The DH 10 hydrolysate has markedly improved foam capacity over that of β-lactoglobulin at all pHs except 8 and 9. This is particularly noticeable in the pH 4–6 range (near the pI). The DH 3 hydrolysate has greatly improved foam stability at pH 4–7 (and especially at pH 5) compared to β-lactoglobulin (Fig. 9.8E). In summary, the data indicate that both the emulsifying and foaming properties of β-lactoglobulin can be improved by selective hydrolysis with trypsin.

Solubility and Functional Properties of Polypeptides Formed by Endoprotease Lys-C. Comparison of solubility and functional properties of endoprotease Lys-C hydrolysate (DH 5) with β-lactoglobulin is shown in Fig. 9.9. Solubility of the DH 5 hydrolysate was significantly less at pH 4 (and a little less at pH 5) than that of β-lactoglobulin (50% vs. 90%; Fig. 9.9A). Solubilities at pHs above and below pH 4–5 were similar for the hydrolysate and β-lactoglobulin. The hydrolysate had significantly higher E.A.I. at pH 6 and above compared to β-lactoglobulin (Fig. 9.9B). At pH 4 and 5 the E.A.I. was smaller for the hydrolysate. The emulsion was stable at pH 5 and 6 for the hydrolysate and not for β-lactoglobulin (Fig. 9.9C). This is caused by a shift in minimum stability at pH 5–6 for β-lactoglobulin to pH 4 for the hydrolysate. An identical effect was found for trypsin hydrolysate (DH 10; Fig. 9.8C). At pH 4 β-lactoglobulin has excellent emulsion stability, in contrast to pH 5 and 6, while the endoprotease Lys-C hydrolysate has little emulsion stability. The foaming capacity was increased somewhat for the hydrolysate vs. β-lactoglobulin from pH 2 to 7 (Fig. 9.9D). Even more important, the foam stability is better at pH 2–8 for the hydrolysate than for β-lactoglobulin (Fig. 9.9E). The greatest improvement is in the pH 4–5 range, although it is significant (two to three times) at pHs 2, 3, 6, and 7. In summary, while there is a decrease in solubility of the endoprotease Lys-C hydrolysate compared to β-lactoglobulin, the emulsifying and foaming properties were improved. Significantly, the emulsion and foam stabilities were greatly improved at pH 5–6 (and pH 4 for foaming also) by the limited specific hydrolysis.

Factors Affecting Emulsion Properties. The data presented in Figs. 9.6–9.9 show that limited specific hydrolysis of β-lactoglobulin with different proteases can improve its emulsifying properties in relation to

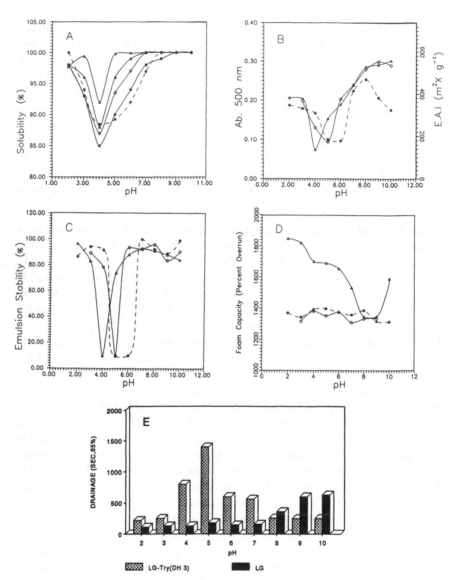

FIG. 9.8   Functional properties of β-lactoglobulin and of trypsin hydroly-sates of β-lactoglobulin. The experimental procedures were as described in Fig. 9.6. Symbols used: •----•, β-lactoglobulin; •——•, DH 1.5; ○——○, DH 3.1; ▲——▲, DH 6.4; △——△, DH 10.5 (A) Solubility. (B) Emulsion activity index (E.A.I.). (C) Emulsion stability. (D) Foam capacity. (E) Foam stability. (See symbols used at bottom of Fig. 9.8E.)

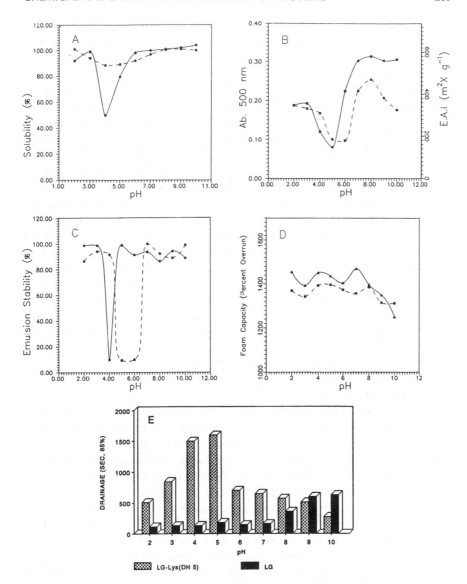

FIG. 9.9 Functional properties of β-lactoglobulin (•---•) and of endoproteinase Lys-C hydrolysate (•—•) DH 5. The experimental procedures were as described in Fig. 9.6. (A) Solubility. (B) Emulsion activity index (E.A.I.). (C) Emulsion stability. (D) Foam capacity. (E) Foam stability. (See symbols used at bottom of Fig. 9.9E.)

those of the intact protein, especially near the isoelectric point. What general conclusions can we draw from these data in relation to factors affecting functional properties? What have we done? Chemically, the changes are two-fold.

There is a decrease in the size of β-lactoglobulin from 18,300 to a mixture of smaller polypeptides ranging from about 13,500 to as low as <1000 (see Table 9.6). This reduction in size did not significantly affect the solubility, especially near the pI of β-lactoglobulin. In two cases [β-Lg Glu-C (DH 5 and 7) and β-Lg Lys-C (DH 5) hydrolysates] there was a significant decrease in solubility near the pI, no change in a third case (β-Lg Arg-C hydrolysate), and a slight increase in solubility only for β-Lg trypsin DH 10 hydrolysate. Therefore, increase in solubility near the pI cannot explain the results.

A second change was in the charge density. At a newly generated amino group there would be one positive charge increase near the pI. A new carboxyl group is also generated at the end of the newly hydrolyzed peptide bond. At pH 4–6 this may be 10 to 90% as the carboxylate ion. This would imply that the pI might shift upward. But in all cases there is an indication that the pI may have shifted downward by as much as one pH unit (compare Figs. 9.6C, 9.7A,B, 9.8A,B, and 9.9A,B). In the case of endoproteases Arg-C, Lys-C, and trypsin hydrolysates, there is one newly formed carboxyl group on arginine and lysine residues at the C-terminal end, both of which already have one positive charge on the side chain. Is this important?

Do the polypeptides have less secondary structure than β-lactoglobulin? Table 9.7 indicates that there is less α-helix structure in all derivatives except for β-Lg Arg C (DH 2) hydrolysate. There is little difference in the effect of hydrolysis on emulsion activity (E.A.I.) and emulsion stability of β-Lg Arg-C (DH 2) hydrolysate vs. that of β-Lg Glu-C (DH 5), β-Lg trypsin (DH 3 and 10), and β-Lg Lys-C (DH 5) hydrolysates. Do the tertiary structures differ between β-lactoglobulin and the hydrolysates? As shown in Table 9.8, the answer is no and yes. β-Lg Arg-C (DH 2) and β-Lg trypsin (DH 3) hydrolysates have about the same number of hydrophobic sites on the surface as β-lactoglobulin. However, in the case of the β-Lg trypsin (DH 10), β-Lg Lys C (DH 5), and β-Lg Glu-C (DH 5 and 7) hydrolysates there is a significant increase in hydrophobic sites on the surface.

Perhaps in this regard, it is instructive to compare β-Lg trypsin (DH 5) and β-Lg trypsin (DH 10) hydrolysates (see Fig. 9.8). β-Lg Trypsin (DH 5) hydrolysate has about the same solubility at pH 4 as β-lactoglobulin and the same E.A.I. at pH 5. However, the emulsion stability is very much better for β-Lg trypsin (DH 5) hydrolysate at pH 6 than for β-lactoglobulin, because the minimum point for emulsion stability has been shifted to pH 4 (rather than ∼5 for β-lactoglobulin). In the case of β-Lg trypsin (DH 10) hydrolysate, it is slightly more soluble (∼6%) than β-Lg trypsin (DH 5) hydrolysate and has a higher emulsion activity (E.A.I.; ∼55%). It has much better emulsion stability at pH 5 than β-Lg trypsin (DH 5) hydrolysate because the minimum emulsion stability is shifted from pH 5 to pH 4.

Is the shift in minimum point more important than the change in emulsion activity? In the case of foam capacity (Fig. 9.8D), β-Lg trypsin (DH 5) hydrolysate and β-lactoglobulin gave about the same results at pH 5. However, the foam capacity is about 22% higher for β-Lg trypsin (DH 10) hydrolysate than for β-Lg trypsin (DH 5) hydrolysate.

Which has the major effect, the 6% increase in solubility or the one pH unit difference in minimum point for emulsion stability? Most literature data indicate that emulsion stability is poor near the pI of a protein and improves at pHs away from the pI. The DLVO theory indicates that emulsion stability results from a balance of repulsive forces (due to charge on the protein molecules, as affected by salt ions in the solution) and attractive forces (van der Waals and hydrophobic forces). Therefore, theory would predict emulsions to be least stable at the pI (all other things being equal) because the net charge on the protein is zero. How does hydrolysis of β-lactoglobulin increase emulsion stability at pH 5–6? By shifting the pI downward, as appears to be the case with our data?

Factors Affecting Foam Stability.   Most literature data indicate that foam stability should be best at the pI of a protein. This is explained by absence of repulsive forces which tend to weaken the strength and thickness of the film (denatured protein) at the interface between air (or gas) bubbles and water, permitting interactions by hydrophobic bonds. How does hydrolysis of β-lactoglobulin help to stabilize the interface between air and water near the pI, as was found for the β-Lg Glu-C (DH 7), β-Lg trypsin (DH 3), and β-Lg Lys-C (DH 5) hydrolysates? Why is the foam stability for β-Lg Arg-C (DH 2) hydrolysate less than for β-lactoglobulin (except at pH 5, where they are about the same)? It is important to note that the foam capacity was about the same for β-Lg Arg-C hydrolysate and β-lactoglobulin at pH 5, and less for β-Lg Arg-C hydrolysate at all other pHs. Perhaps the hydrophobic interactions are too great between molecules of the β-Lg Arg C hydrolysate?

Are there other factors that account for the differences between the functional properties of β-lactoglobulin and its enzymatic hydrolysates? Is β-lactoglobulin a monomer or a dimer in emulsions and foams? According to Timasheff and Townend (1969), β-lactoglobulin is a monomer below pH 3 and above pH 8 and a dimer at pH 3–8. At pH 3.8–5.1, low temperature, and high protein concentration it forms an octamer.

Would β-lactoglobulin behave differently if it were a monomer vs. a dimer? In examples to be discussed below, Kato (1991) showed that the polymeric forms of ovomucin and ovalbumin give better foaming properties than lower molecular weight units. On the other hand, the emulsifying properties of ovomucin showed a good correlation with the surface hydrophobicity, but not with the degree of polymerization (Kato et al., 1987a). Kato (1991) reported that the emulsifying properties were not affected by the degree of polymerization.

Diffusion Rates of Proteins and Polypeptides to Surfaces

Above, we asked whether the increased rates of diffusion of polypeptides formed by enzymatic hydrolysis of β-lactoglobulin could explain why the enzymatic hydrolysates had better emulsifying and foaming properties than β-lactoglobulin, especially near the pI (pH 4–6). There are two aspects to this question. One is the diffusion rate from the bulk solution during formation of the oil/water (emulsions) or air/water (foams) interfaces. The rates of diffusion (k_1) and rearrangement (k_2) are measured as one investigates the rate of approach to equilibrium of the surface tension. As shown in Table 9.11, the rate constants (k_1) for diffusion of the polypeptides to the water/air interface and rearrangement are similar to those for β-lactoglobulin. k_1 for β-Lg Glu-C (DH 5) hydrolysate is larger than k_1 for β-lactoglobulin, while k_1 values for β-Lg Arg-C (DH 2) hydrolysate and β-Lg Lys-C (DH 5) hydrolysate

TABLE 9.11   First-Order Rate Constants for Adsorption and Rearrangement (k_1) and Rearrangement (k_2) of β-Lactoglobulin and β-Lactoglobulin Hydrolysates at the Air/Water Interface at pH 5.0[a]

Protein		k_1 (min^{-1})	k_2 (min^{-1})
β-Lactoglobulin		0.092	0.026
β-Lg Arg-C	(DH 2)	0.056	0.011
β-Lg Glu-C	(DH 5)	0.17	0.007
	(DH 7)	0.095	0.008
β-Lg Lys-C	(DH 5)	0.046 (0.10)[b]	0.007
β-Lg Trypsin	(DH 10)	0.084	0.008

[a]Determined in 20 mM citrate buffer, pH 5.0, at 23°C, by the Whilhelmy plate method (Graham and Phillips, 1979a,b) with a Cahn electrobalance.
[b]At pH 4, the maximum rate constant.

are smaller. It is interesting that k_1 at pH 4 is larger than at pH 5 for β-Lg Lys-C (DH 5) hydrolysate, since the emulsion stability is minima at pH 4 (Fig. 9.9). The rate constants for rearrangement (k_2) are significantly smaller for the hydrolysates than for β-lactoglobulin. We are not sure of the significance of these differences, although the hydrophobicity and hydrophilicity properties would certainly be important.

Effect of Protein-Protein Interaction on Functional Properties

β-Lactoglobulin forms a dimer of 36,000 MW within the pH range where we studied the emulsion and foaming properties of this protein and its enzymatic hydrolysates. Timasheff and Townsend (1969) reported that β-lactoglobulin forms tightly associated dimers, via hydrophobic binding between two molecules, at pH 3–8. In addition, oxidation can lead to disulfide bond formation between the molecules. β-Lactoglobulin used in our experiments chromatographed as an 18,300 monomer, indicating there were no disulfide bonds.

Does dimer formation give better or worse emulsion and foaming properties than the monomeric form of β-lactoglobulin? Do the hydrophobic interactions involved in dimerization decrease important hydrophobic interactions in emulsions and foams? We did not test these questions in our work. Kato (1991) studied the polymerization effect using ovomucin and ovalbumin. Ovomucin, a glycoprotein found in egg white, is one of the best foam-producing proteins known. Soluble ovomucin is a polymer (8,300 kDa) consisting of α- and β-ovomucin subunits. These subunits can be selectively dissociated by reduction of disulfide bonds or by sonication to give smaller polymers of 1100 and 540 kDa (Hayakawa and Sato, 1976). Kato (1991) found that the higher the degree of polymerization of ovomucin, the better the foaming capacity (Fig. 9.10A). However, the emulsifying activity of ovomucin does not depend on the degree of polymerization, but rather correlates with the surface hydrophobicity (Kato, 1991).

Kato (1991) also investigated the effect of the degree of polymerization of ovalbumin on foaming and emulsion properties. Ovalbumin forms aggregates on heating at 80°C at pH 7.4. Figure 9.10B shows the effect of molecular weight of the aggregate on the foaming capacity. The foaming capacity correlated positively with the increase in molecular weight from 6.0, 7.8, and 9.5 million Da. However, the degree of polymerization did not affect emulsion activity.

Therefore, based on Kato's results, one would predict that the dimeric β-lactoglobulin should have better foaming capacity than does monomeric β-lactoglobulin. This effect of size on foaming capacity is presum-

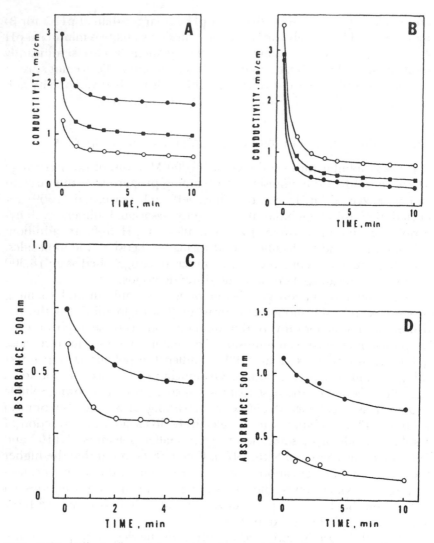

FIG. 9.10 Effect of protein-protein interaction and charge repulsion on functional properties of proteins. (A) Foaming properties of native (•), sonicated (■) and reduced (○) ovomucins. (B) Foaming properties of the soluble aggregates of ovalbumin heated to 80°C and then cooled at room temperature for 0 (•), 3 (■), and 6 (○) hr. (C) Effect of sialidase treatment on the emulsifying properties of ovomucin. •, untreated; ○, treated with sialidase. (D) Effect of phosphatase treatment on the emulsifying properties of phosvitin. • untreated, ○ treated with phosphatase. (A) and (B) (*From Kato, 1991.*) (C) (*From Kato et al., 1987a.*) (D) (*From Kato et al., 1987b.*)

ably the result of a greater coverage of the air/water interface per unit of protein structure. In this regard, we might have anticipated that the enzyme-derived polypeptides of β-lactoglobulin would give poorer foaming capacity. Perhaps this is compensated for in a positive fashion by other factors.

Effect of Charge Repulsion on Emulsion and Foaming Properties

Above, we indicated that there would be an increase in charge density of the polypeptides derived from the enzymatic hydrolysis of β-lactoglobulin (based on smaller peptides). The effect of shift in the apparent pI (based on point of minimum emulsion stabilization) appears to indicate that some charge repulsion is essential near the pI of β-lactoglobulin and of the hydrolysates.

Kato et al. (1987a,b) systematically studied the effect of charge repulsion, using ovomucin and phosvitin, before and after removal of some of the anionic groups. Ovomucin has about 10% sialic acid as part of the glyco groups. The terminal sialic acid residues can be removed readily by sialidase with no effect on the peptide bonds. Figure 9.10C shows that the emulsifying properties of ovomucin are greatly decreased by removal of the anionic sialic acid groups. Kato et al. (1987a) showed that the enzymatically modified ovomucin had better foaming properties.

Phosvitin is an egg white protein of 35,500 MW with about 10% by weight (120 residues) phosphate groups covalently bound to serine residues as organophosphate. As shown in Fig. 9.10D, removal of some of the phosphate groups markedly decreased the emulsion activity (Kato et al., 1987b). These results show how important charged groups are near the pI of proteins. Therefore, charged groups enhance emulsion properties, while charged groups decrease foaming properties near the pI.

Deamidation of Proteins to Change Charged Properties

Deamidation of the asparagine and glutamine residues are known to affect the charged properties of proteins near the pI. A calculation of effect of deamidation on the pI of a theoretical protein is shown in Table 9.12. The observed effects on emulsion and foaming properties by deamidating proteins (Wu et al., 1976; Matsudomi et al., 1986; Kato et al., 1987c) are similar to those found for ovomucin and phosvitin above (Kato et al., 1987a,b). Deamidation of wheat gluten gave much better emulsion properties (Matsudomi et al., 1986). Because of the marked

TABLE 9.12   Effect of Deamidation of Asparagine and Glutamine
Residues on pI of Hypothetical Protein

Ammonium groups[a]	Carboxyl groups[b]	Amide groups	pI
10	8	10	8.4
10	9	9	8.1
10	10	8	7.0
10	11	7	5.0
10	12	6	4.7
10	14	4	4.4
10	16	2	4.2
10	18	0	4.1

[a]$-NH_3^+$ + $-NH_2$; assumed average pK_{a2} = 9.0.
[b]$-COOH$ + COO^-; assumed average pK_{a1} = 4.0.

effect of deamidation on functional properties, we have included a brief description of how proteins can be deamidated.

Proteins contain both aspartic and glutamic acid residues, as well as the amidated derivatives (asparagine and glutamine residues). Plant proteins contain a substantial percentage of asparagine and glutamine residues. For example, in soybean protein and wheat gluten, the % amide N is 1.617 (Shih, 1990a) and 0.98 (Matsudomi et al., 1986) g/100 g protein, respectively. For soybean protein, the 1.617 g N/100 g protein is equivalent to 14.8 g of glutamine residues or 13.2 g of asparagine residues per 100 g of protein. For wheat gluten, the 0.98 g N/100 g protein is equivalent to 8.97 g of glutamine residues or 7.99 g of asparagine residues per 100 g of protein.

Acid- or Base-Catalyzed Deamidation.   Deamidation of proteins occurs below pH 3 (acid catalyzed) and above pH 8 (base catalyzed) at rates dependent upon the H^+ or OH^- concentration, temperature, adjacent amino acid residue (R), and predominate pathway. The reaction is complex as it involves competing reactions as shown in Fig. 9.11. Reactions 1 and 2 are competing reactions. Reaction 1 involves nucleophilic attack of the peptide bond N on the carbonyl group C of the β- or γ-amide group leading to a cyclic intermediate (II) and NH_3. This reaction is more facile for peptide bound asparagine than for glutamine, explaining why deamidation of asparagine is faster than for glutamine (Robinson et al., 1973). Alternatively, deamidation can occur by direct protonation of the carbonyl group O with H^+ or nucleophilic attack of OH^- on

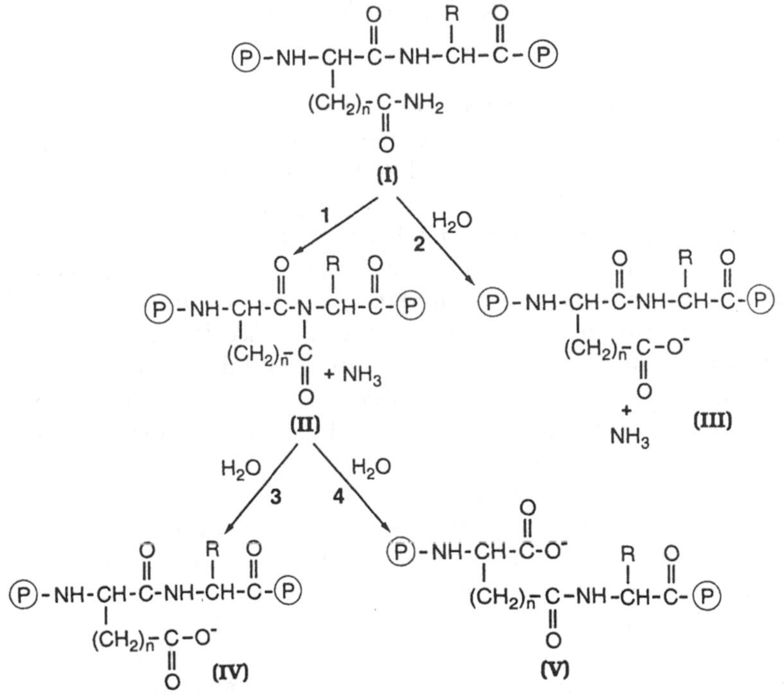

FIG. 9.11 Nonenzymatic deamidation of proteins. See text for details.

the carbonyl group C, leading to expulsion of NH_3, without a cyclic intermediate (III) (Step 2). Step 2 is the preferred pathway for deamidation of peptide-bound glutamine. Hydrolysis of the cyclic intermediate (II), catalyzed by H^+ or OH^-, can lead to either the normal α-carbonyl–linked peptide (IV) or the β- or γ-carbonyl–linked peptide (V) (Sondheimer and Holley, 1954; Murray and Clarke, 1984; Aswad, 1984; Meinwald et al., 1986). Pathway 3 is the predominant pathway (Sondheimer and Holley, 1954; Battersby and Robinson, 1955).

The normal α-carbonyl peptide bond will also be susceptible to direct hydrolysis by H^+ or OH^-, dependent on pH and temperature. The use of large alkylsulfuric and alkylsulfonic acids, arylsulfonic acids, phosphate or bicarbonate ions as catalysts favored deamidation vs. peptide bond hydrolysis (Shih, 1987, 1990c; Shih and Kalmar, 1987). Approximately 40% deamidation, with 1–4% peptide bond hydrolysis, occurred. The product had better solubility, water-binding capacity, foam expansion, emulsion capacity, and viscosity compared to the unmodified product (Shih, 1987; Shih and Kalmar, 1987). The process has been patented (Shih, 1989).

Enzyme-Catalyzed Deamidation. Kato et al. (1987a,b) reported that several proteins could be deamidated by the proteases papain, pronase E, or chymotrypsin at pH 10 and 20°C. The reported deamidation was about 20%, with 0–8% proteolysis. Subsequently, Kato et al. (1989) reported that chymotrypsin immobilized on glass also deamidated several proteins to 5–10% at pH 10 and 20°C. However, Shih (1990a,b) more recently showed that the deamidation is due to nonenzymatic OH^--catalyzed hydrolysis at pH 10, which can be accelerated by phosphate and bicarbonate ions in the alkaline buffers (Shih, 1990a).

Deamidation of glutamine is catalyzed by two nonproteolytic enzymes. These are transglutaminase (EC 2.3.2.13) and peptidoglutaminase (L-glutamine amidohydrolase, EC 3.5.1.2).

Transglutaminase is responsible, in vivo, for the cross-linking of soluble fibrin, formed by highly specific proteolysis of fibrinogen by thrombin, as the terminal step in the blood clotting cascade of proteolytic steps leading to blood clotting (insoluble cross-linked fibrin). The reaction catalyzed by transglutaminase is shown in Fig. 9.12. The enzyme specifically recognizes L-glutamine residues of proteins (I) and uses other amine compounds [the ϵ-amino group of lysine residue of the same protein (intrachain) or of other proteins (interchain)] as nucleophiles to attack the γ-carbonyl C and displace the $-NH_2$ as NH_3 (III), thereby cross-linking the protein(s). In the absence of a good nucleophile, the enzyme can also use water in a nucleophilic attack to hydrolyze the glutaminyl residue to a glutamyl residue (II) and NH_3.

When transglutaminase is used for deamidating the γ-glutamine residue, the transamination reaction must be minimized by blocking the amino groups (Mycek and Waelsch, 1960; Bercovici et al., 1987). The blocking can best be done with citraconic anhydride prior to treatment with transglutaminase; after the enzyme treatment, the citraconyl group is hydrolyzed at pH 3.3 (4 hr, 37°C) (Brinegar and Kinsella, 1980; Motoki et al., 1986). The cost of the treatment of protein with citraconic anhydride and removal of the citraconyl group at the end, plus required FDA approval, probably precludes the commercial use of transglutaminase for deamidating food proteins. Transglutaminase is being studied as a means of cross-linking proteins to make them more useful as edible packaging materials (Krochta et al., personal communication).

Peptidoglutaminase (PGase) is a hydrolase, with minimal transpeptidation reaction (unlike transglutaminase). Two isoenzymes were re-

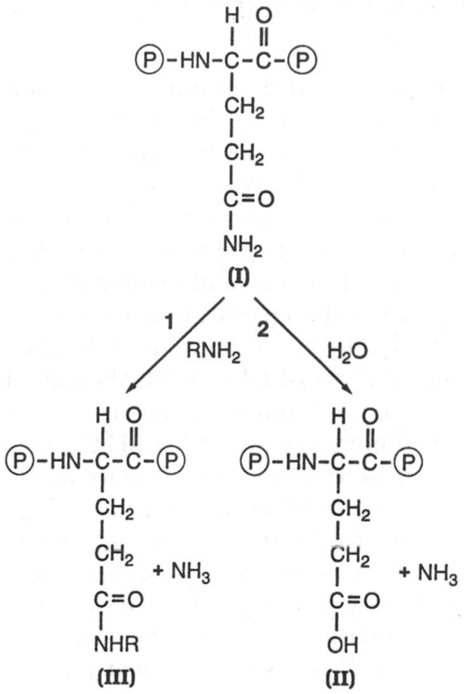

FIG. 9.12 Transglutaminase-catalyzed deamidation of glutamine residues in proteins. See text for details.

ported in the soil bacterium *Bacillus circulans* by Kikuchi et al. (1971). PGase I catalyzes the deamidation of the C-terminal glutamine residue in small peptides while PGase II catalyzes the deamidation of glutamine residues within the peptide chain. More recently, Gill et al. (1985) reported low levels of deamidation of casein and whey proteins by PGase II. Hamada et al. (1988) found that partially purified PGase (mixture of PGases I and II) rapidly deamidated glutamine residues in soy protein hydrolysates but showed very little activity on soy protein. Heating of the protein prior to adding PGase increased the rate of deamidation (Hamada and Marshall, 1988). Hamada (1991) has applied the methodology to several plant and animal proteins, and Hamada and Marshall (1989) showed that PGase treatment of soy proteins increased the solubility (at pH 4–7 and at alkaline pH), emulsifying activity, and stability and foaming power but did not improve the foam stability.

Effect of Stability of Secondary and Tertiary Structures on Functional Properties

We found that most of the β-sheet and β-bend motifs of β-lactoglobulin were preserved in the enzymatic hydrolysates (see Table 9.7). However, there was a decrease in α-helical motifs after protease treatment, except for β-Lg Arg-C (DH 2) hydrolysate.

Kato and Yutani (1988) and Kato (1991) have examined the effect of stability of proteins on emulsion and foam properties, using genetically engineered tryptophan synthases. Replacement of the Glu_{49} residue with other amino acid residues changed the thermal stabilities of the subunits as shown in Table 9.13. The lower the value of ΔG, the lower the stability of the subunit. Replacement of Glu_{49} with Ala and Thr did not significantly change ΔG at pH 7.0, but it did at pH 9. Substitution with Gly decreased ΔG at pH 7 but increased ΔG at pH 9 relative to the wild type (Glu_{49}). Similar results were found with replacement of Glu_{49} with Lys. Phe and Ile had the largest effects on ΔG, stabilizing the subunits.

Table 9.14 shows the correlation of Glu_{49} replacement in tryptophan synthase subunits with surface tension, foaming capacity, foam stability, and emulsifying activity. An increase in ΔG, indicating increased stability of the protein, increased the surface tension at the air/water interface at both pH 7 and 9 (Fig. 9.13A). Note the linear relationship between ΔG and surface tension. As ΔG increases, there is a linear decrease in foam-

TABLE 9.13 Values of Gibbs Free Energy (ΔG) for Unfolding of Seven α-Subunits of Tryptophan Synthase Substituted at Position 49 by Site-Directed Mutagenesis

Residue at position 49	ΔG	
	pH 7.0	pH 9.0
Glu (wild)	8.8 ± 0.1	4.9 ± 0.3
Gly	7.1 ± 0.1	6.4 ± 0.1
Ala	8.5 ± 0.2	6.8 ± 0.2
Ile	16.8 ± 0.5	10.0 ± 0.3
Phe	11.2 ± 0.2	8.3 ± 0.2
Lys	7.9 ± 0.5	7.5 ± 0.9
Thr	8.8 ± 0.2	7.0 ± 0.8

Source: Kato and Yutani, 1988.

TABLE 9.14 Correlation Coefficients of Surface Properties at pH 9 with the ΔG of Unfolding of Tryptophan Synthase α-Subunits

Surface properties	r
Surface tension	$0.90\ (p < 0.01)$
Foaming power	$-0.93\ (p < 0.01)$
Foam stability	$-0.95\ (p < 0.01)$
Emulsifying activity	$-0.83\ (p < 0.05)$

Source: Kato, 1991.

ing capacity (Fig. 9.13B,C). There is less effect at pH 7 than at pH 9. Increasing protein stability (increase in ΔG) also linearly decreased foam stability (Fig. 9.13D). There is much less effect of ΔG on foam stability at pH 7 than at pH 9. At pH 9 foam stability decreased almost by a factor of 3 on replacing Glu_{49} with Ile. There is also a decrease in emulsion activity as ΔG increased (Fig. 9.13E). These very nice results clearly indicate that changes in stability of a protein have substantial effects on foaming and emulsion properties. In both functionalities, increasing stability of tryptophan synthase had a negative effect on the properties (Table 9.14). We conclude that the positive effects of enzymatic hydrolysis of β-lactoglobulin on foaming and emulsifying properties near the pI may be primarily due to decreasing the stability of β-lactoglobulin.

CONCLUSIONS

From the data presented in this chapter based on our own research on the effect of enzymatic hydrolysis of β-lactoglobulin, and on other researchers results, we conclude that emulsion activity is affected by surface hydrophobicity and protein stability and that emulsion stability is affected by electrostatic repulsive forces near the pI. Foaming capacity is affected primarily by protein stability, while foaming stability is affected primarily by protein-protein interaction, which is modulated negatively by charge repulsion.

We conclude that limited specific enzymatic hydrolysis is an effective way of improving the emulsifying and foaming properties of proteins, especially near the pI. We used the mixture of polypeptides formed in our studies on enzymatic hydrolysis of β-lactoglobulin. It will be interesting, and essential, to separate the polypeptides (by gel filtration chroma-

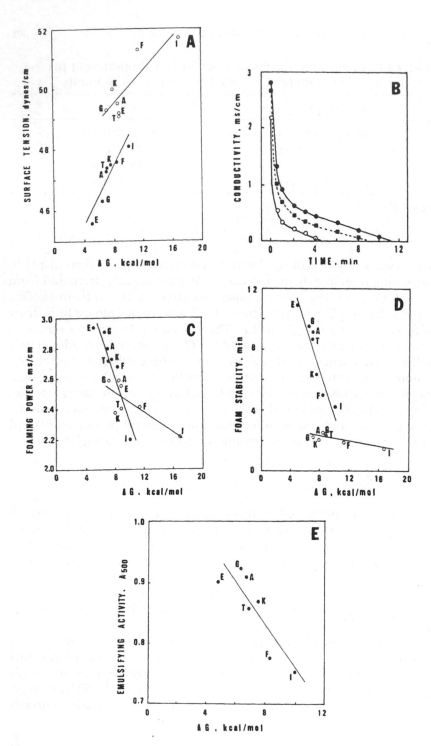

302

tography for example) so as to individually study their effects. Much more data on the relation of structure to functionality will be gained.

Finally, we suggest that genetic engineering has much to offer in the way of designing functionality into proteins. In the near future it will be possible to use transgenic plants and animals to produce designed proteins not only as enzymes and hormones but also as food proteins with better functional and nutritional properties.

ACKNOWLEDGMENT

The authors thank the California Dairy Research Center for financial support of this research. This research is a part of the requirements for the Ph.D. degree in Agricultural and Environmental Chemistry for Fakhrieh Vojdani.

REFERENCES

Adler-Nissen, J. 1986a. *Enzymatic Hydrolysis of Food Proteins.* Elsevier Applied Science Publishers Ltd., London.

Adler-Nissen, J. 1986b. Relationship of structure to taste of peptides and peptide mixtures. In *Protein Tailoring for Food and Medical Uses.* R. E.

FIG. 9.13   Effect of protein stability on functional properties of proteins. (A) Relationship between $\triangle$G of unfolding and surface tension of tryptophan synthases substituted at position 49 by recombinant DNA engineering. Measurements were made at pH 7 (o) and 9 (•). A, E, F, G, I, K, and T are data points for mutated proteins at position 49 (A, alanine; E, glutamic acid (wild type; F, phenylalanine; G, glycine; K, lysine, T, threonine). (B) Foaming properties of mutants of tryptophan synthase substituted at position 49. o, substituted with isoleucine; ■; substituted with glycine; • wild type. (glutamic acid). (C) Relationship between foaming capacity (power) and $\triangle$G of unfolding of mutant tryptophan synthases substituted at position 49. Experiments at pH 7 (o) and 9 (•). See A for explanation of A–T. (D) Relationship between foam stability and $\triangle$G of unfolding of mutant tryptophan synthases substituted at position 49. Experiments at pH 7 (o) and 9 (•). See A for explanation of A–T. (E) Relationship of emulsion activity and $\triangle$G of unfolding of mutant tryptophan synthases substituted at position 49. Experiments at pH 9. See A for explanation of A–T. (A) (C) (D) and (E) (*From Kato and Yutani, 1988.*) (B) (*From Kato, 1991.*)

Feeney and J. R. Whitaker (Ed.), pp. 97–122. Marcel Dekker, Inc., New York.

Aswad, D. W. 1984. Stoichiometric methylation of porcine adrenocorticotropin by protein carboxyl methyltransferase requires deamidation of asparagine 25. *J. Biol. Chem.* 259: 10714–10721.

Battersby, R. and Robinson, J. C. 1955. Studies on the specific chemical fission of peptide links. I. Rearrangements of aspartyl and glutamyl peptides. *J. Chem. Soc.* 1955: 259–269.

Bercovici, D., Gaertner, H. F., and Puigserver, A. J. 1987. Transglutaminase-catalyzed incorporation of lysine oligomers into casein. *J. Agric. Food Chem.* 35: 301–304.

Bingham, E. W. and Farrell, H. M., Jr. 1974. Casein kinase from the golgi apparatus of lactating mammary gland. *J. Biol. Chem.* 249: 3647–3651.

Brake, J. M. and Wold, F. 1962. Carboxymethylation of yeast enolase. *Biochemistry* 1: 386–391.

Braunitzer, G., Chen, R., Schrank, R., and Stangl, A. 1973. The sequential analysis of β-lactoglobulin. *Hoppe-Seyler's Z. Physiol. Chem.* 354: 867–878.

Brinegar, A. C. and Kinsella, J. E. 1980. Reversible modification of lysine in soybeans, using citraconic anhydride: Characterization of physical and chemical changes in soy protein isolate, the 7S globulin, and lipoxygenase. *J. Agric. Food Chem.* 28: 818–824.

Buttkus, H., Clark, J. R., and Feeney, R. E. 1965. Chemical modifications of amino groups of transferrins: Ovotransferrin, human serum transferrin and human lactotransferrin. *Biochemistry* 4: 998–1005.

Drapeau, G. R. 1976. Protease from *Staphylococcus aureus*. *Methods Enzymol.* 45: 469–475.

Elder, J. H. and Alexander, S. 1989. Endoglycosidases from *Flavobacterium meningosepticum:* Application to biological problems. *Methods Enzymol.* 179: 505–518.

Feeney, R. E. and Whitaker, J. R. 1988. Importance of cross-linking reactions in proteins. *Adv. Cereal Sci. Technol.* IX: 21–46.

Fields, R. 1972. The rapid determination of amino groups with TNBS. *Methods Enzymol.* 25B: 464–468.

Gill, B. P., O'Shaughnessey, A. J., Henderson, P., and Headon, D. R. 1985. An assessment of potential of peptidoglutaminase I and II in modifying the charge characteristics of casein and whey proteins. *Ir. J. Food Sci. Technol.* 9: 33–41.

Ginsburg, V. (Ed.). 1989. Complex carbohydrates, Part F. *Methods Enzymol.* 179: 1–639.

Graham, D. E. and Phillips, M. C. 1979a. Proteins at liquid surfaces. I. Kinetics of adsorption and surface denaturation. *J. Colloid. Interface Sci.* 70: 403–414.

Graham, D. E. and Phillips, M. C. 1979b. Proteins at liquid interfaces. II. Adsorption isotherms. *J. Colloid. Interface Sci.* 70: 415–426.

Graham, D. E. and Phillips, M. C. 1979c. Proteins at liquid interfaces. III. Molecular structures of adsorbed films. *J. Colloid Interface Sci.* 70: 427–439.

Grossberg, A. L. and Pressman, D. 1963. Effect of acetylation on the active site of several antihapten antibodies: Further evidence for the presence of tyrosine in each site. *Biochemistry* 2: 90–96.

Hamada, J. S. 1991. Peptidoglutaminase deamidation of proteins for improved food use. *J. Am. Oil Chemists' Soc.* 68: 459—462.

Hamada, J. S. and Marshall, W. E. 1988. Enhancement of peptidoglutaminase deamidation of soy proteins by heat and/or proteolysis. *J. Food Sci.* 53: 1132–1134, 1149.

Hamada, J. S. and Marshall, W. E. 1989. Preparation and functional properties of enzymatically deamidated soy proteins. *J. Food Sci.* 54: 598–601, 635.

Hamada, J. S., Shih, F. F., Frank, A. W., and Marshall, W. E. 1988. Deamidation of soy peptides and proteins by *Bacillus circulans* peptidoglutaminase. *J. Food Sci.* 53: 671–672.

Hayakawa, S. and Nakai, S. 1985. Relationships of hydrophobicity and net charge to the solubility of milk and soy proteins. *J. Food Sci.* 50: 486–491.

Hayakawa, S. and Sato, Y. 1976. Studies on the dissociation of the soluble ovomucin by sonication. *Agric. Biol. Chem.* 40: 2397–2404.

Hoare, D. G. and Koshland, D. E. 1967. A method for the quantitative modification and estimation of carboxylic acid groups in proteins. *J. Biol. Chem.* 242: 2447–2453.

Hunter, T. and Cooper, J. A. 1985. Protein-tyrosine kinases. *Ann. Rev. Biochem.* 54: 897–930.

Hunter, M. J. and Ludwig, M. L. 1962. The reaction of imidoesters with proteins and related small molecules. *J. Am. Chem. Soc.* 84: 3491–3504.

Jentoft, N. and Dearborn, D. G. 1979. Labelling of proteins by reductive methylation using sodium cyanoborohydride. *J. Biol. Chem.* 254: 4359–4365.

Kato, A. 1991. Significance of macromolecular interaction and stability in functional properties of food proteins. In *Interaction of Food Proteins*, N. Parris and R. Barford (Ed.), pp. 13–24. ACS Symposium Series 454, American Chemical Society, Washington, DC.

Kato, A. and Nakai, S. 1980. Hydrophobicity determined by a fluorescent probe method and its correlation with surface properties of proteins. *Biochim. Biophys. Acta* 624: 13–20.

Kato, A. and Yutani, K. 1988. Correlation of surface properties with conformational stabilities of wild-type and six mutant tryptophan synthase α-subunits substituted at the same position. *Protein Eng.* 2: 153–156.

Kato, A., Miyachi, N., Matsudomi, N., and Kobayashi, K. 1987a. The role of sialic acid in the functional properties of ovomucin. *Agric. Biol. Chem.* 51: 641–645.

Kato, A., Miyazaki, S., Kawamoto, A., and Kobayashi, K. 1987b. Effect of phosphate residues on the excellent emulsifying properties of phosphoglycoprotein phosvitin. *Agric. Biol. Chem.* 51: 2989–2994.

Kato, A., Tanaka, A., Lee, Y., Matsudomi, N., and Kobayashi, Y. 1987c. Effects of deamidation with chymotrypsin at pH 10 on the functional properties of proteins. *J. Agric. Food Chem.* 35: 285–288.

Kato, A., Tanaka, A., Matsudomi, N., and Kobayashi, K. 1987d. Deamidation of food proteins by proteases in alkaline pH. *J. Agric. Food Chem.* 35: 224–227.

Kato, A., Lee, Y., and Kobayashi, K. 1989. Deamidation and functional properties of food proteins by the treatment with immobilized chymotrypsin at alkaline pH. *J. Food Sci.* 54: 1345–1347, 1371–1372.

Kikuchi, M., Hayashida, H., Nakano, E., and Sakahuchi, K. 1971. Peptidoglutaminase: Enzymes for selective deamidation of γ-amide of peptide-bound glutamine. *Biochemistry* 10: 1222–1229.

Kinsella, J. E. 1982. Protein structure and functional properties: Emulsification and flavor binding effects. In *Food Protein Determination: Mechanisms and Functionality*, J. P. Cherry (Ed.), pp. 301–326. ACS Symposium Series 206, American Chemical Society, Washington, DC.

Kinsella, J. E. and Phillips, L. C. 1989. Structure-function relationships in food proteins, films and foaming behavior. In *Food Proteins*, J. E. Kinsella and W. G. Soucie (Ed.), pp. 52–77. American Oil Chemists Society, Champaign, IL.

Krebs, E. G. 1986. The enzymology of control of phosphorylation. In *Control by Phosphorylation*, P. D. Boyer and E. G. Krebs (Ed.), pp. 3–20. The Enzymes, 3rd ed., Vol. XVII, Academic Press, Inc., Orlando.

Lowry, O. H., Rosebrough, N. J., Farr, A. L., and Randall, R. J. 1951. Protein measurement with folin phenol reagent. *J. Biol. Chem.* 193: 265–275.

Matheis, G. and Whitaker, J. R. 1984a. Modification of proteins by polyphenol oxidase and peroxidase and their products. *J. Food Biochem.* 8: 137–162.

Matheis, G. and Whitaker, J. R. 1984b. Peroxidase-catalyzed crosslinking of proteins. *J. Protein Chem.* 3: 35–48.

Matheis, G., Penner, M. H., Feeney, R. E., and Whitaker, J. R. 1983. Phosphorylation of casein and lysozyme by phosphorus oxychloride. *J. Agric. Food Chem.* 31: 379–387.

Matsudomi, N., Tanaka, A., Kato, A., and Kobayashi, K. 1986. Functional properties of deamidated gluten by treating with chymotrypsin. *Agric. Biol. Chem.* 50: 1989–1994.

Means, G. E. and Feeney, R. E. 1968. Reductive alkylation of amino groups in proteins. *Biochemistry* 7: 1366–1371.

Meinwald, Y. C., Stinson, E. R., and Scheraga, A. 1986. Deamidation of the asparaginyl-glycyl sequence. *Int. J. Peptide Protein Res.* 28: 79–84.

Melchior, W. B. and Fahrney, D. 1970. Ethoxyformylation of proteins. Reaction of ethoxyformic anhydride with α-chymotrypsin, pepsin and pancreatic ribonuclease at pH 4. *Biochemistry* 9: 251–258.

Motoki, M., Seguro, K., Nio, N., and Takinami, K. 1986. Glutamine-specific deamidation of α_{s1}-casein by transglutaminase. *Agric. Biol. Chem.* 50: 3025–3030.

Murray, E. D. and Clarke, S. 1984. Synthetic peptide substrates for the erythrocyte protein carbonyl methyltransferase. *J. Biol. Chem.* 259: 10722–10732.

Mycek, M. J. and Waelsch, H. 1960. The enzymatic deamidation of proteins. *J. Biol. Chem.* 235: 3513–3517.

Pearce, K. N. and Kinsella, J. E. 1978. Emulsifying properties of proteins: Evaluation of a turbidimetric technique. *J. Agric. Food Chem.* 26: 716–723.

Peters, K. and Richards, F. M. 1977. Chemical crosslinking: Reagents and problems in studies of membrane structure. *Ann. Rev. Biochem.* 46: 523–551.

Riordan, J. E. 1973. Functional arginine residues in carboxypeptidase-modification with butanedione. *Biochemistry* 12: 3915–3923.

Robinson, A. B., Scotchler, J. W. and McKerrow, J. H. 1973. Rates of nonenzymatic deamidation of glutaminyl and asparaginyl residues in pentapeptides. *J. Am. Chem. Soc.* 95: 8156–8159.

Shih, F. F. 1987. Deamidation of protein in a soy extract by ion-exchange resin catalysis. *J. Food Sci.* 52: 1529–1531.

Shih, F. F. 1989. Partially deamidated oilseed proteins and process for the preparation thereof. U.S. Patent 4,824,940, April 25, 1989.

Shih, F. F. 1990a. Deamidation during treatment of soybean with protease. *J. Food Sci.* 55: 127–129, 132.

Shih, F. F. 1990b. Deamidation studies on selected food proteins. *J. Am. Oil Chem. Soc.* 67: 675–677.

Shih, F. F. and Kalmar, A. D. 1987. SDS-catalyzed deamidation of oilseed proteins. *J. Agric. Food Chem.* 35: 672–675.

Sklar, L. A., Hudson, B. S., and Simoni, R. D. 1977. Conjugated polyene fatty acids as fluorescent probes: Binding to bovine serum albumin. *Biochemistry* 16: 5100–5108.

Sondheimer, E. and Holley, R. W. 1954. Imides form asparagine and glutamine. *J. Am. Chem. Soc.* 76: 2467–2470.

Spande, T. F. and Witkop, B. 1967. Tryptophan involvement in the function of enzymes and protein hormones as determined by selective oxidation with N-bromosuccinimide. *Methods Enzymol.* 11: 506–521.

Stauffer, C. E. and Etson, D. 1969. The effect on subtilisin activity of oxidizing a methionine residue. *J. Biol. Chem.* 244: 5333–5338.

Swaisgood, H. E. and Horton, H. R. 1989. Immobilized enzymes as processing aids or analytical tools. In *Biocatalysis in Agricultural Biotechnology,* J. R. Whitaker and P. Sonnet (Ed.), pp. 242–261. ACS Symposium Series 389, American Chemical Society, Washington, DC.

Timasheff, S. N. and Townend, R. 1969. β-Lactoglobulin as a model of subunit enzymes. *Protides Biol. Fluids, Proc. Colloq.* 16: 33–40.

Trimble, R. B., Trumbly, R. J. and Maley, F. 1987. Endo-β-N-acetylglucosaminidase H from *Streptomyces plicatus. Methods Enzymol.* 138: 763–770.

Uy, R. and Wold, F. 1980. Posttranslational chemical modification of proteins. In *Chemical Deterioration of Proteins,* J. R. Whitaker and M. Fujimaki (Ed.), pp. 49–62. ACS Symposium Series 123. American Chemical Society, Washington, DC.

Vodjani, F. 1992. Effects of specific enzymatic and chemical modification on functional properties of proteins: α-Lactalbumin and β-lactoglobulin. Ph.D. thesis, University of California, Davis.

Wallace, C. J. A. and Harris, D. E. 1984. The preparation of fully N-ε-acetimidylated cytochrome c. *Biochem. J.* 217: 589–594.

Whitaker, J. R. 1977. Enzymatic modification of proteins applicable to foods. In *Improvement Through Chemical and Enzymatic Modification,*

R. E. Feeney and J. R. Whitaker (Ed.), pp. 95–155. Advances in Chemistry Series 160, American Chemical Society, Washington, D.C.

Whitaker, J. R. and Puigserver, A. J. 1982. Fundamentals and applications of enzymatic modifications of proteins: An overview. In *Modification of Proteins: Food, Nutritional and Pharmacological Aspects*, R. E. Feeney and J. R. Whitaker (Ed.), pp. 57–87. Advances in Chemistry Series 198, American Chemistry Society, Washington, D.C.

Whitney, R. M., Brunner, J. R., Ebner, K. E., Farrell, H. M., Josephson, R. V., Moor, C. V., and Swaisgood, H. E. 1976. Nomenclature of the proteins of cow's milk: Fourth revision. *J. Dairy Sci.* 59: 795–815.

Wu, C. H., Nakai, S., and Powrie, W. D. 1976. Preparation and properties of acid-solubilized gluten. *J. Agric. Food Chem.* 24: 504–510.

Yamasaki, R. B., Vega, A., and Feeney, R. E. 1980. Modification of available arginine residues in proteins by *p*-hydroxyphenylglyoxal. *Anal. Biochem.* 109: 32–40.

Yang, J. T., Wu, C. C., and Martinez, H. M. 1986. Calculation of protein conformation from circular dichroism. *Methods Enzymol.* 130: 208–269.

10
Functionality of Soy Proteins

Khee Choon Rhee

Texas A&M University
College Station, Texas

INTRODUCTION

Soybeans have been an integral part of the diet of people in the Far East for more than 5000 years. However, the history of soy protein products—flours, concentrates, isolates, and their derivatives—is relatively short. In early years, soy protein products were mainly used to meet nutritional needs, but more recently they have been used primarily for their unique functional characteristics. Today, many thousands of tons of concentrated forms of functional soy protein ingredients and products are used by the food industry, feed manufacturers, and other nonfood, nonfeed industries in a variety of applications. This chapter will describe the various types of soy protein ingredients and products available commercially, the methods of their preparation, their physicochemical and functional characteristics, and the current and possible functional applications of these ingredients.

Soy protein ingredients have become an important protein source, not only for human food, but also for animal feed and industrial applications.

Although soy protein is used in animal feed, human food use and industrial applications have increased steadily over the past several years. Soy protein ingredients are important in food and industrial applications principally because of their functional characteristics.

Functional characteristics are generally dependent on intrinsic physicochemical properties, which are affected greatly by the methods used for their manufacture, storage, handling, and use. It is, therefore, very important to clearly understand the detailed conditions of manufacturing of soy protein ingredients as they affect the composition, structure, and functional properties of the component proteins.

MANUFACTURING PROCEDURES FOR SOY PROTEIN INGREDIENTS

Various manufacturing procedures for the basic soy protein ingredients, i.e., soy flours and grits, soy protein concentrates, soy protein isolates, and textured soy proteins, are briefly reviewed below.

Soy Flours and Grits

Soy flours and grits are produced by grinding and screening soybean flakes either before or after the oil is removed (Fulmer, 1988). Protein contents of these products are between 40 and 54%. Soy flours and grits are the least refined forms of soy protein ingredients used for human consumption and may vary in particle size distribution, fat content, and degree of protein denaturation. Depending on the type and level of oil present, soy flours and grits are commonly divided into full-fat flours, high-enzyme flours, defatted flours, defatted grits, and lecithinated/refatted flours.

Full-Fat Soy Flours.   Full-fat soy flours are becoming increasingly important in food. With steadily improving production technology and intensive training in applications, full-fat soy flours are finding wide application as a food or food ingredient. Full-fat flours are the least refined soy protein ingredients, containing about 40% protein. They are prepared by grinding dehulled cotyledons to specific size (Heiser and Trentleman, 1988). They are produced primarily in Europe and Asia for the baking industry and production of soymilks.

High-Enzyme Flours.   High-enzyme flours are prepared by grinding defatted flakes with minimum heat denaturation to maintain high nitrogen solubility index (NSI) and usually contain 52–54% protein. They are used in baking applications to increase mixing tolerance and bleaching in bread and as starting raw materials for preparation of functional concentrates and isolates.

Defatted Soy Flour.   Defatted flours, which contain 52–54% protein, are prepared by finely grinding defatted flakes to pass through No. 100 U.S. Standard Screen size. Controlled moist heat treatment is used to produce white (85–90 NSI), cooked (20–60 NSI), and toasted (<20 NSI) grits. Defatted flours are used in a variety of food applications where a wide range of protein solubilities is required.

Defatted Grits.   Defatted grits are similar to defatted flours, except a screen size between No. 10 and No. 80 is used. These products are generally used in ground meat systems and bakery products.

Lecithinated/Refatted Flours.   About 0.5–30% lecithin or vegetable oil is blended with defatted flours. These products are usually used to improve water dispersibility and emulsifying capacity in baking applications.

Soy Protein Concentrates

Soy protein concentrates are prepared from dehulled and defatted soybean flakes by removing most of the water-soluble, nonprotein constituents (Berry, 1988). These products usually contain at least 70% protein on a moisture-free basis. There are three basic processes for preparation of soy protein concentrates: acid leaching at pH 4.5, extraction with 70–90% aqueous ethanol, and extraction with water after heat denaturation of the protein with moist heat. In a more recently developed process, "functional concentrates" are prepared by subjecting a low water-soluble soy protein concentrate to heat treatment by steam injection or jet cooking to increase the solubility and functionality. Homogenization further improves the functionality. The functional concentrates act like emulsifiers and emulsion stabilizers, bind water and fat, and offer special adhesive properties similar to those of isolates.

Soy Protein Isolates

Soy protein isolates are the most refined soy protein ingredients. They are prepared from dehulled and defatted flakes by removing most of the nonprotein components and usually contain not less than 90% protein on a moisture-free basis (Johnson and Kikuchi, 1988).

Isoelectric Soy Protein Isolate.   Soy protein is first extracted from defatted flakes with mild alkaline water at pH 8.5–9.0 followed by centrifugation to remove the insoluble residue. The liquid extract is then adjusted to pH 4.5 to precipitate most of the proteins, the precipitated curd separated by centrifugation, washed with water, and finally spray dried to obtain an isoelectric (or acid) isolate. This unique form of isolate is frequently used in infant formulas and nutritional applications.

Neutralized Soy Protein Isolates.   More commonly, the isoelectric isolate is neutralized with alkali to form Na-, K-, or Ca-proteinates to make them more soluble and functional. These forms of isolates are used in a wide variety of meat and dairy applications where emulsification, emulsion stabilization, water and fat absorption, adhesive, and fiber forming properties are desired.

Textured Soy Proteins

Textured soy proteins (TSP) are prepared to impart a structure (i.e., fiber or chunk) for use as food ingredients. They are frequently made to resemble meat, seafood, or poultry in structure and appearance when hydrated.

Textured Soy Flours and Soy Protein Concentrates.   These products are prepared by thermoplastic extrusion or steam texturization of soy flours or alcohol or heat-denatured concentrates. Compositions of these products are similar to the corresponding raw materials. Types of foods using these products successfully include fibrous foods, ground meat products, poultry, and seafoods.

Structured Soy Protein Concentrates.   Soy protein concentrates are extruded into different shapes and sizes, and used in poultry, meats, and seafoods.

Structured Soy Protein Isolates.   These types of products are prepared by extruding soy protein isolate as above or by extruding a solution of soy protein isolate into an acid-salt bath to coagulate the protein into fibers. The fibers are then put together with binders to form fiber tows or fiber bundles. These products can be used in poultry and seafoods and as food analogs.

Specialty Soy Protein Ingredients

Partially hydrolyzed soy protein ingredients are produced by controlled partial hydrolysis of soy proteins with proteolytic enzymes (e.g., pepsin, papain, bromelain) to reduce the molecular size of proteins to large peptides (3000–5000 daltons). This enzyme treatment improves acid solubility and whipping properties substantially. Fully hydrolyzed proteins [i.e., hydrolyzed soy protein (HSP), soy sauce, etc.] are prepared from soy grits usually by acid hydrolysis to produce flavoring agents.

BASIC PHYSICOCHEMICAL PROPERTIES OF SOY PROTEIN

Classification

Soy proteins were initially classified according to their sedimentation velocity into 2S, 7S, 11S, and 15S fractions (Naismith, 1955; Wolf and Briggs, 1956). The 2S fraction contains trypsin inhibitors and cytochrome and constitutes about 8% of the protein. The 7S fraction contains globulins and enzymes (lipoxygenase and amylase) and constitutes about 35% of the protein. The 11S fraction is considered to be a single protein and constitutes about 52% of the total protein (Wolf et al., 1961; Wolf and Cowan, 1975). The 15S fraction is a polymer form of 11S fraction and constitutes about 5% of the protein. The major soy globulins were also classified into glycinin, α-, β-, and γ-conglycinin based on their different immunological responses (Catsimpoolas, 1969b). The 11S fraction is believed to be identical to glycinin and the globulin portion of the 7S fraction to conglycinin.

Molecular Structure

Most of the structural studies of soy proteins were conducted on the 11S fraction (or glycinin) because it is a single protein, constitutes more than half of the protein in soybean, and can easily be prepared in a relatively pure form. There are three kinds of acidic (A_1, A_2, and A_3) and three kinds of basic (B_1, B_2, and B_3) subunits, with different molecular weights, in the 11S protein molecule (Catsimpoolas, 1969b; Bradley et al., 1975; Yamaguchi et al., 1979). Monomeric glycinin consists of six subunits: three acidic and three basic. The acidic and basic subunits alternate in the same layer and are held together by hydrophobic and disulfide bonds. The dimer has been considered to consist of two identical monomer layers held together by hydrophilic bonds—electrostatic and/or hydrogen bonding—and has a molecular weight of approximately 360,000 (Catsimpoolas et al., 1967, 1971b; Catsimpoolas, 1969b, 1970; Bradley et al., 1975; Draper and Catsimpoolas, 1977). However, other researchers reported that there are four to six kinds of acidic subunits and four to five kinds of basic subunits (Kitamura and Shibasaki, 1975; Kitamura et al., 1976; Mori et al., 1979; Staswick et al., 1981), which have been attributed largely to varietal differences (Wolf, 1976). The optical rotary dispersion studies indicated that 11S protein has 5.2% α-helix, 34.8% β-structure, and 60% random coil (Koshiyama, 1972).

Amino Acid Profile

The exact amino acid sequence of soy proteins has not yet been determined, but the amino acid compositions of acidic and basic subunits of the 11S fraction are known (Catsimpoolas et al., 1971; Ochiai-Yanagi et al., 1977). Acidic subunits have higher contents of glutamic acid, proline, and cysteine than basic subunits and have a molecular weight of around 37,000. Basic subunits have higher hydrophobic amino acid (alanine, leucine, phenylalanine, tyrosine, and valine) contents than acidic subunits and have a molecular weight of around 20,000. The 11S protein contains approximately 48 moles of cysteine per mole of protein (Shavarts and Vaintraub, 1967; Catsimpoolas et al., 1969). The 11S protein has three kinds of N-terminal amino acids—glycine, leucine, and phenylalanine—with 8:2:2 molar ratios (Mitsuda et al., 1965; Catsimpoolas et al., 1967)

Solubility

Solubility of soy proteins can be affected by many factors, i.e., varietal difference, storage conditions, relative humidity, types and concentration of ions, freezing and thawing, pH, and temperature. The 11S protein fraction has an isoelectric point of pH 4.64; however, protein solubilities at this pH in buffer solutions vary considerably depending upon the ionic strength and temperature (Koshiyama, 1972; Eldridge and Wolf, 1967). Also, the 11S protein is more sensitive to lowering of pH than the 7S protein (Thanh and Shibasaki, 1976; Wolf and Sly, 1967).

FUNCTIONAL APPLICATIONS OF SOY PROTEIN INGREDIENTS

Soy protein ingredients must possess appropriate functional properties for intended applications and consumer acceptability. These are intrinsic physicochemical characteristics that affect the behavior of proteins in food systems during processing, manufacturing, storage, and preparation (Kinsella, 1979; Morr, 1990; Peng et al., 1984; Snyder and Kwon, 1987). Because functional properties are affected by composition, structure, and conformation of ingredient proteins as well as by the manner in which they are used, systematic elucidation of the physicochemical properties of component proteins as well as the surrounding conditions is critical for understanding the mechanisms of particular functional traits. Some of the important functional applications of soy protein ingredients are briefly reviewed below.

Dairylike Products

A number of dairy analogs have been developed from soybeans and soy protein ingredients to lower the cost, improve nutritional value, reduce allergic responses, alleviate lactose intolerance, and improve functional characteristics. These products include imitation milks (soymilks), imitation cheeses, nondairy frozen desserts, ice creams, yogurts, creams, coffee whiteners, and soymilk drinks. These products are very popular in Asian and Southeast Asian countries, but they are not yet produced in any significant volume in the United States. However, soy protein ingredients offer considerable potential in the manufacture of dairy-type products.

Meat/Fish Products

The use of soy protein ingredients in processed meat systems is increasing due to changing attitudes of consumers, processors, and regulatory agencies. Soy protein ingredients are used as partial replacements of meat, binders, flavor enhancers, emulsifiers, brine ingredients, and meat analogs and contribute to nutrition, flavor, and critical functional properties. Most of the current meat applications of soy protein ingredients are in the area of processed (comminuted and coarsely ground) meat products. The quality of whole meat products can also be improved by injecting soy protein brine to tenderize and reduce cooking losses.

Comminuted (Emulsified) Meat Products. Comminuted meat products containing soy protein ingredients have excellent texture, flavor, and appearance. Typically, about 1–4% of soy protein ingredients are used in comminuted meat products on a prehydrated basis depending on the ingredients used and the final products desired, resulting in substantial savings without reducing nutritional or eating quality (Kolar et al., 1985; Kinsella et al., 1985).

Soy protein isolates and neutralized soy protein concentrates are used in finely ground (chopped) meats (e.g., bologna and frankfurters) for their moisture and fat-binding, emulsifying, and emulsion-stabilizing properties. These properties make these types of soy protein ingredients ideal for use in processed meat products, both fine and coarse emulsions (i.e., loaves, patties, and sausages).

Coarsely Ground (Chopped) Meats. Textured soy proteins (TSP) are the ingredients of choice for coarsely ground (chopped) meat (e.g., chili, meatballs, meat patties, meat sauces, pizza toppings, Salisbury steaks) applications. Primary functions of textured soy proteins in these types of products are to give structure to the product, reduce cooking losses, and extend freshness and keeping quality. If properly used, the resulting products will be more moist, contain more protein and less fat, and maintain the quality longer. In general, use of up to about 20% hydrated textured soy protein ingredients produces acceptable texture, flavor, and overall quality in coarsely ground meat products.

Whole Muscle Meats. It is possible to incorporate soy protein isolate or functional soy protein concentrate solutions into whole muscles or large pieces of muscle tissues (e.g., fish, ham, poultry, roast beef). Usually, brine solutions of these protein ingredients are injected or massaged directly or tumbled into the muscle using conventional cured meat technology. This treatment typically increases the weight of products by 20–40% over untreated meats. The appearance of the resulting products is normal, with improved firmness, enhanced slicing characteristics, and less weepage under vacuum packaging (Kolar et al., 1985). Soy protein isolates can also be used as binders (adhesives or glues) in formed products. The resulting products have excellent appearance, good texture, and acceptable flavor and maintain juiciness after cooking.

Canned Meats. Various types of soy protein ingredients are used in retorted meat products to absorb juices and reduce fat separation during canning, resulting in firmer products (e.g., chili, hot snacks, meatballs, meat loaf mixes, meat pie fillings, minced hams, sloppy joes, soups, taco fillings, tamales). Relatively high levels of textured soy flours and textured and structured soy protein concentrates can also be used in retorted meat products (e.g., corned beef-type products, stews) with proper adjustment of fat content to maintain the quality attributes such as flavor.

Poultry Products. Processed products have become the fastest growing section of the poultry market. Soy protein concentrates and isolates are playing a key roll in making high-quality processed poultry products. They are increasingly used to bind meat cuts and trimmings in nuggets, patties, poultry rolls, and pressed loaves. Poultry breasts pumped with slurries of soy protein isolate, salt, and flavors are also becoming popular (Kolar et al., 1985).

Seafood Products. In the United States, the use of oilseed protein products in seafoods has been rather limited when compared with red meats but is growing steadily. However, soy protein ingredients have been used in a variety of traditional Japanese fish-based products (e.g., agekama, chikuwa, kamaboko) for centuries. Most of these products are based on surimi (a minced fish flesh), a portion of which, usually 1–3%, can be replaced with soy protein isolates without affecting the traditional quality. Also, textured soy protein ingredients can be used in seafood products at a level of up to about 8% on a prehydrated basis. Hydrated textured soy proteins can also be used in preparing fish blocks, fish cakes, and fish patties as binders, taking advantage of their water absorption and retention properties.

Analog Products. Complete meat analogs (e.g., bacon bits and crumbles, breakfast sausages, ham and ham crumbles, sliced beef, turkey) have been produced from soy protein ingredients and marketed for several years.

Institutional Feeding. In 1983, the U.S. Department of Agriculture issued permission to use all forms of soy protein ingredients, in either dry or partially hydrated forms, as partial replacement for meat, poultry, and seafoods, especially for school lunch programs and military uses.

Confectionery Products

A small but significant food use for specially processed soy protein ingredients is as aerating and whipping agents. Partially hydrolyzed soy proteins possess good foaming and foam-stabilizing properties, which make them suitable for use as aerating agents. They can be used alone or in combination with egg albumen or whole egg to improve whipping properties—whipping rate and whip stability. These modified soy protein ingredients are also used in coffee whiteners, vegetable creams, confections, and desserts. Confections containing modified soy protein ingredients are less sticky and show better handling properties.

Bakery Products

Soy protein ingredients are used in a variety of bakery products for various functional and nutritional reasons.

White Breads. It is common to replace about 3–5% of the wheat flour with soy flour to make white breads without altering the formula, except the need for more water. Soy flours are commonly used in bakery products, primarily for economic reasons, as replacers for more expensive milk and nonfat dry milk. The enzyme-active soy flours are used to bleach wheat flour and impart flavor to bread (Smith and Circle, 1978). Some of the benefits of using soy flours in bread include improved water absorption, better dough handling, tenderizing effects, firmer body, resiliency, improved crust color, and better keeping quality (maintain freshness longer).

Specialty Breads.   Specialty breads, with 13–14% protein contents, can be made by incorporating soy protein ingredients into bread formulas, usually with additional emulsifiers and vital wheat gluten, to improve the nutritional value. Soy protein–fortified wheat flours have been used throughout the world in mass feeding programs and school lunch programs for the past 18 years (Dubois and Hoover, 1981).

Cakes and Cake Mixes.   A variety of soy protein ingredients can be added to various mixes (e.g., bread, buns, pancakes, waffles) at a level of 2–15% to improve emulsification of fats and other ingredients, resulting in more uniform, smoother, more pliable, and less sticky doughs. The finished baked products have improved crust color and texture and maintain freshness longer. Lecithinated and/or refatted soy flours, at a level of 3–5%, are often used to make sponge and pound cakes, taking advantage of their excellent emulsification properties. Similar applications are found in cookies, crackers, biscuits, pancakes, and sweet pastries.

Doughnuts.   Better quality doughnuts can be made by incorporating 3–5% soy protein ingredients into the dough. Soy protein–containing doughnuts contain fewer calories and remain fresher longer due to less fat absorption during deep-fat frying and the natural antioxidant activity in the soy protein ingredients. The use of lightly lecithinated soy flours can also significantly reduce the amount of egg yolk used in doughnuts.

Pasta Products.   Pasta products can also be fortified with soy protein ingredients to improve the nutritional value. The fortified pasta products have been accepted by the U.S. military, government feeding programs, and the national school lunch program (Soy Protein Council, 1987).

Breakfast Cereals.   The use of soy protein ingredients in various breakfast cereals has increased considerably to improve the nutritive value and quality of the products.

Nonfood Applications

The current interest in soy protein ingredients for nonfood, industrial applications is based mainly on those properties that enable the use of soy proteins as ingredients to impart favorable, desirable, and acceptable changes in structure, texture, and composition to finished products at an attractive price. In many cases, soy protein ingredients replace more expensive ingredients such as casein without detracting from finished product quality. Among areas in which there has been increasing interest in the use of soy protein ingredients are pharmaceutical, textile sizing, and adhesive and bonding materials. Soy proteins usually need to be modified to be usable for specific applications (Rhee and Kim, 1992).

CONCLUSIONS

Undoubtedly, soy protein ingredients have made a significant impact in the food industry. Thus far, however, commercial success has fallen short of expectations due to many unfavorable factors—less than desirable organoleptic properties, variable product quality, and lack of functionality. To fully realize their potential as food and industrial ingredients, the organoleptic and functional properties must be fully understood and standardized soy protein ingredients with uniform, reliable and specific functional properties must be produced for specific applications.

REFERENCES

Berry, K. E. 1988. Preparation of soy protein concentrate products and their application in food systems. In *Proceedings of the World Congress on Vegetable Protein Utilization in Human Foods and Animal Feedstuffs*, T. H. Applewhite (Ed.), pp. 62–65. American Oil Chemists Society, Champaign, IL.

Bradley, R. A., Atkinson, D., Hauser, H., Oldani, D., Green, J. P., and Stubbs, J. M. 1975. The structure, physical and chemical properties of soybean protein glycinin. *Biochem. Biophys. Acta* 412: 214.

Catsimpoolas, N. 1969a. A note on the proposal of an immunochemical system of reference and nomenclature for the major soybean globulins. *Cereal Chem.* 46: 369.

Catsimpoolas, N. 1969b. Isolation of glycinin subunits by isoelectric focusing in urea-mercaptoethanol. *FEBS Lett.* 4: 259.

Catsimpoolas, N. 1970. A note on dissimilar subunits present in dissociated glycinin. *Cereal Chem.* 47: 70.

Catsimpoolas, N., Campbell, T. G., and Keyer, E. W. 1969. Association-dissociation phenomena in glycinin. *Arch. Biochem. Biophys.* 131: 577.

Catsimpoolas, N., Kenny, J. A., Meyer, E. W., and Szuhaj, B. F. 1971.

Molecular weight and amino acid composition of glycinin subunits. *J. Sci. Food Agric.* 22: 448.

Catsimpoolas, N., Rogers, D. A., Circle, S. J., and Meyer, E. W. 1967. Purification and structural studies of the 11S component of soybean proteins. *Cereal Chem.* 44: 631.

Draper, M and Catsimpoolas, N. 1977. Isolation of the acidic and basic subunits of glycinin. *Phytochemistry* 16: 25.

Dubois, D. K. and Hoover, W. J. 1981. Soya protein products in cereal grain foods. *J. Am. Oil Chem. Soc.* 58: 343–346.

Eldridge, A. C. and Wolf, W. J. 1967. Purification of the 11S component of soybean protein. *Cereal Chem.* 44: 645.

Fulmer, R. W. 1988. The preparation and properties of defatted soy flours and their uses. In *Proceedings of the World Congress on Vegetable Protein Utilization in Human Foods and Animal Feedstuffs.* T. H. Applewhite (Ed.), pp. 55–61. American Oil Chemists Society, Champaign, IL.

Heiser, J. and Trentleman, T. 1988. Full-fat soya products—manufacturing and uses in foodstuffs. In *Proceedings of the World Congress on Vegetable Protein Utilization in Human Foods and Animal Feedstuffs*, T. H. Applewhite (Ed.), pp. 52–54. American Oil Chemists Society, Champaign, IL.

Johnson, D. W. and Kikuchi, S. 1988. Processing for producing soy protein isolates. In *Proceedings of the World Congress on Vegetable Protein Utilization in Human Foods and Animal Feedstuffs*, T. H. Applewhite (Ed.), pp. 66–77. American Oil Chemists Society, Champaign, IL.

Kinsella, J. E. 1979. Functional properties of soy proteins. *J. Am. Oil Chem. Soc.* 56: 242–258.

Kinsella, J. E., Damodaran, S., and German, B. 1985. Physicochemical and functional properties of oilseed proteins with emphasis on soy proteins. In *New Protein Foods*, A. A. Altshul and H. L. Wilcke (Ed.). Academic Press, Yew York.

Kitamura, K. and Shibasaki, K. 1975. Isolation and some physicochemical properties of the acidic subunits of soybean 11S globulin. *Agric. Biol. Chem.* 39: 945.

Kitamura, K., Takagi, T., and Shibaski, K. 1976. Subunit structure of soybean 11S globulin. *Agric. Biol. Chem.* 41: 351.

Kolar, C. W., Richert, S. H., Decker, C. D., Steinke, F. H., and Vander Zaden, R. J. 1985. Isolated soy protein. In *New Protein Foods*, A. A. Altshul and H. L. Wilcke (Ed.), pp. 259-299. Academic Press, New York.

Koshiyama, I. 1972. Purification and physico-chemical properties of 11S globulin in soybean seeds. *Int. J. Peptide Protein Res.* 4: 167.

Mitsuda, H., Kusano, T., and Hasegawa, K. 1965. Purification of the 11S component of soybean proteins. *Agric. Biol. Chem.* 29: 7.

Mori, T., Utsumi, S., and Inada, H. 1979. Interaction involving disulfide bridges between subunits of soybean seed globulin and between subunits of soybean and sesame globulin. *Agric. Biol. Chem.* 43: 2, 317.

Morr, C. V. 1990. Functionality of oilseed and legume protein. *J. Am. Oil Chem. Soc.* 67: 265–271.

Naismith, W. E. F. 1955. Ultracentrifuge studies on soya bean protein. *Biochem. Biophys. Acta* 16: 203.

Ochiai-Yanagi, S., Takagi, T., Kitamura, K. Tajima, M., and Watanabe, T. 1977. Reevaluation of the subunit molecular weights of soybean 11S globulin. *Agric. Biol. Chem.* 41: 647.

Peng, I. C., Quass, D. W., Dayton, W. R., and Allen, C. E. 1984. The physicochemical and functional properties of soybean 11S globulin—a review. *Cereal Chem.* 61: 480–490.

Rhee, K. C. and Kim K. H. 1992. Prospects for industrial uses of plant proteins. *INFORM* 3(9): 1044–1054.

Shavarts, V. S. and Vaintraub, I. A. 1967. Isolation of the 11S component of soya bean protein and determination of its amino acid composition by an automatic chromatopolarographic method. *Biochemistry* (USSR) 32: 135.

Smith, A. K. and Circle, S. J. 1978. *Soybeans: Chemistry and Technology.* AVI Publishing Co., Westport, CT.

Snyder H. E. and Kwon, T. W. 1987. Functional properties of soy proteins. In *Soybean Utilization,* pp. 163–186. Van Nostrand Reinhold Co., New York.

Soy Protein Council. 1987. *Soy Protein Products—Characteristics, Nutritional Aspects and Utilization.* Soy Protein Council, Washington, DC.

Staswick, P. E., Hermodson, M. A., and Nielsen, N. C. 1981. Identification of the acidic and basic subunit complexes of glycinin. *J. Biol. Chem.* 256: 8, 752.

Thanh, V. H. and Shibasaki, K. 1976. Major proteins of soybean seeds. A straightforward fractionation and their characterization. *J. Agric. Food Chem.* 24: 1, 117.

Wolf, W. J. 1976. Chemistry and technology of soybeans. In *Advances in Cereal Science and Technology,* Vol. 1, 1st ed., Y. Pomeranz (Ed.), p. 325. American Association of Cereal Chemists, Inc., St. Paul, MN.

Wolf, W. J., Babcock, G. E., and Smith, A. K. 1961. Ultracentrifugal differences in soybean protein composition. *Nature* 191: 1, 395.

Wolf, W. J. and Briggs, D. R. 1956. Ultracentrifugal investigation of the effect of neutral salts on the extraction of soybean proteins. *Arch. Biochem. Biophys.* 63: 40.

Wolf, W. J. and Cowan, J. C. 1975. *Soybeans as a Food Source.* rev. ed. CRC Press, Cleveland, OH.

Wolf, W. J. and Sly, D. A. 1967. Cryoprecipitation of 11S component of soybean protein. *Cereal Chem.* 44: 653.

Yamaguchi, F., Ono, H., Kamata, Y., and Shibasaki, K. 1979. Acetylation of amino groups and its effect on the structure of soybean glycinin. *Agric. Biol. Chem.* 43: 1,309.

11
Whey Protein Functionality

Arun Kilara

The Pennsylvania State University
University Park, Pennsylvania

TYPES OF PRODUCTS AND DEFINITIONS

Whey is the by product of cheese and casein manufacture. Therefore, the primary emphasis is on production of cheese and casein of consistent quality attributes and attributes of whey resulting from these processes vary considerably. The only certainty with whey protein products is that their functionality will vary from batch to batch and manufacturer to manufacturer. Therefore, an understanding of the types of products and their manufacture may help in rationalizing the observed variabilities in functionalities of whey products. Typical compositions of wheys are given (Table 11.1).

Dry Whey

Whey obtained is concentrated and spray dried. Sometimes fat may be centrifuged off prior to concentration. Processes such as reverse os-

TABLE 11.1   Composition (%) of Different Types of Whey from Cow's Milk

	Rennet	Lactic	Mixed	Sweet	Acid
Dry matter	7.08	6.58	7.05	7.00	6.50
Lipids	0.51	0.09	0.34	0.20	0.04
Lactose	5.18	4.53	5.05	4.90	4.40
Total nitrogen	0.15	0.12	0.15	0.13	0.11
Lactic and citric acids	0.16	0.78	0.32	0.20	0.05
Ash	0.53	0.07	0.47	0.50	0.80

Source: Fevrier and Bourdin, 1977; Morr, 1984.

mosis may also be used to partially concentrate the whey but are not economical for use without vacuum evaporation.

When whey is dried the moisture content of sweet whey powder approaches 4.6%, while that of acid whey is 3.9% (Morr, 1984). Because of their low nitrogen contents, whey powders are not regarded as rich sources of functional proteins.

"Partial" Whey Products

Whey from which lactose is removed by ultrafiltration is often called "partially" delactosed whey powder. Such products are lower in lactose by about 33% when compared to whey powder. The ash content is still high. Therefore, a process called electrodialysis can be employed to reduce the ash content and the subsequent product is called "partially delactosed, partially demineralized" whey powder. Alternately, whey may only be demineralized and not delactosed, and in this case the product is called "partially demineralized whey powder." Composition of these products was given by Batchelder (1986) (Table 11.2). These "partial" products find uses in infant formulas and dry mixes, but their protein contents were too low, rendering them functionally ineffective.

Whey Protein Concentrates

Products containing more than 35% protein on a dry basis are called whey protein concentrates. A number of different techniques have been developed to concentrate the proteins. Morr (1986) classified these techniques as laboratory and commercial processes. Laboratory processes include those based on differential solubility (e.g., polyphosphate/carboxymethyl cellulose complexes), polyethylene glycol precipitation, pH-temperature precipitation, demineralization, and chromatographic techniques (ion exchange and molecular size exclusion). Laboratory processes were either cost prohibitive for commercial scale-up or produced nonfunctional products. Compositions of some of the laboratory-processed samples are given in Table 11.3.

TABLE 11.2 Composition (%) of "Partial" Whey Products Compared to Whole Whey Powder

Constituent	Whey powder	50% Reduced mineral whey powder	Delactosed whey powder	55% Delactosed, demineralized whey powder
Protein	12.5	13.3	28.0	30.2
Lactose	74.0	77.2	48.0	54.0
Fat	1.0	1.0	1.5	1.8
Ash	8.0	4.0	16.0	7.0
Moisture	4.5	4.5	2.5	2.5
Organic milk salts	0.8	0.4	6.0	4.5
Potassium 1.9	0.5	4.5	1.2	—
Phosphate	0.75	0.60	4.2	2.6
Chloride	0.15	0.12	3.2	0.6
Calcium	1.6	0.2	1.1	0.8
Magnesium	2.0	1.6	0.3	0.2
Sodium	0.55	0.40	1.6	0.7

Source: Batchelder, 1986.

TABLE 11.3 Composition (%) of Laboratory Process Whey Protein Concentrates

Process	Protein	Lactose	Ash	Fat
Metaphosphate complex	55.7	13.0	13.7	5.3
CMC complex	66.0	20.1	8.0	1.2
Iron complex	99.8	0.8	54.0	0.6
Sephadex	41.9	24.9	11.5	0.8

Source: Morr et al., 1973.

TABLE 11.4   Composition of Whey Protein Concentrate Powders

	Whey protein concentrates protein (%)			
Constituent	35	50	65	80
Moisture	4.6	4.3	4.2	4.0
Crude protein	36.2	52.1	63.0	81.0
True protein	29.7	40.9	59.4	75.0
Lactose	46.5	30.9	21.1	3.5
Fat	2.1	3.7	5.6	7.2
Ash	7.8	6.4	3.9	3.1

Source: Glover, 1985.

Commercial products with 35, 50, or 80% protein on a dry basis are routinely available. Commercial processes rely on ultrafiltration and diafiltration. Composition of such products is provided in Table 11.4. Mangino (1992) observed that as the protein content of the products increases so do manufacturing costs, fat content, protein:ash ratios, and protein:lactose ratios. Morr (1986) reported that concentration of whey by ultrafiltration to 4% protein in the retentate followed by evaporation and spray drying results in a 35% whey protein concentrate. If the retentate containing 4% protein is diafiltered to 16% protein and then evaporated and spray-dried, a 50–75% protein whey protein concentrate results.

Whey Protein Isolates

The protein content of whey protein isolates is 90% or higher. Ion exchange processes are utilized for such fractionation. The Vistec process uses a cellulose-based cation exchanger, and lactose and nonprotein nitrogen components are not adsorbed on the ion exchanger. Adsorbed proteins are released by pH adjustment to >5.5. Eluted proteins are concentrated by ultrafiltration, evaporated, and spray-dried (Marshall, 1982; Palmer, 1981). Other ion exchange processes involving Spherosil-S and Spherosil-QMA (Rhône-Poulenc) were reported to result in whey protein concentrates with high ash contents and poor solubilities (Nichols and Morr, 1985; Barker and Morr, 1986).

Specialized Fractions

Presently, attempts have been successful at producing fractions enriched in α-lactalbumin. Such fractions are useful in infant formulas.

Heating whey to >55°C at pH 4–4.5 leads to aggregation of α-lactalbumin, while β-lactoglobulin is unaffected (Pearce, 1987). The denaturation of α-lactalbumin is reversible. Morr et al. (1968) described a process in which cheese whey is concentrated to remove lactose and ash. Pearce (1992) suggests using ultrafiltration for this purpose. The concentrate is adjusted to pH 4.1–4.3 and is heated to 63–65°C for 3 min to precipitate proteins. Precipitated proteins are collected by centrifugation at 4000 × g, and this fraction is lactalbumin. The supernatant fraction is enriched in β-lactoglobulin. Amundsen et al. (1982) proposed a method in which β-lactoglobulin was precipitated and α-lactalbumin remained soluble. In the process delineated by these Wisconsin researchers, whey is concentrated by ultrafiltration, pH adjusted, demineralized, and centrifuged to obtain the β-lactoglobulin–enriched fraction. While the β-lactoglobulin fraction so obtained was pure, the α-lactalbumin fraction was not. Modler and Jones (1987) compared processes developed by Pearce (1987), Fauquant et al. (1985), and Maubois et al. (1987) and pointed out a number of similarities. The end product compositions of the products were also similar.

FACTORS AFFECTING COMPOSITION

From the previous section it is evident that the composition of products depends upon the methods employed to reduce lactose and ash contents. Morr and Foegeding (1990) conducted a study in which compositions of eight commercial whey protein concentrates and three whey protein isolates were reported (Table 11.5). These samples originated in Denmark, the United States, the United Kingdom, the former West Germany, New Zealand, and Ireland. Upon further analysis the types of minerals in the ash were found to vary, as did the proportions of individual proteins in the samples tested. These data are in agreement with previous studies by deWit et al. (1986, 1988). There can be several reasons for the observed differences in compositions.

Seasonal Changes

In contrast to the United States, milk production is highly seasonal in some countries. Calving in Australia, New Zealand, and Ireland occurs in the spring (Zadow, 1986). Seasonal changes in milk composition translate to differences in cheeses made. Hence, it can be anticipated that composition of whey can be similarly altered (Regester and Smithers, 1991). Whey protein concentrates manufactured in the last trimester of lactation gave consistently lower α-lactalbumin contents. The β-lacto-

TABLE 11.5 Compositions (%) of Whey Protein Isolate and Whey Protein Concentrate Samples from the United States, Europe, and New Zealand

Sample code	Protein	Moisture	Ash	Lactose	Total lipids	Phospho-lipids
1	88.9	5.57	1.37	ND	0.64	0.22
2	92.7	2.40	2.15	0.42	0.39	0.11
3	91.6	3.28	1.94	0.46	0.67	0.31
4	76.6	4.14	6.04	3.14	6.07	1.34
5	73.8	4.70	2.56	2.13	7.38	1.54
6	72.5	5.59	5.35	5.22	3.88	0.80
7	73.0	5.22	2.52	2.56	5.97	1.33
8	72.0	5.76	4.99	5.75	4.13	1.08
9	72.5	5.92	2.97	3.78	3.30	1.18
10	73.5	6.01	4.70	4.84	4.64	1.49

Source: Morr and Foegeding, 1990.

globulin contents were, however, increased during the same period. Casein content of whey protein concentrates during the first quarter of lactation was higher. There were no observed differences in the glycomacropeptide contents during the same period.

Roeper (1971) observed that protein contents of acid and rennet wheys increased from midlactation to the end of lactation. Koning et al. (1974) were the first to observe that variations in protein contents of whey were a direct result of the increases in β-lactoglobulin. Matthews (1978) studied compositions of sulfuric and lactic acid whey protein concentrates and observed that the protein content increased as lactation progressed.

Lipid Composition

The presence of residual lipids in whey protein products has been adequately and consistently demonstrated (Tables 11.1–11.4). It is extremely difficult to completely remove all lipid from whey. Yet, minimizing lipid content greatly improves process efficiencies in the manufacture of whey protein concentrates. Removal of lipids from whey increases flux rates through ultrafiltration membranes (deWit and de Boer, 1975). Centrifugation is a common unit operation for lipid removal. In whey, lipids exist as lipoprotein complexes, and the density of these complexes is not disparate enough from the aqueous phase as to render them easily separable by centrifugation (Houlihan and Hirst, 1987). As the protein content of the whey increases, so does the lipid content (Glover, 1985). Houlihan and Hirst (1987) also report that the composition of lipids associated with proteins in whey protein concentrates differs considerably from bulk lipid in milk. Not only were whey protein concentrates enriched in phospholipids and milk fat globule membrane materials significant amounts of mono- and di-glycerides and free fatty acids were also observed (Vaghela and Kilara, unpublished data). Prior work by Joseph and Mangino (1988a,b) has shown the presence of milk fat globule membrane material to be associated with the phospholipid fraction in whey protein concentrate. deWit et al. (1986) reported the composition of lipids in whey protein concentrates (Table 11.6).

Rinn et al. (1990) evaluated nine physical pretreatments to whey, which are shown in Table 11.7. The treatments were accorded to whey after calcium chloride addition (31.6 mg/L), pH adjustment to 7.3 followed by rapid heating to 55°C and holding at this temperature for 8 min. The most effective pretreatment for removal of phospholipids involved 0.6 μm microfiltration without clarification or a single pass clarification. However, these authors observe that clarification prior to microfiltration was necessary to improve the throughput of the microfiltration system.

Heat treatment of milk used for cheese or casein manufacture has been speculated to alter functionality of whey protein concentrates. Also, pasteurization of whey prior to membrane processing may damage the proteins in whey. Heating may also be used to pasteurize ultrafiltered

TABLE 11.6   Lipids in Commercial Whey Protein Concentrates

	Milk fat content (%)	Triglyc- erides	Diglyc- erides	Free fatty acids	Phospho- lipids
UF WPC	5.0	51	6	15	28
UF diafiltered WPC	6.7	48	15	10	27
Acid UF WPC	6.4	53	13	7	27
Defatted WPC	0.5	62	20	0	18
Spherosil QMA WPC	1.6	45	9	3.4	12
Vistec WPC	1.7	37	10	38	15
Demineralized de- lactosed WPC	3.3	43	12	20	25

Source: deWit et al., 1986.

Table 11.7   Nine Physical Pretreatment Conditions Accorded to Whey

Gravity settling	Clarification	Microfiltration
−	−	
−	Single pass	1.0 μm
−	Single pass	—
+	—	—
−	Single pass	0.6 μm
−	20-min recycle	0.6 μm
−	30-min recycle	0.6 μm
−	Single pass	0.6 μm

Source: Rinn et al., 1990.

retentates. Morr (1987) reported that pasteurization of retentates resulted in significant protein denaturation, as evidenced by loss in solubility of whey protein concentrates. Mangino et al. (1987) observed changes in functionality due to pasteurization treatments given to whey. The specific effects were dependent on the particular functional property in question. In general, Mangino et al. (1988) observed that heating retentates to 72°C decreased hydrophobicity and adversely affected functionality but that heating to 64°C did not.

FUNCTIONALITY OF WHEY PRODUCTS

Functional properties of whey proteins encompass those physicochemical attributes of a protein that make it useful in food products. Discussion in the above two sections points out that many factors influence the composition of protein concentrates. Further, a number of different types of whey products are available in the marketplace. Morr (1979) observes that for any food protein ingredient to be useful, it must be free from toxic and antinutritional factors, free of off-flavors and off-colors, compatible with other ingredients and processes, readily available at an affordable price, and serve a function in the product. Functional properties of proteins can be studied in model systems, model food systems, and in "real" foods. The complexity of evaluating functionality increases from model systems to "real" foods. Further, functionality testing in model systems has not been standardized. Lack of standardization presents challenges in comparing results between laboratories and sometimes within a laboratory. Many empirical approaches exist for functionality testing. In this section, general concepts of terms used in functionality testing will be discussed, followed by effects of processing pretreatments on functionality and use of whey proteins in foods, and finally prediction of functionality based on physicochemical measurements will be presented.

Common Functional Attributes Investigated

Solubility. Macromolecules are not truly soluble in the same manner as low molecular weight solutes. However, amino acids in protein chains interact with water, and proteins can be suspended in water. This property is often used as an indicator of whey protein denaturation. Protein solubility is often affected by temperature, pH, and the presence of other solutes and salts, and the values for solubility obtained are particularly influenced by the procedures used to solubilize the protein sample. Proteins are least soluble in the pH range close to their isoelectric point, but whey proteins are soluble at these pH values. The wide range of pH values over which whey proteins are soluble make them ideal for use in a variety of products. The protein composition of sweet and acid whey was presented by Pearce (1990) (Table 11.8).

The two main proteins in whey are β-lactoglobulin and α-lactalbumin. Properties of these proteins are therefore presumed to be important in the overall functionality of whey proteins. Increases in the temperature of solubilization can be detrimental to the native conformations of proteins. Therefore, increasing temperatures result in decreased solubility.

Mangino et al. (1987) observed that pasteurization of ultrafiltered retentates decreased solubilities. They also observed that β-lactoglobulin content of whey protein concentrates decreased as the retentates were pasteurized. Morr and Foegeding (1990) analyzed solubilities of several commercial whey protein concentrates and observed that solubilities at

TABLE 11.8 Individual Proteins in Sweet and Acid Whey

Protein	Acid whey	Sweet whey
β-Lactoglobulin	54	45
α-Lactalbumin	23	18
Casein-derived peptides	2	20
Serum albumin	6	5
Immunoglobulins	6	5
Phospholipid-protein complexes	5	5
Enzymes	2	2

Source: Pearce, 1990.

pH 3, 4.5, and 7 were good and that whey protein isolates were more soluble at any given pH than whey protein concentrates. These results indicate that manufacturers of whey protein products are cognizant of the need to minimize thermal exposure of proteins during processing.

Delaney (1976) reported that solubility of whey protein concentrate was affected by ionic strength in some cases. Protein solubility for meta-phosphate-complexed whey protein concentrate was highly pH dependent, but for all other samples tested it was independent of pH. Commercially electrodialyzed whey protein concentrate showed lower solubility, presumably due to higher heat treatments received by these samples in the manufacturing process. Ionic strength had little effect on protein solubility at pH 8.4 but markedly decreased solubility at pH 3. Under acidic pH conditions solubility decreased as ionic strength increased. Morr et al. (1973) reported the effects of pH on solubility at various pH values (Table 11.9). Solubility of protein concentrates prepared from carboxymethylcellulose complex and metaphosphate complex are highly pH dependent, but other samples were not affected. Solubility is an important attribute for other functional properties. Solubilization of the protein has to occur prior to rearrangements at air/water and oil/water interfaces. Likewise, gelation depends on protein-protein interactions in an aqueous environment.

Water-Protein Interactions. The structure and conformation of proteins is dependent on water. Hydrogen bonding between amino acid residues and water, ion dipole and dipole-dipole interactions are all important in protein-water interactions. Apart from these molecular interactions between protein and water, physicochemical forces (such

TABLE 11.9 Solubility of Whey Protein Concentrates as Affected by pH and Method of Manufacture

Process	Solubility (%) at pH			
	2.0	4.0	6.0	8.0
Metaphosphate complex	31.1	24.0	73.1	85.9
Electrodialysis	90.4	92.2	91.1	93.6
Ultrafiltration	80.0	77.9	77.3	76.7
Sephadex	88.0	89.5	90.2	91.2
Dialysis	79.4	91.2	—	81.2
Carboxymethylcellulose complex	21.2	57.7	80.9	86.2

Source: Morr et al., 1973.

as adsorption) may also cause water-protein interactions. Water can be contained in capillaries or physically entrapped in particles of proteins. As protein insolubility increases, these interactions become dominant. If proteins are soluble they do not bind water as readily. Whey proteins are soluble and do not bind water unless the proteins are denatured. Hence, lactalbumin (heat-denatured whey protein) absorbs more water than undenatured whey protein. In many applications, denaturation of proteins is achieved by heat, pH, and ions used singly or in combination. Morr (1989) provided values for protein hydration by different methods (Table 11.10).

Viscosity is another property that results from water-protein interactions. deWit (1989) has extensively discussed theoretical aspects of viscosity of proteins in solution. Rennet whey at pH 4.5, 25°C, has very low viscosity and electroviscous effects caused by pH changes. deWit (1989) states that above 65°C relative viscosity of whey increases and an even greater increase can be observed if whey is heated to >85°C. Between 65 and 85°C, whey protein denaturation is thought to occur, and at temperatures >85°C protein aggregation causes further viscosity increases.

Foaming.   Behavior of proteins at the air/water interface are of particular importance in many foods. Proteins have to diffuse rapidly to the interface and then undergo molecular rearrangements. Theoretical discussion of the behavior of proteins at air/water interfaces are covered elsewhere in this book (see Chapter 6). Devilbiss et al. (1975) reported that heating whey protein resulted in better foaming. This suggests that

TABLE **11.10**   Hydration Values for Some Milk Proteins

Method	Protein	Water of hydration (g/g)
Bet monolayer		
	Casein	0.050
	β-Lactoglobulin	0.067
Equilibrium moisture at $a_{w\ 0.92}$		
	Casein	0.400
	β-Lactoglobulin	0.320
Nuclear magnetic resonance		
	Casein	0.390
	β-Lactoglobulin	Not determined

Source: Morr, 1989.

partial denaturation of the protein facilitates molecular rearrangements conducive to rigid, high-viscosity surface film formation. This confirmed work of Richert et al. (1974), who had reported that heating whey protein concentrate to 55–60°C led to an improvement in foaming properties of whey proteins. It has also been observed that cooling whey protein solutions to <4°C reduces foaming (Richert et al., 1974; Haggett, 1976). If the solutions were heated to <55°C, foaming properties improved. Haggettt (1976) speculated that β-lactoglobulin may be involved in this temperature-dependent phenomenon. It has been observed by Liao and Mangino (1987) that in whey protein concentrates from acid whey, hydrophobicity and sulfhydryl content were predictive of the foaming properties. These findings paralleled results with commercial whey protein concentrates reported by Peltonen-Shalaby and Mangino (1986). Native conformation of β-lactoglobulin strongly correlated with foaming performance (Kim et al., 1981). Pasteurization of retentates drastically reduces foaming properties of whey protein concentrates (Mangino et al., 1987). In a study of 11 commercial whey protein isolates and concentrates, Morr and Foegeding (1990) report variable foaming properties. Foaming experiments were conducted after pretreatment at two temperatures (25 and 55°C) and at three pH values (4.5, 7.0, and 9.5). Results for pH 7.0 experiments are shown in Table 11.11. Regardless of pretreatment, some samples did not form stable foams even though the foam expansion was good.

Regester et al. (1992) report that whey from late lactation milk resulted in increased foam capacity and decreased foam stability. The slight improvement in foam capacity was attributed to (1) an increase in β-lactoglobulin in late lactation milk and (2) reductions in ash, phosphorus, and lactose contents of whey protein concentrates made with late-lactation milk. In general, studies with dilute solutions of whey protein concentrates indicate that they do not form stable foams in model systems. Conditions applicable to model systems differ considerably from model food systems and real formulations.

Emulsification.   Behavior of proteins at the oil/water interface are of interest in foods. Emulsions can be liquids, semi-solids, or solids, and standardized methods to study emulsion properties do not exist. Energy is applied to disperse one phase into another of two normally immiscible phases.

If the dispersed phase is oil and the continuous phase is water, then an oil-in-water emulsion results. On the other hand, if the continuous phase is oil and the dispersed phase is aqueous, a water-in-oil emulsion is obtained. In addition to work being performed, an energy barrier is

TABLE 11.11 Foam Expansion (Overrun %) and Foam Stability (% Drainage) for 11 Commercial Whey Protein Concentrate Samples at pH 7.0

	Pretreatment, 25°C		Pretreatment, 55°C	
Sample code	Foam expansion (%)	Foam stability (% drainage)	Foam expansion (%)	Foam stability (% drainage)
1	873	44	862	52
2	1556	24	896	39
3	804	56	1066	45
4	435	100	814	55
5	476	100	728	100
6	388	100	541	79
7	580	100	686	100
8	353	100	624	74
9	448	73	858	49
10	690	69	782	82
11	676	74	685	98

Samples were preheated for 30 min.
Source: Morr and Foegeding, 1990.

necessary to prevent the dispersed phase from coalescing. If the density of the two phases are different, separation of the phases occurs sooner. Emulsifiers or amphiphilic compounds can delay or prevent coalescence. Proteins are amphiphiles but are not as effective as low molecular weight emulsifiers in preventing coalescence. Proteins do retard creaming more effectively than low molecular weight amphiphiles. Details on the theory of emulsification are provided elsewhere in this book (see Chapter 5).

Use of whey proteins as emulsifiers is limited. Pearce and Kinsella (1978) observed that as whey protein concentration increased from 0.5 to 5% and dispersed phase volume was kept constant at 25%, droplet size of the emulsion decreased. After more than 15 passes in a piston homogenizer, droplet size decrease was dramatic. This could also be due to the slower rate of adsorption of protein at the interfaces as shown by Tornberg and Hermannson (1977). Factors affecting whey protein emulsions include pH and ionic strength. Around their isoelectric point, whey proteins form poor, unstable emulsions (de Wit, 1989). Mangino et al. (1987) observed no adverse effects of pasteurization of milk or whey on emulsification, but pasteurization of the retentate greatly reduced the emulsion capacity. Denaturation of proteins caused by the heat treatment was speculated to result in the observed loss of emulsion properties. Further studies by Liao and Mangino (1987) have led these workers to conclude that emulsion capacity of whey proteins does not vary. Even though emulsion capacity and stability are important attributes in many food products, scant information is available on reliable methods to assess this important functionality.

Gelation.   Gelation is the result of a balance between polymer molecules interacting with other polymer molecules and the interaction between polymer molecules and the solvent in such a manner that a network results. This network is capable of holding large amounts of water and other materials. Gelation differs from coagulation. Coagulation is primarily a chain-chain interaction, and coagula are not capable of holding water as are gels. Protein gelation (see Chapters 7 and 13) is a two-step process. The first step involves denaturation and unfolding, while the second step involves rearrangement and aggregation. Good gels form if the first step is faster than the second step.

Varunsatian et al. (1983) studied the effects of calcium, magnesium, and sodium ions on the thermal aggregation of whey protein concentrates. At pH > 8 chloride salts of the cations increased the rate of aggregation, and calcium was the most effective cation. In the presence of divalent cations, denaturation temperatures were considerably lower but increased slightly in the presence of sodium chloride. β-Lactoglobulin was sensitive to heat in the presence of calcium ions. This study did not report on gelation but can provide some insights into gelation.

Later Rector et al. (1991) reported that dry heat treatment of dialyzed whey protein concentrate decreased the hardness of gels. Polymerization of β-lactoglobulin by non−sulfhydryl-mediated reactions was implicated in the loss of gel strength. The time and temperature conditions employed in this study were unrealistic. Whey protein concentrates were stored at 80°C for 7 days, which does not mimic any storage or processing condition in food processing.

Kohnhorst and Mangino (1985) report that protein hydrophobicity (measured as alkane binding) and calcium content are important predictors of gel strength in whey protein gels. Mangino et al. (1987) also demonstrated that pasteurizing milk used in cheese making affected the ability of whey protein to gel at pH 6.5 but not at pH 8. Heating ultrafiltration retentates significantly reduced gel strength. Pasteurization of whey did not significantly affect gel strength. Rector et al. (1989) report that whey protein isolates heated at 90°C for 15 min at pH 6.5−8.5 and protein concentrations 9−10.5% form reversible gels. The melting temperatures of gels at pH 8 ranged from 24.5 to 57.8°C. The maximum enthalpy of formation was -858 calories per mole of cross-links, and a maximum storage modulus of 240 dynes/cm^2 was obtained following holding at 8°C for 7 hr.

Rinn et al. (1990) report that whey protein concentrates prepared by microfiltration through 0.6-μm pores exhibited "superior" gels at 4 and 5% protein concentration. In comparison, conventionally produced whey protein concentrates were needed at 9% concentrations to form nonpourable gels. Morr and Foegeding (1990) reported gelling properties of 11 commercial whey protein concentrate and isolate samples. Stable nonpourable gels were obtained from all samples at pH 4.5. Formation of gels at pH 6 and 7.5 necessitated higher protein concentrations in most instances, and one whey protein isolate was observed not to gel even at 14% concentrations in distilled water. However, the addition of sodium chloride reduced the concentration needed to form stable gels below 10% protein for this sample. In the same study rheological properties of 10% protein gels were studied when gels were formed in 0.1 M sodium chloride. Generally, shear stress was observed to increase with increases in pH. Shear strain values varied considerably, and no consistent pH effect was observed.

Regester et al. (1992) studied seasonal variation in gelation of whey proteins. They report that gel firmness increased as lactation progressed. Whey protein concentrates prepared during the last month of lactation gave firmest gels, while whey proteins prepared during the second month of lactation resulted in the weakest gels.

Studies with model systems provide insights into the physicochemical behavior of proteins but may not accurately reflect its performance in a food. This is because foods have other components and additives and processing procedures may induce complex interactions between proteins and these additives.

Model Food Systems to Evaluate Whey Protein Functionality

A model food system is one in which the food formulation and processing accorded to the protein approximates industrial processes. A model food system is formulated and processed to emphasize differences in whey protein functionality. Results from model food systems, although not directly transposable to foods, are considered more realistic than model systems. Thus, model food systems provide an intermediate step between model systems and real foods. Often food formulations and processing conditions vary from processor to processor. Model food systems permit evaluation of functionality without the added variables of differences between manufacturers conditions. The inherent variability of whey protein products has been discussed above. In model food systems more than one functional property may be involved. For example, a whipped topping involves emulsification, foaming, and water absorption, while a pudding involves emulsification, water binding, gelation, and elasticity properties.

Harper (1984) outlined the process for developing model food systems in which functionality of whey proteins can be investigated. The process begins by evaluating a number of formulations for the food of interest. It is suggested that as many formulations as possible be reviewed in order to choose a representative formula containing ingredients commonly found in most recipes. In order to simplify the process it is advisable to use only one additive of its kind. For example, if a combination of hydrocolloids are used, try to choose the one that is deemed more functional. After choosing the formulation, evaluate how this formulation can be processed on a laboratory bench scale. Using a valid statistical procedure, determine variables that affect the manufacture of this formulation. Select quantitative, objective measures that indicate quality of the finished product. The next step is more tedious and requires statistically designed experimentation to ascertain ingredient responses to food characteristics being evaluated. These steps will ultimately lead to a formulation that differentiates whey products that vary in their functionality as tested by model food systems. The aim of the above exercise is to develop a model food system that is sensitive enough to pick up differences between proteins with respect to combined functional attributes.

Harper (1984) also provided some notes of caution. Seemingly "insignificant" factors can cause variations in observations. For example, 10% whey protein concentrates from two different sources were whipped in plastic, glass, and stainless steel bowls. One sample whipped well in all three bowls, whereas the other did not whip well in plastic but did so in stainless steel. Although seemingly trivial, bowl type was also shown to be of importance in both the magnitude and rate of whipping. Because of the need to verify all details of the procedure, the development of an appropriate "model" system can take 3–9 months. Harper (1984) has suggested several model food systems to evaluate functionalities of whey protein products. In each instance formulation, processing procedure and evaluation were provided. These are discussed below. The appropriate references to consult for these food systems are Harper (1984) and Harper et al. (1980).

"No-Time" Bread Dough. The key functionality involves water absorption, adhesion, and elasticity. The formula suggested is 130 g flour,

60% flour weight as water, 2.7% yeast, 2% salt, 2% beef tallow, 100 ppm ascorbic acid, 30 ppm potassium bromate, 100 ppm malt flour, and protein source to be evaluated to obtain 1% protein in the final product. High-heat nonfat dry milk is used as the reference protein additive. Ingredients that are dry are blended and mixed together for 30 sec. Then water is added and mixed in a dough mixer and the dough is developed for 3–5 min. Finished dough temperature should be 28–30°C. The dough is divided into pieces within 5–10 min and further allowed to rest for 8–10 min. The dough pieces are placed in pans and proofed to a suitable height, a procedure that takes 45–60 min. The proofed loaves are baked at 225°C for an appropriate period. Evaluation of loaves involves cooling prior to measuring loaf volume, evaluating crumb and crust characteristics. After 24 hr the loaf is also cut and scored for color, flavor, and texture.

Cake. Key characteristics needed are water binding, cohesion, elasticity, emulsification, and foaming. The cake formulation recommended by Harper et al. (1980) uses 53 g cake flour, 60 g sugar, 1.5 g baking powder, 0.7 g cream of tartar, 22 g shortening, 1.9 g glycerol monostearate, 17 g whole egg, milk protein to give 10% protein, and water to make a 212-g batch. The flour, salt, sugar, baking powder, and cream of tartar are dry blended, followed by shortening, eggs, water, and protein. Blending is continued until a smooth batter is obtained. The batter is placed in a pan and baked at 180°C to desired doneness. The cake is cooled, its volume measured, and then it is stored overnight. The following morning the cake is cut and evaluated for texture. Visual inspection of thin slices using a microcomparator is also recommended. This procedure allows for the comparison of air cell size in the cakes.

Salad Dressing. Important attributes in a salad dressing are emulsification and viscosity. The model food system suggested by Harper et al. (1980) is 40% salad oil (soybean oil), 12% sugar, 8% 100-grain vinegar, 5% egg yolk solids, 2% salt, 0.2% mustard flour, 0.1% xanthan gum, water to make 100%. Egg yolk solids serve as the reference material, and whey proteins are substituted for egg yolk solids to achieve a 2% concentration in the final product. This substitution is on a mass basis and not on a protein basis. The procedure for making the salad dressing begins with adding water and salt in a mixing bowl. Xanthan gum and sugar are dry mixed and added to the water and salt under agitation in a manner as to minimize clumping. Mix at 150 rpm and add vinegar. Reduce mixing speed to 90 rpm and add egg yolk solids. Increase mixing speed to 150 rpm and continue mixing until smooth. This step may take 1–2 min. Slow down the mixer to 120 rpm and add oil at a constant rate.

As the amount of oil added increases, increase the mixing speed until 210 rpm is reached. Maintain 210 rpm for 2 min after all the oil has been added. Homogenize the mixture by passing it through a piston homogenizer. Store dressing at 4°C for 24 hr prior to evaluation. The salad dressing so obtained is a creamy style product. Evaluation consists of viscosity measurement, stability to creaming, and microscopic examination. Temper sample to 20°C prior to viscosity determination using a Brookfield viscometer at 5, 10, 50 rpm. To test for stability to creaming add 40 mL dressing to a 200-mL beaker followed by 0.5 mL of 1% Sudan V in acetone-ethanol (1:1). Add 15 mL of dye dressing mixture to a conical centrifuge tube and centrifuge at a predetermined rpm. Record amount of free fat layer (red layer) volume, emulsion layer (white) volume, and serum phase (blue layer) volume. Microscopic examination of dressing involves phosphine and acridine orange as stains for fat and protein and provides an idea of fat-to-protein dispersion in the emulsion.

Coffee Whitener.   For a protein to function in a coffee whitener it should emulsify, be opaque (light scattering), and not feather. The coffee whitener formulation contains 10% liquefied, hardened coconut oil containing 5% sodium stearoyl—2 lactylate and ATMOS-150 (mono- and diglycerides) in a 1:1 ratio—protein to obtain 1.8% concentration, 5% sugar, 0.05% xanthan gum, and water to make 100%. Sodium caseinate is the reference protein. Coffee whitener is prepared by making a 0.15% xanthan gum dispersion in water followed by warming and hydrating the gum. Use 100 mL of this solution in a 400-mL beaker. Add sugar and mix. Hydrate protein in 50 mL water and add to beaker. Add oil emulsifier to water and bring volume to desired level with water. Heat the mixture to 60° C and blend in a blender for 2 min at high speed. Homogenize in a piston homogenizer at 1000 psi. Evaluation is conducted for stability of whitener to feathering and stability of the emulsion. For feathering stability hot coffee at 90°C pH 5.4 is used, to which 20 mL of the whitener is added and allowed to stand for 2 min. Visually inspect for feathering and oiling off. Stability of emulsion is evaluated in a manner similar to salad dressing except that centrifugation is at 2500 rpm for 10 min.

Starch Pudding.   Puddings are formed by a combination of emulsification, water binding, gelation, and viscosity. Generally skim milk is used as the source of protein, and starch is an important contributor to gelation. The starch pudding formulation provided by Harper et al. (1980) uses 200 mL of coffee whitener described above, 10 g sugar, and 10 g pregelatinized cross-linked starch. The coffee whitener is heated in a double boiler to 60°C, and sugar is stirred in. Continue stirring and add starch and heat to 80°C. Place pudding in a closed container for 24 hr prior to evaluation. Evaluation is conducted for organoleptic qualities of flavor and mouthfeel. A standardized procedure needs to be developed for this subjective test. Freeze-thaw stability is another important attribute. Subject the pudding to −20°C freezing followed by thawing to room temperature. Repeat the procedure three times and record the change in viscosity. Viscosity is measured using a Brookfield viscometer using a variety of speeds and times.

Whipped Topping. Whipped toppings involve emulsification, foaming, and water absorption, and sodium caseinate is commonly used as the protein source. Harper et al. (1980) suggest a formulation consisting of 30% coconut oil, 6% protein, 7% sugar, 0.2% stabilizer (carrageenan, guar gum, and carboxymethyl cellulose 1:1:3), 0.05% emulsifier (Tween 60, Span 80), and water to make 100%. The stabilizers are suspended in 20 mL water at 80°C. The protein is hydrated in water at pH 6.0. Combine gums and proteins and mix for 2 min at 150 rpm. Add sugar and mix for 1 min. Add emulsifier to coconut oil and heat to 60°C. Slowly add oil-emulsifier mixture to gum protein blend and continue mixing at low speed. Then mix at 200 rpm for 2 min. Warm to 50°C in a water bath, homogenize twice in a hand homogenizer, and cool rapidly using an ice bath. This liquid whipped topping mix is aged for 24 hr at 4°C. Evaluation begins by chilling the mixing bowl and then whipping the mix in a Sunbeam Mix Master at #2 speed for 30 sec followed by whipping at speed #8. Record whipping time to maximum overrun and whip stability. Stability is measured as time to lose one half of the liquid originally used in the whipping.

In the model coffee whitener system, Harper (1984) reported that for emulsion volume index the following were important in decreasing order of significance-emulsifier, gum, protein, phosphate protein interaction, emulsifier-phosphate interaction. Emulsion volume index is a measure of emulsion capacity. For serum separation—a measure of emulsion instability—the following parameters were important in decreasing order of significance: emulsifier, gum, emulsifier-gum interaction, protein, protein-phosphate interaction. These observations illustrate the importance of specifying the objectives of functionality testing.

CORRELATIONS BETWEEN COMPOSITION AND FUNCTIONALITY

A number of researchers have pointed out the necessity of understanding whey protein functionality based on structural properties (Kinsella,

1984; Mangino, 1984; Morr, 1984; Schmidt et al., 1984). A few studies have determined the relationship of structure of food proteins to their activity or functionality (de Wit and de Boer, 1975; Nakai and Li Chan, 1985; Townsend and Nakai, 1983; Voutsinas et al. 1983). Attempts have been made to characterize and relate the composition and physicochemical and functional properties of acid whey protein concentrates (Liao and Mangino, 1987; Mangino et al., 1985; Peltonen-Shalaby and Mangino, 1986)

Surface hydrophobicity determined with *cis*-paranaric acid has been related to protein functionality in emulsions (Kato and Nakai, 1983), foaming (Townsend and Nakai, 1983), and gelation properties (Voutsinas et al., 1983) of a number of food proteins. Mangino et al. (1985) have reported that heptane binding by whey protein concentrates is related to gel strength, emulsion capacity, and creaming rates for coffee whitener and salad dressing model foods. Peltonen-Shalaby and Mangino (1986) have reported that ash content of whey products was highly correlated with foaming properties.

Mangino et al. (1987) studied effects of heat treatments on whey protein functionality. They reported that pasteurization of milk used for cheese making resulted in whey protein concentrates that whipped to a higher overrun and formed more stable foams than other treatments. Pasteurization of milk highly and negatively correlated with alkane binding, neutral lipid, and gel strength at pH 6.5 and positively correlated with foam stability and phospholipid content. For example, up to 98% of the variations in results could be attributed to pasteurization of milk. In another study (Mangino et al., 1988), surface hydrophobicity of retentates was significantly correlated to surface hydrophobicity of the whey protein concentrates. Surface hydrophobicity of retentates was highly correlated with overrun, solubility, gel strength, and cake height.

Kilara and Mangino (1991) studied 12 whey protein concentrate samples for solubility at pH 7.0 using three different methods. Thermal properties of the samples were determined using a differential scanning calorimeter. Regardless of the procedure used for determining solubility, onset temperature and enthalpy of denaturation were highly correlated to this property. Therefore it was proposed that thermal properties could serve as predictors of solubility.

Liao and Mangino (1987) reported that for foaming, heptane binding, protein solubility, and calcium contents were important for overrun development. For foam stability, however, disulfide content, heptane binding, and residual lipids were important predictors. In a high-fat, whipped-topping model exposed hydrophobicity was the most important predictor of overrun development, and exposed hydrophobicity, disulfide content, and potassium all positively correlated with whipped topping overrun.

Surface hydrophobicity of retentates was significantly correlated to surface hydrophobicity of whey protein concentrates (Mangino et al., 1988). Surface hydrophobicity of retentates was significantly correlated with overrun, solubility, gel strength, and cake height of whey protein concentrates. Kohnhorst and Mangino (1985) determined the composition of several membrane-processed whey protein concentrates. In addition, physicochemical properties of these samples were also determined. Using statistical procedures, models were developed to predict gel strength. Compositional and physical factors determined were gel strength, processing pH, ash, total and free sulfyhydryls, protein distribution, fat, calcium, phosphorus, protein hydrophobicity, and soluble protein. The best model was one in which protein hydrophobicity (as measured by alkane binding) and calcium content were both included. In this study sulfhydryls did not vary enough to be of use in predicting gel strength.

Patel et al. (1990) studied compositional factors of whey protein protein concentrates that may affect thermal properties. Denaturation enthalpy was positively correlated with β-lactoglobulin and protein contents and negatively correlated with bound fat, membrane protein, and membrane-associated lipid components. The denaturation temperature correlated positively with phospholipid content, and the onset temperature of denaturation correlated positively with iron content. A plausible mechanism explaining the involvement of iron could not be formulated.

Foaming, emulsifying, and physicochemical properties of processed Cheddar cheese–type whey protein concentrates was studied to understand the relationship between structure and function (Patel and Kilara, 1990). The concentrates were prepared from cheese whey obtained from skim milk, whole milk, and buttermilk-enriched skim milk. In comparison with the other whey protein concentrates, concentrates prepared from skim milk whey had lower surface hydrophobicity, and concentrates prepared from buttermilk enriched skim milk whey had lower solubility. Whey protein concentrates prepared from whole milk whey had poor foaming and emulsifying properties. Among the compositional and physicochemical attributes studied, free fat, bound fat, ash, calcium, denaturation enthalpy, and denaturation temperature were related to solubility, foaming, and emulsifying properties. In general, free fat and bound fat were negatively related with foaming, and emulsifying properties, whereas ash calcium and denaturation enthalpy were positively re-

TABLE 11.12 Regression Models for Functional Properties of Whey Proteins

| Functional property | Independent variable | | Regression coefficient | t-ratio | R^2 ($p < 0.05$) |
	No.	Predictor			
Solubility	1	Constant	54	3.56	0.64
		Denaturation enthalpy	13.7	2.64	
	2	Constant	34.9	3.45	0.92
		Denaturation enthalpy	13.5	3.25	
		Calcium	43.1	4.78	
Foaming capacity	1	Constant	−32.36	−4.18	0.84
		Calcium	91.07	3.87	
	2	Constant	−42.43	−3.03	0.87
		Calcium	91.07	3.87	
		Surface hydrophobicity	2.7	0.88	
Foam stability	1	Constant	−76.6	−3.68	0.79
		Calcium	−17.5	3.86	
	2	Constant	−12.2	−5.49	0.93
		Calcium	17.3	5.91	
		Denaturation enthalpy	16	2.57	
Emulsifying capacity	1	Constant	21.1	70.55	0.54
		Free fat	−0.26	−2.17	
	2	Constant	−56.18	−2.84	0.93
		Free fat	−0.36	−5.85	
		Onset temperature	1.11	3.91	

Source: Patel and Kilara, 1990.

lated to foaming and emulsifying properties. Regression models were developed for predicting solubility, foaming, and emulsifying properties from the various physicochemical properties studied (Table 11.12).

The area of developing models to predict functionality is in its infancy and has shown promise in many instances. This concept should be explored in greater detail and may prove to be a boon in developing new protein sources in a shorter period of time.

USES OF WHEY PROTEIN PRODUCTS IN FOODS

Marshall and Harper (1987) cite a number of foods in which whey protein concentrates are used (Table 11.13). The array of foods docu-

TABLE 11.13 Use of Whey Protein Concentrates in Foods

Food category	Foods
Beverages	Acid-clear, acid-turbid, neutral, chocolate drink, cultured beverages
Baked goods	Biscuits, cakes, bread, donuts, cream fillings, cream icings, cake fillings, baked custard, whipped topping, meringue
Candy/Confectionery	Caramels, milk chocolates, puddings
Dairy and analogs	Coffee whitener, yogurt, ice cream, imitation cream cheese, imitation milk, sherbet, frozen yogurt, infant formula, cream desserts
Sauces	Gravies, cream sauces, flavored sauces
Pasta	Macaroni, noodles, variously shaped pasta
Meat	Hot dogs, meat analogs, meat extenders, meat loaf, sausages
Other	Cereals, canned refried beans, egg white replacers, egg yolk replacers, potato flakes, tortillas

Source: Marshall and Harper, 1987.

mented by these authors is impressive. deWit (1989) delineated a combination of food uses, processing steps required, and the desired composition (Table 11.14).

It has been estimated that 600,000 tons of high-quality whey proteins are produced globally per year. In the United States 70% of the processed whey solids are used in human foods. In Europe 50% of the whey solids are converted to whey powder (500,000 tons), delactosed whey powder (180,000 tons), lactose (200,000 tons), demineralized whey powder (50,000 tons), and various whey products (35,000 tons) (de Wit, 1989). Reliable statistics and market values are extremely hard to collate. Estimates about potential markets are often speculative. For example, fortification of soft drinks with whey protein was estimated by Teixeira (1982). In this estimate it was assumed that only 1% of the soft drinks sold in the United States were fortified to 3% protein using 35% protein whey protein concentrate. The annual requirement for such products alone would be 40,800 tons of whey protein concentrate per year. Clearly in the intervening 11 years this scenario has not materialized.

Beverages hold a lot of promise for utilizing whey protein concentrates. Suggested traditional uses for whey products in beverages include fortifying fruit juices, soft drinks, and dairy-based beverages. Several

TABLE 11.14 Relationship of Type of Food in Which Whey Protein Products Are Used to Functionality and Protein Concentration

Functional attribute	Protein conc. (%)	Food use
Solubility	Varied	Beverages
Whippability	Varied	Confectionery
Emulsification	Varied	Convenience foods
	>35	Desserts
Viscosity	Varied	
Gelation	>65	New products
Fat/Water binding	>85	Meat products
Heat setting	>65	Baked goods
Nonallergenicity	Varied	Infant formulas

Source: deWit, 1989.

functionalities may be involved including solubility, emulsification, viscosity, turbidity (opacity), and nutritional quality. Whey products for such applications would need to be bland, pH stable, low in cost, and in some instances heat stable. An example of a commercial beverage based on whey is Yoohoo, a shelf-stable chocolate-flavored drink. Harper (1985) has described a model system for the evaluation of functionality of whey protein concentrates in soft drinks. Drinks fortified to 2% level protein using 75% whey protein concentrate showed ring formation around the neck of bottles. The rings resulted from protein rising in the bottles. The amount of ring formed was a function of the thermal heat treatment received by the samples. Infant formulas are not traditional beverages, but these products use many specific fractions of whey proteins.

deWit (1989) cites uses of whey solids in confectionery products and points out that condensed whey and electrodialyzed whey can serve as cost-reducing ingredients in low-quality caramels. Too much lactose causes sandiness and graininess in toffees. Defatted whey protein concentrates can serve as egg white replacers in frappé and meringue.

Whey products are used in numerous convenience foods such as creamed soups, white sauces, and gravies. In some instance, whey products serve as whiteners or opacifiers. Dehydrated soups, gravy mixes, and sauces commonly contain whey solids.

Numerous desserts may contain whey products. These desserts can either be aerated products like whipped toppings and ice cream or jellied products such as puddings, fillings, and cheeselike products. In the ice cream industry whey powder is used as an inexpensive substitute for nonfat dry milk. More recently, however, it has been demonstrated that whey protein products do serve a role as an emulsifier. Goff et al. (1989) tested the performance of several milk protein preparations in an ice cream system. Enhanced levels of whey proteins at the fat globule interface were arrived at either by selective homogenization or by enhancement of whey protein:casein ratio in the mix by the addition of whey protein concentrates. Whey protein concentrates contributed to a reduction in the fat water interfacial tension. Mix viscosity increased slightly, the ice cream was drier, and more fat had been destabilized. If 95% whey protein Isolate was used the desirable characteristics of ice cream were enhanced.

Various whey products are used in the manufacture of a number of dairy foods, e.g., yogurt, quarg, Ricotta cheese, cream cheese, and cream dips. Functional properties of importance in such food applications are water holding, emulsification, aeration, and gelation.

In the meat product category, whey protein concentrates are used to prevent or minimize shrinkage of processed meats during cooking, to retain juiciness, and to act as an emulsifier. Thus, whey protein products are used in hams, frankfurters, and luncheon meats. In attempts to reduce the fat content of these products, whey proteins may provide more juiciness in the absence of fat.

The use of whey protein products in baked goods is thwarted by the unique structure and function of gluten. In addition whey protein products contain loaf volume depressants. The net result is that quality characteristics of baked goods is seriously diminished by the addition of whey products.

Infant formulas are generally based on cow's milk. Composition of cow's milk differs considerably from human milk. Human milk contains less casein and more whey protein. Human milk also contains α-lactalbumin and is devoid of β-lactoglobulin. Therefore attempts to "humanize" cow's milk have included the addition of whey protein concentrates and, more recently, fractionated whey proteins.

CONCLUSIONS

This chapter has attempted to outline the types of whey products available in the market. The effect of compositional and other factors on characteristics of the whey products were examined followed by the functionality of such products. The relationships between structure and functionality and some promising prognosticators of functionality were also discussed, and finally the use of whey products in various foods was surveyed.

Whey proteins have been considered as functional and nutritional ingredients in many foods. Roadblocks to the increased use of whey products include the lack of standardized methods to evaluate functionality in foods. The functionality of whey products of different manufacturers and different batches of a given manufacturer further complicates the use of whey products. As a general observation, in the United States food industry whey powder was used as a source of inexpensive solids and these powders contain high lactose levels. Insolubility of lactose posed numerous problems in food formulations. Therefore, partially delactosed and partially demineralized whey powders resulted. Next on the evolutionary ladder was 35% whey protein concentrate. Use of this product shifted the focus from inexpensive ingredient to functionality. Currently higher-protein powders and specialized fractions are gaining more acceptance. In some specialized applications such as sports beverages and diet drinks, higher levels of protein were required and opened markets for whey protein isolates. If the problems outlined in this chapter are successfully addressed, use of whey protein concentrates and isolates can be further enhanced.

REFERENCES

Amundsen, C. H., Watanawanichakon, S., and Hill, C. G. 1982. Production of enriched protein fractions of β-lactoglobulin and α-lactalbumin from cheese whey. *J. Food Proc. Preserv.* 6: 55–71.

Barker, C. M. and Morr, C. V. 1986. Composition and properties of Spherosil-QMA whey protein concentrate. *J. Food Sci.* 51: 919–922.

Batchelder, B. T. 1986. *Electrodialysis Applications in Whey Processing.* Bull. IDF 212, Ch. 13, pp. 84–90. International Dairy Federation, Brussels, Belgium.

Delaney, R. A. M. 1976. Composition, properties and uses of whey protein concentrates. *J. Soc. Dairy Technol.* 29: 91–101.

Devilbiss, E. D., Holsinger, V. H., Posti, L. P., and Pallansch, M. J. 1975. Properties of whey protein concentrate foams. *Food Technol.* 28: 40–48.

deWit, J. N. 1989. Functional properties of whey proteins. In *Developments in Dairy Chemistry-4*, P. F. Fox (Ed.), pp. 285–322. Elsevier Applied Science, New York.

deWit, J. N. and de Boer, R. 1975. Ultrafiltration of cheese whey and some functional properties of the resulting whey protein concentrates. *Neth. Milk Dairy J.* 29: 128–140.

deWit, J. N., Hontelex-Backs, E., and Adamse, M. 1988. Evaluation of functional properties of whey protein concentrates and isolates. 3. Functional properties in aqueous solutions. *Neth. Milk Dairy J.* 42: 155–168.

deWit, J. N., Klarenbeek, G., and Adamse, M. 1986. Evaluation of functional properties of whey protein concentrates and whey protein isolates. 2. Effects of processing history and composition. *Neth. Milk Dairy J.* 40: 41–56.

Fauquant, J., Pierre, A., and Brule, G. 1985. Clarification de lactoserum acide de caseinerie. *Techn. Lait. Marketing* 1003: 37–41.

Fevrier, C. and Bourdin, C. 1977. Utilisation du lactoserum et des produits lactoses par la porcins. In *Colloque: Lactoserum, une Richesse Alimentaire*, pp. 129–174. Association pour la Promotion Industrie-Agriculture, 35 Rue du General Foy, 75008 Paris.

Glover, F. A. 1985. *Ultrafiltration and Reverse Osmosis for the Dairy Industry.* Technical Bull. No. 5. National Institute for Research in Dairying, Reading, UK.

Goff, H. D., Kinsella, J. E., and Jordan, W. K. 1989. Influence of various milk protein isolates on ice cream emulsion stability. *J. Dairy Sci.* 72: 385–397.

Haggett, T. O. R. 1976. The effects of refrigerated storage on whipping properties of whey protein concentrates. *N. Z. J. Dairy Sci. Technol.* 11: 275–277.

Harper, W. J. 1984. Model food system approaches for evaluating whey protein functionality. *J. Dairy Sci.* 67: 2745–2756.

Harper, W. J. 1985. Whey protein utilization in soft drinks. Paper presented at the Australia, New Zealand, United States Collaborative Conference on Whey Protein Utilization, Auckland, New Zealand, Oct. 29–31.

Harper, W. J., Peltonen, R., and Hayes, J. 1980. Model food systems yield clearer utility evaluations of whey proteins. *Food Prod. Dev.* 14(10): 52–56.

Hermansson, A. M. 1975. Functional properties of proteins for foods. *J. Texture Stud.* 5: 425–435.

Houlihan, A. V. and Hirst, P. A. 1987. Estimation of milk membrane fraction from whey protein concentrates. Paper presented at Australia, New Zealand, United States Collaborative Conference on Whey Protein Utilization, Columbus, OH, June 17–19.

Hsu, K. H. and Fennema, O. 1989. Changes in the functionality of dry whey protein concentrate during storage. *J. Dairy Sci.* 72: 829–837.

Joseph, M. S. B. and Mangino, M. E. 1988a. Contribution of milk fat globule membrane proteins to the effective hydrophobicity of whey protein concentrates. *Austr. J. Dairy Technol.* 43: 6–8.

Joseph, M. S. B. and Mangino, M. E. 1988b. The effects of milk fat globule membrane protein in the foaming and gelation properties of β-lactoglobulin solutions and whey protein concentrates. *Austr. J. Dairy Technol.* 43: 9–11.

Kato, A. and Nakai, S. 1980. Hydrophobicity determined by a fluorescence probe method and its correlation with surface properties of proteins. *Biochim. Biophys. Acta* 624: 13–20.

Kilara, A. and Mangino, M. E. 1991. Relationship of solubility of whey protein concentrates to thermal properties determined by differential scanning calorimetry. *J. Food Sci.* 56: 1448–1449.

Kim, Y. A., Chism, G. W., and Mangino, M. E. 1987. Determination of the β-lactoglobulin, α-lactalbumin and bovine serum albumin content of whey protein concentrates and their relationship to protein functionality. *J. Food Sci.* 52: 124–127.

Kinsella, J. E. 1984. Milk proteins: Physicochemical and functional properties. *CRC Crit. Rev. Food Sci. Nutr.* 21: 197.

Kohnhorst, A. L. and Mangino, M. E. 1985. Prediction of the strength of whey protein gels based on composition. *J. Food Sci.* 50: 1403–1405.

Koning, P. J., de Koops, J., and van Rooijen, P. J. 1974. some features of heat stability of concentrated milk. III. Seasonal effects on the amount of casein, individual whey proteins and NPN and their relation to variations in heat stability. *Neth. Milk Dairy J.* 28: 186–202.

Liao, S. Y. and Mangino, M. E. 1987. Characterization of the composition, physicochemical and functional properties of acid whey protein concentrates. *J. Food Sci.* 552: 1033–1037.

Mangino, M. E. 1984. Physicochemical aspects of whey protein functionality. *J. Dairy Sci.* 67: 2711.

Mangino, M. E. 1992. Properties of whey protein concentrates. In *Lactose and Whey Processing*, J. G. Zadow and R. J. Pearce (Ed.), pp. 231–270. Elsevier Applied Science, London.

Mangino, M. E., Fritsch, D. A., Liao, S. Y., Fayerman, A. M., and Harper, W. J. 1985. The binding of *n*-alkanes as a predictor of whey protein functionality. *N. Z. J. Dairy Sci. Technol.* 20: 103–107.

Mangino, M. E., Huffman, L. E., and Regester, G. O. 1988. Changes in hydrophobicity and functionality of whey during the processing of whey protein concentrates. *J. Food Sci.* 53: 1684–1686, 1693.

Mangino, M. E., Liao, S. W., Harper, W. J., Morr, C. V., and Zadow, J. G. 1987. The effects of heating during processing on the functionality of whey protein concentrates. *J. Food Sci.*52: 1522–1524.

Marshall, K. R. 1982. Industrial isolation of milk proteins: Whey proteins. In *Developments in Dairy Chemistry-1*, P. F. Fox (Ed.), pp. 339–374. Applied Science Publishers, New York.

Marshall, K. R. and Harper, W. J. 1987. Whey protein concentrates. IDF 338/13, pp. 21–32. International Dairy Federation, Brussels, Belgium.

Matthews, M. E. 1978. Composition of rennet, sulphuric and lactic whey protein concentrates. *N. Z. J. Dairy Sci. Technol.* 13: 149–156.

Maubois, M. L., Pierre, A., Fauquant, J., and Piot, M. 1987. *Industrial Fractionation of Main Whey Proteins.* Bull. Int. Dairy Fed. 212, pp. 154–159. International Dairy Federation, Brussels, Belgium.

Modler, H. W. and Jones, J. D. 1987. Selected processes to improve the functionality of dairy ingredients. *Food Technol.* 41(10): 114–117.

Morr, C. V. 1984. Production and uses of milk proteins in food. *Food Technol.* 38(7): 39–48.

Morr, C. V. 1986. *Fractionation and Modification of Whey Protein in the U.S.* Bull. IDF 212, Ch. 22, pp. 145–149. International Dairy Federation, Brussels, Belgium.

Morr, C. V. 1987. Effect of HTST pasteurization of milk, cheese whey and cheese whey UF retentate upon composition, physicochemical and functional properties of whey protein concentrates. *J. Food Sci.* 52: 312–317.

Morr, C. V. 1979. Functionality of whey protein products. *N.Z. J. Dairy Sci. Technol.* 14: 185–194.

Morr, C. V. 1989. Beneficial and adverse effects of water-protein interactions in selected dairy products. *J. Dairy Sci.* 72: 575–580.

Morr, C. V., Coulter, S. T., and Jenness, R. 1958. Comparison of column and centrifugal Sephadex methods for fractionating whey and skim milk systems. *J. Dairy Sci.* 51: 1155–1160.

Morr, C. V. and Foegeding, E. A. 1990. Composition and functionality of commercial whey and milk protein concentrates: A status report. *Food Technol.* 44(4): 100–112.

Morr, C. V., Swenson, P. E., and Richter, R. L. 1973. Functional characteristics of whey protein concentrates. *J. Food Sci.* 38: 324–330.

Nakai, S. and Li Chan, E. 1985. Structure modification and functionality of proteins: Quantitative structure-activity relationship approach. *J. Dairy Sci.* 68: 2763.

Nichols, J. A. and Morr, C. V. 1985. Spherosil-S ion exchange process for preparing whey protein concentrates. *J. Food Sci.* 50: 610–614.

Palmer, D. E. 1981. Recovery of protein from food factory wastes by ion exchange. In *Food Proteins*, P. F. Fox and J. J. Condon (Ed.), pp. 219–240. Applied Science Publishers, New York.

Patel, M. T. and Kilara, A. 1990. Studies on whey protein concentrates. 2. Foaming and emulsifying properties and their relationship with physicochemical properties. *J. Dairy Sci.* 73: 2731–2740.

Patel, M. T., Kilara, A., Huffman, L. M., Hewitt, S. A., and Houlihan, A. V. 1990. Studies on whey protein concentrates. 1. Compositional and thermal properties. *J. Dairy Sci.* 1439–1449.

Pearce, R. J. 1987. Fractionation of whey proteins. Trends in Whey Utilization. IDF Bull. 212. International Dairy Federation, Brussels, Belgium.

Pearce, R. J. 1990. *Thermal Denaturation of Whey Proteins.* IDF Bull. 238, pp. 17–23. International Dairy Federation, Brussels, Belgium.

Pearce, R. J. 1992. Whey protein processing. In *Whey and Lactose Processing*, J. G. Zadow (Ed.), pp. 271–316. Elsevier Applied Science, London.

Pearce, K. N. and Kinsella, J. E. 1978. Emulsifying properties of proteins: Evaluation of a turbidimetric technique. *J. Agric. Food Chem.* 26: 716–723.

Peltonen-Shalaby, R. and Mangino, M. E. 1986. Factors that affect the emulsifying and foaming properties of whey protein concentrates. *J. Food Sci.* 51: 103–105.

Rector, D., Matusdomi, N., and Kinsella, J. E. 1991. Changes in gelling behavior of whey protein isolate and β-lactoglobulin during storage: Possible mechanism(s). *J. Food Sci.* 56: 782–788.

Rector, D. J., Kella, N. K., and Kinsella, J. E. 1989. Reversible gelation of whey proteins: Melting thermodynamics and viscoelastic behavior. *J. Texture Stud.* 20: 457–471.

Regester, G. O. and Smithers, G. W. 1991. Seasonal changes in the β-lactoglobulin, α-lactalbumin, glycomacropeptide and casein content of whey protein concentrates. *J. Dairy Sci.* 74: 796–802.

Regester, G. O., Smithers, G. W., Mangino, M. E., and Pearce, R. J. 1992. Seasonal changes in the physical and functional properties of whey protein concentrates. *J. Dairy Sci.* 75: 2928–2936.

Richert, S. H., Morr, C. V., and Cooney, C. M. 1974. Effect of heat and other factors upon foaming properties of whey protein concentrates. *J. Food Sci.* 39: 42–48.

Rinn, J. C., Morr, C. V., Seo, A., and Surak, J. G. 1990. Evaluation of nine semi pilot scale whey protein pre-treatment modifications for providing whey protein concentrates. *J. Food Sci.* 55: 510–515.

Roeper, J. 1971. Seasonal changes in composition of rennet and acid wheys. *N.Z. Dairy Tech.* 6: 112–114.

Schmidt, R. H., Packard, V. S., and Morris, H. A. 1984. Effect of processing on whey protein functionality. *J. Dairy Sci.* 67: 2723.

Teixeira, A. A. 1982. Whey protein concentrate market potential and out-look. Paper presented at Cornell University and New York State Energy Research & Development Authority Meeting, April 7, Ithaca, NY.

Tornberg, E. and Hermannson, A. M. 1977. Functional characterization of protein stabilized emulsions. Effects of processing. *J. Food Sci.* 42: 468–472.

Townsend, A. and Nakai, S. 1983. Relationship between hydrophobicity and foaming characteristics of food proteins. *J. Food Sci.* 48: 588.

Varunsatian, S., Watanabe, K., Hayakawa, S., and Nakamura, R. 1983. Effects of Ca^{2+}, Mg^{2+}, and Na^+ on heat aggregation of whey protein concentrates. *J. Food Sci.* 48: 42–46, 70.

Voutsinas, L. P., Cheung, B., and Nakai, S. 1983. Relationship of hydrophobicity to emulsifying properties of heat denatured proteins. *J. Food Sci.* 48: 26.

Zadow, J. G. 1986. Utilization of milk components: Whey. In *Modern Dairy Technology*, Vol. 1, R. K. Robinson (Ed.), pp. 273–316. Elsevier Applied Science Publishers, London.

12

Color as a Functional Property of Proteins

James C. Acton and Paul L. Dawson

Clemson University
Clemson, South Carolina

INTRODUCTION

Proteins are only indirectly involved in the color of foods with respect to absorption of light energy in the 380- to 760-nm-wavelength range. The three roles that proteins do have that affect the color properties of foods are: (1) light-scattering effects resulting in product lightness (or whiteness), (2) participation in browning and enzymatic reactions, and (3) serving as a carrier/ligand for prosthetic chromaphores such as in myoglobin, flavoproteins, and cytochromes.

Instrumental methods of measuring color, including whiteness, must be related to our sensory perception of a product's color arising from visual stimuli. For visual human perception and for instrumental measurements, the existing or defined energy emission of the light source utilized for sample illumination and the sample's characteristics influencing incident light reflection or transmission are important to color specification. Hunter (1975) and MacAdam (1985) provide excellent discussions on light sources, types of objects, detectors of light energy, and the integration of physical measurements of light-object interactions with the human's psychological perception of color. From such integration, psychophysical color specification through defined numerical assignments has been achieved. The CIE (Commission Internationale de l'Eclairage) in 1931 incorporated spectral aspects of illuminants with the specification of spectrum colors in terms of tristimulus values (X, Y, Z). The CIE XYZ system defines color by the additive mixture of the X, Y, and Z primaries (red, green, and blue counterparts, respectively) that are required to match the color of a mixture as viewed by "standard observers" (humans) under defined illumination and viewing conditions.

Several physical methods of measuring color consist of the eventual definition of a point in the three-dimensional space of various color solids or spaces, although other nonspatial physical and visual techniques exist (Hunter, 1975; Clydesdale, 1976). Today's spectrophotometers with absorption and/or reflection detectors and tristimulus colorimeters with filters for factoring mixes of the primary wavelengths of incident light (allowing X, Y, Z approximation) and its detection by reflection are used with rapid data reduction to yield tristimulus values of many color solids. Hunter (1975) has traced the lineage of the most widely used color spaces. Two of these used frequently for foods are the Hunter L, a, b solid (Fig. 12.1) and the CIE LAB space (Fig. 12.2). The relationships of these and other color scales to instrumentation and to each other are given by Hunter (1975) and Clydesdale (1978). In each color space, the respective tristimulus values serve as the space coordinates and are related to the 1931 CIE tristimulus values as shown in Table 12.1. The CIE Y, x, y color solid (not shown) developed from the X, Y, Z tristimulus values is not frequently used to measure color or color differences of products due to major inadequacies (Hunter, 1975). In various parts of the CIE Y, x, y color space, spacing between colors is not uniform and does not correspond to differences as seen by the human eye. The Hunter L, a, b and CIE LAB color spaces give reasonably uniform measures of color difference.

Three measured or calculated sample attributes derived from the spatial coordinates of these color solids relate remarkably well to visual color evaluations and have been the basis for acceptance of psychophysical terms that define a sample's color. These attributes from the Hunter L, a, b solid are shown in Fig. 12.3. They are *hue*, the angular specification for color perceived as red, blue, green, and so forth; *saturation*, a calculated chromatic unit associated with color perception expressing the

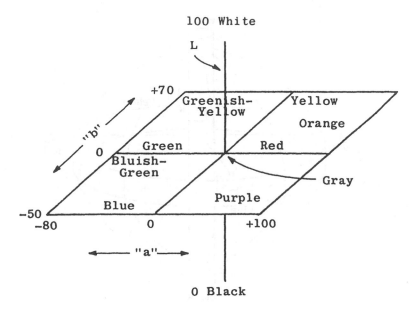

FIG. 12.1 Schematic of the Hunter L, a, b color solid. The vertical "L" scale denotes lightness-to-darkness in 100 to 0 units, respectively. The "a" scale represents redness (+ a) versus greeness (− a), and the "b" scale represents yellowness (+ b) versus blueness (− b). The L, a, and b values are referred to as tristimulus values for the Hunter L, a, b solid. (*With permission of Hunter Associates Laboratory, Inc., Reston, VA.*)

amount of hue departure from gray of the same lightness; and *lightness*, a scaled proportion of the light reflected from or transmitted by the sample relating to the achromatic white-to-gray-to-black appearance for the sample. For the CIE LAB color space, saturation is termed *chroma*. The psychophysical color attributes of the CIE LAB color space are shown in Fig. 12.4. Correlates of the psychophysical color attributes of the CIE Y, x, y color space are indicated in Table 12.1, although there is no direct conversion method between hue and saturation of the Hunter L, a, b solid and dominant wavelength and purity, respectively, of the CIE Y, x, y space.

 In either color solid system, tristimulus values and color attribute values, when quantitated, are illuminant dependent. Therefore, the exact dimensional boundaries and the location of the sample within a color solid are dependent on the spectral characteristics of the light source.

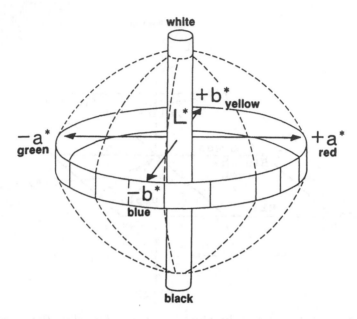

FIG. 12.2   Schematic of the CIE (1976) LAB color space. The vertical "L"
scale denotes lightness-to-darkness in 100 to 0 units, respectively. The "a*"
scale represents redness (+a*) versus greenness (−a*), and the "b*" scale
represents yellowness (+b*) versus blueness (−b*). The L*, a*, and b* values
are referred to as tristimulus values for the CIE LAB space. (*Adapted from
Precise Color Communication© 1989. Minolta Corporation, NJ. With permission.*)

LIGHT-SCATTERING EFFECTS

Light scattering is essentially a nonabsorbent interaction of light with a
sample and is a physical characteristic for proteins in suspension when
existing in particulate dimensions from approximately 0.001 to 1 μm in
the greatest dimension. Scattering results in a reflection of the light from
the particle surface throughout a range of angles. Since most protein
particles, when existing in colloidal dimensions, are not smooth-surfaced,
the scattered light is diffusely reflected.

Without absorption occurring, the reflected light has the appearance
characteristic of the light source, generally white, although a bluish color
may be evident at 90° from the incident light path due to scatter dependence on wavelength.

TABLE 12.1 Relationship of Hunter L, a, b and CIE LAB Color Spaces to CIE XYZ Tristimulus Values and Their Associated Psychophysical Color Attributes

Color space	Tristimulus value relationships to the CIE XYZ values[a]	Psychophysical color attributes[b]
Hunter L,a,b	$L = 10 (Y)^{1/2}$ $a = 17.5 (X_\% - Y)/Y^{1/2}$ $b = 7.0 (Y - Z_\%)/Y^{1/2}$	Lightness = L Saturation = $(a^2 + b^2)^{1/2}$ Hue = $\theta = \tan^{-1}b/a$
CIE (1976) LAB	$L^* = 25 (100Y/Y_0)^{1/2} - 16$, where $1 \leq Y \leq 100$ $a^* = 500 [(X/X_0)^{1/2} - (Y/Y_0)^{1/2}]$ $b^* = 200 [(Y/Y_0)^{1/2} - (Z/Z_0)^{1/2}]$	Lightness = L* Chroma = $(a^{*2} + b^{*2})^{1/2}$ Hue = $\theta = \tan^{-1}b^*/a^*$
CIE (1931) Y, x, y	$Y = \%Y$ $x = X/(X + Y + Z)$ $y = Y/(X + Y + Z)$	Lightness correlate is "luminosity", %Y Saturation correlate is "purity" (1.0 to 0.0) Hue correlate is dominant wavelength, λ_{nm}

[a] For Hunter L,a,b, $X_\%$, $Y_\%$, and $Z_\%$ are tristimulus values expressed as a percent of X, Y, and Z of a standard white surface (Hunter, 1975). For CIE 1976 L*a*b*, X_0, Y_0, and Z_0 are tristimulus values of a normal white object-color stimulus (Wyszecki, 1974). For standard illuminants, $Y_\%$ and Y_0, respectively, equal 100.
[b] Only correlate attributes are indicated for the CIE (1931) Y,x,y color space.

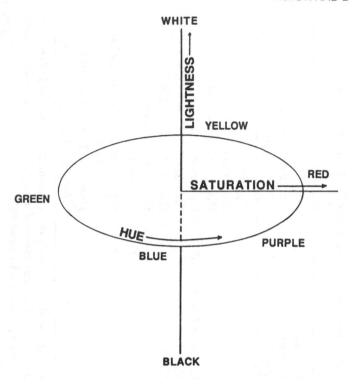

FIG. 12.3   Schematic of psychophysical color space for attributes derived from tristimulus values of the Hunter L, a, b color space (see Table 12.1). *(Adapted from* The Measurement of Appearance, *R. S. Hunter, Copyright© 1975 of John Wiley & Sons, NY. Reprinted with permission of John Wiley & Sons, Inc.)*

The intensity of light that is scattered (I_s) is related to the incident light (I_o) by the inverse fourth power of the wavelength of light ($I_s/I_o =$ constant/λ^4). This relationship is known as Rayleigh scattering (R_o). By this relationship to wavelength, violet light at 400 nm is scattered approximately 10 times more than red light at 710 nm. The amount of scattered light has also been shown to depend on the effective diameter of dispersed particles (Fig. 12.5), having a maximum effect at a diameter of approximately 0.1 μm about equivalent to one-fourth the wavelength of violet light (400 nm). When the scattering particles are smaller than the wavelength of light, Rayleigh scattering produces blue colors, whereas when the particles are larger than the wavelength of light, Mie scattering is produced, which usually results in a white sample appearance (Nassau, 1983).

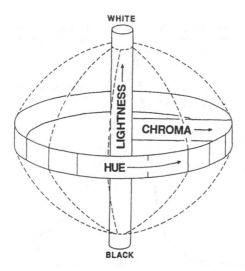

FIG. 12.4   Schematic of psychophysical color space for attributes derived from tristimulus values of the CIE (1976) LAB color space (see Table 12.1). (*Adapted from* Precise Color Communication© *1989, Minolta Corporation, NJ. With permission.*)

    There are two processing examples that can be used to demonstrate the light-scattering effect of proteins on the "whiteness" or "lightening" of foods. First, when discrete aggregates of proteins are initially being formed during precipitation from solution by acidification, or when initially aggregating in solution during heat-induced gelation, the particles attain light-scattering dimensions. When the larger structural protein matrices are formed, incident light is reflected. Respective products of these two protein matrices are casein curds formed in cottage cheese production and myofibrillar protein gels formed in manufacture of surimi-derived seafood products. Second, when protein fractions or isolates are prepared as powders following dehydration and milling, product whiteness due to scatter would increase with particle sizes between 1.0 and 0.1 μm in diameter. Of course, particle sizes larger than 1.0 μm also reflect light, acting more as solid matter or opaque objects rather than as colloidal particles. Sodium caseinate used in coffee whiteners provides some whitening power secondary to its emulsification and flavor effects; the objective in dehydration of the whitener is to have powder particles approximately 125–150 μm in diameter, with entrapped fat globules at about 1 μm (Knightly, 1969).

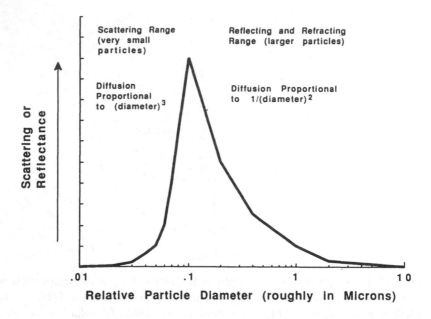

FIG. 12.5   The approximate relationship between the diameter of spherical suspended particles and the intensity of light scattering. The relationship assumes a constant weight of particulate material in a thin layer of suspension. (*With permission of Hunter Associates Laboratory, Inc. Reston, VA.*)

Surfaces of foods may be opaque, translucent, or transparent and have varying degrees of gloss and light penetration below the surface. Plane surfaces of foods are rare. Many foods present matte or fibrous finishes with variable moisture contents at their surface. Internal scattering of light due to refraction and reflection, in addition to absorption, for translucent (turbid) and moist-surfaced opaque food materials present problems in color measurement (Clydesdale, 1972). The major problem was termed "masking" by Little and MacKinney (1969) due to loss of definitive spectral reflectance characteristics for pigment absorption in tuna samples. The heme-based pigments of tuna were considered to exist within a proteinaceous, achromatic muscle fiber matrix. Francis and Clydesdale (1975) presented the Kubelka-Munk theory and its application for materials that both scatter and absorb part of the incident light energy (at any wavelength). Kubelka-Munk analysis for colorant concentration provides a solution to colorimetric measures for problem foods because the methodology takes into account reflection, transmission, and absorption of light from subsurface layers. A scattering coefficient (S) and an absorption coefficient (K) are related to the amount of light scattered and absorbed, respectively, per unit length of travel by the incident light. Using reflectance spectrophotometry, the K/S ratio has been extensively applied in the measurement of fresh meat color (see Francis and Clydesdale, 1975; Hunt, 1980; Kropf et al., 1984). The achromatic muscle fiber matrix is composed of interlinked proteins, likely having particulate sizes of colloidal dimensions and larger opaque surfaces. It is likely that Rayleigh and Mie scattering from particle (fibrous) surfaces within the matrix are components of the Kubelka-Munk scattering coefficient.

ROLE OF PROTEINS IN BROWNING AND ENZYMATIC REACTIONS

There are four types of browning reactions in foods—enzymatic, Maillard, ascorbic acid oxidation, and carmelization—three of which are nonenzymatic. Of these four reactions, only enzymatic and Maillard browning involves proteins.

Enzymatic Reactions

Enzymes can be defined as complex globular protein catalysts that accelerate reaction rates by factors of 10^{12}–10^{20}.

Phenolases are enzymes that catalyze the oxidation of phenols. Those isolated from food sources are oligomers and contain one copper prosthetic group. Phenolases cause the undesirable browning on the cut surfaces of light-colored fruits and vegetables (bananas, apples, potatoes) and desirable characteristic color development of teas, ciders, cocoa, and to a lesser extent raisins, figs, dates, and prunes. The mechanism (Fig. 12.6) starts with a substrate of a phenolic compound (often tyrosine), and the phenolase (tyrosinase) catalyzes a two-step reaction (hydroxylation and oxidation) to form dopamine (DOPA)-quinone. The reaction is continued nonenzymatically via oxidation and polymerization of indole 5,6-quinone to form brown melanin pigments. Melanins will also combine with other proteins to form complexes.

Lipoxygenases are a diverse group of enzymes that oxygenate polyunsaturated fatty acids (or their esters and acylglycerols). These enzymes cause the undesirable destruction of chlorophyll and the green color of

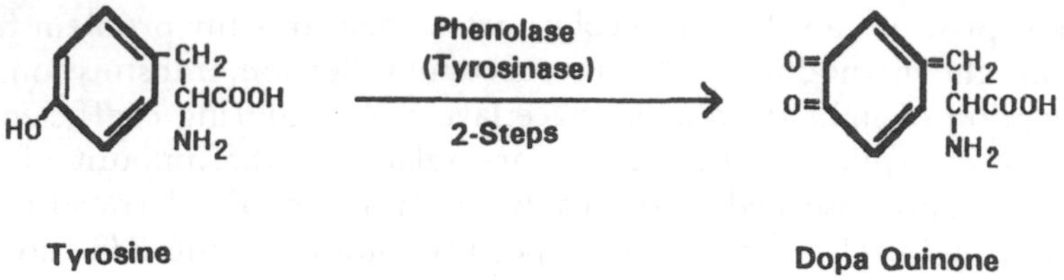

FIG. 12.6 The reaction catalyzed by phenolase that leads to browning. Dopa quinone is transformed nonenzymatically via oxidation and polymerization to form brown pigments.

vegetables, the destruction of carotenoids and the loss of color in pasta and animal feed, the destruction of color additives, and the loss of fish skin pigmentation. One use of lipoxygenases in food processing is the bleaching of hard wheat flour via carotenoid destruction. Lipoxygenases have been implicated in some contribution to the color of meat via prostaglandin synthase. The mechanism of lipoxygenases is to oxygenate the unsaturated fatty acid between the *cis,cis*-1,4-pentadiene double-bond system located between the 6th and 10th carbon from the methyl terminus of the fatty acid (Fig. 12.7). The products are optically active hydroperoxides.

Peroxidases are a common group of enzymes found in plants and milk and usually contain a heme prosthetic group. Horseradish peroxidase has been most extensively researched. Milk is a popular source for peroxidase. Peroxidases catalyze the reaction of an organic hydrogen peroxide with an electron donor (often ascorbic acid, phenols, amines). The reaction is:

$$ROOH + AH_2 \rightarrow H_2O + ROH + A$$

where ROOH is the hydrogen peroxide or an organic peroxide and AH is the electron donor (ascorbic acid, phenols, amines, etc.). The oxidation product is often highly colored, which is sometimes used in colorimetric analyses; however, peroxidase color effects in food are usually associated with the destruction of pigments such as carotenoids and anthocyanins.

Other specific pigment-degrading enzymes such as anthocyanases and chlorophyllases exist in plants and can accelerate the destruction of these pigments in foods.

Maillard Browning Reaction

Maillard browning was first discussed in the writings of French biochemist Louis-Camille Maillard between 1912 and 1916. This nonenzy-

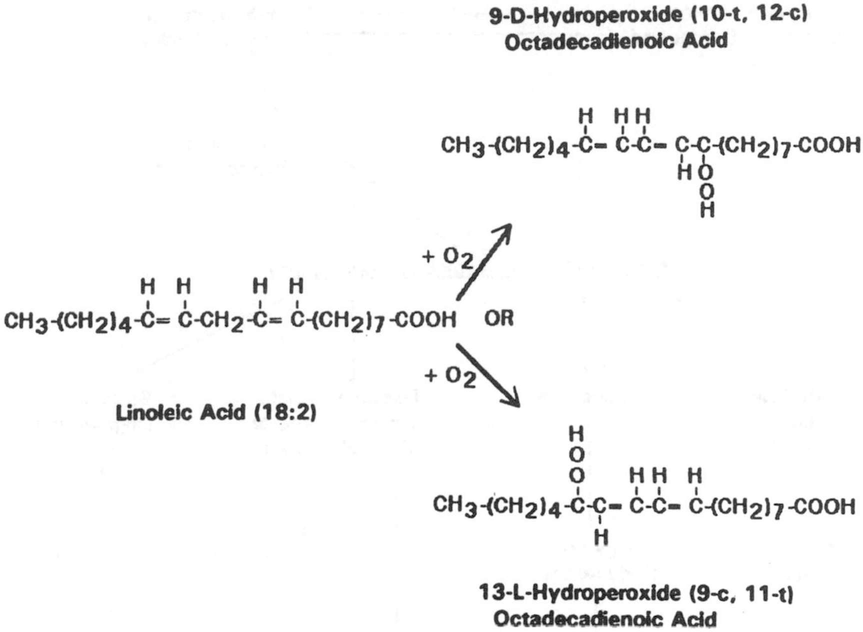

9-D-Hydroperoxide (10-t, 12-c)
Octadecadienoic Acid

$$CH_3\text{-}(CH_2)_4\text{-}\overset{H}{\underset{}{C}}=\overset{H}{\underset{}{C}}\text{-}\overset{H}{\underset{}{C}}\text{-}\underset{\overset{|}{\underset{\overset{O}{\underset{\overset{|}{H}}{}}}{O}}}{\overset{|}{C}}\text{-}C\text{-}(CH_2)_7\text{-}COOH$$

$+ O_2 \nearrow$

$$CH_3\text{-}(CH_2)_4\text{-}\overset{H}{\underset{}{C}}=\overset{H}{\underset{}{C}}\text{-}CH_2\text{-}\overset{H}{\underset{}{C}}=\overset{H}{\underset{}{C}}\text{-}(CH_2)_7\text{-}COOH \quad OR$$

Linoleic Acid (18:2)

$+ O_2 \searrow$

$$CH_3\text{-}(CH_2)_4\text{-}\underset{\overset{|}{H}}{\overset{\overset{H}{\underset{\overset{O}{\underset{\overset{O}{\underset{|}{}}}}{}}}{C}}\text{-}C\text{-}\overset{H}{\underset{}{C}}=\overset{H}{\underset{}{C}}\text{-}\overset{H}{\underset{}{C}}\text{-}(CH_2)_7\text{-}COOH$$

13-L-Hydroperoxide (9-c, 11-t)
Octadecadienoic Acid

FIG. 12.7   The reaction catalyzed by lipoxygenase on linoleic acid. Lipoxygenases can also catalyze reactions of oxygen with esters and acylglycerols resulting in the destruction of pigments.

matic reaction is the condensation of an amino group with a reducing group (often a reducing sugar) resulting in the formation of compounds which ultimately polymerize to form brown pigments (melaniodins). Hodge (1967) reported that sugar-amine browning reactions proceed through at least three stages involving seven different types of chemical reactions. These include an initial stage (colorless with no near-UV absorption) in which the carbonyl-amine condensation and Amadori rearrangement occur; the intermediate stage (colorless or yellow with strong UV absorption) where sugar dehydration, fragmentation, and amino acid degradation occur; and a final stage (colored) that includes aldol condensation, aldehyde-amine polymerization, and formation of heterocyclic nitrogen compounds (Hodge, 1953) (Fig. 12.8). Color development via the Maillard reaction during food processing can be either desirable or undesirable. Browning in meat and bread is often positive, while in milk and dehydrated food it is considered a negative reaction.

These reactions most often occur between sugars and amines, however, much evidence indicates that peroxide formation in oxidizing lipids fosters protein-lipid reactions similar to those leading to Maillard pig-

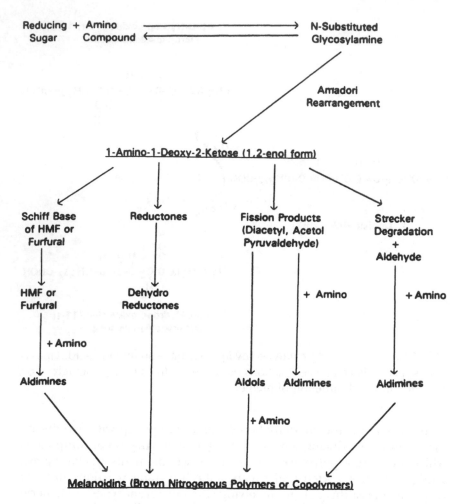

FIG. 12.8   The Maillard reaction pathway. Amines are required to initiate and/or propagate the reaction; however, either reducing sugars or peroxides from oxidizing lipids can react to form brown pigments.

ments (Lundberg, 1961; Schlutz et al., 1962). Hydroperoxide radicals and carbonyl compounds produced from oxidized lipids form Schiff bases with free amine groups from proteins, which are then converted to brown pigments. As a side note, the degree of unsaturation will increase the rate of browning since this degree influences the rate of oxidation and certain phospholipids will participate in browning as they contain both unsaturated fatty acids and an amine group. Brown copolymers (Tappel, 1955) and brown complexes (Narayan and Kummerow, 1963) have been observed from the emulsification of oxidized lipids with aqueous protein solutions. Pokorny et al. (1973a,b) showed that oxidizing lipid peroxides release carbonyl compounds that can readily enter the Maillard pathway forming colorless or slightly colored intermediate compounds. Intermediate-moisture meat products are particularly susceptible to Maillard browning via the lipid-protein pathway when heat is utilized in processing. Heat is usually required for this pathway as the Q_{10} for lipid oxidation is much less than for enzymatic browning. Maillard intermediates are well known for their antioxidative properties, thus heated-lipid oxidation model systems in the absence of Maillard browning reactions (without protein) will spoil via oxidative rancidity (Labuza et al., 1972; Chou and Labuza, 1974). While in a heated system such as is the case with intermediate moisture meats, Maillard intermediates may form thus preventing rancidity but resulting in browning (Brockman, 1973). The lipid-protein browning reaction pathway was illustrated by Obanu et al. (1976a,b) in that browning was the quality-limiting factor during the storage of cook-soak equilibrated intermediate moisture meat. Smoked meat color development also involves the carbonyl-amine reaction. The carbonyls (aldehydes, ketones) in the smoke and the amino compounds in the meat react with no reducing sugar present. Furthermore, the volatile acids deposited on the meat surface from the smoke may facilitate the release of amines from the meat.

Maillard browning more commonly involves the sugar-amine rather than the lipid-amine precursor. The free reducing sugar reacts with a free amine group (NH_2) in the first step leading to the development of numerous brown pigments via the Maillard reaction (Fig. 12.8). Browning will occur in many food systems where a reducing sugar and amine are in close proximity, especially during heating. The Maillard reaction is the principal browning mechanism for grilled meat surfaces and for the interior and surface browning of deep fat fried meats. Browning of batter and breaded foods is sometimes enhanced by the addition of reducing sugars such as dextrose, maltose, or lactose to the batter. The effect of free amine and reducing sugar concentrations is shown by how the absorption at 375 nm and degree of browning increases with the amount of reducing sugar present in the muscle filtrates from the surface of cooked pork loins (Table 12.2).

Dehydration of egg, especially egg albumen, will result in the development of brown pigments via the Maillard reaction. The egg contains 1% carbohydrate, much of which is glucose. During heating this will react with free amines to form melanoidins and discolor the egg product. Glucose was first commercially removed via natural bacterial fermenta-

TABLE 12.2 Reducing Sugar Concentration in Muscle Filtrates and the Development of Brown Coloration in Filtrate Extracts and Deep Fat Fried Pork Loins

Reducing sugars (mg/100 g)	Free-NH$_2$ nitrogen (%)	Absorption at 375 nm	"Browning" rank	
			OD	DFF
57	0.21	0.090	1	2
113	0.28	0.100	2	1
210	0.26	0.142	3	4
387	0.27	0.221	4	3
436	0.24	0.290	5	5
752	0.19	0.480	6	6

OD = oven dried filtrate; DFF = deep fat fried pork loin.
Source: Adapted from Pearson et al., 1962.

tion by the Chinese during the early 1900s. Today glucose is removed from egg albumen prior to drying enzymatically or by yeast fermentation.

PROTEINS WITH PROSTHETIC GROUPS

Myoglobin of Muscle Tissue

Color attributes of meat and meat products are most dependent on the chemical reactions of the pigments myoglobin and hemoglobin of the muscle cell and residual blood, respectively. Myoglobin is a water-soluble protein localized in muscle cells with the function of binding oxygen for use in aerobic metabolism. The myoglobin molecule consists of globin, the protein moiety, and a prosthetic nonprotein heme group composed of a central iron (Fe) atom covalently bonded with four nitrogens of the tetrapyrrole ring of protoporphyrin IX. Of two additional bonding positions of the Fe atom, the fifth coordinate position is occupied by histidine of globin for myoglobin pigments in their native, undenatured state (Fig. 12.9). It is the sixth coordinate position through which Fe bonding with various ligands occurs. The structure shown in Fig. 12.9 is for oxymyoglobin. The specific ligand being bound will also depend

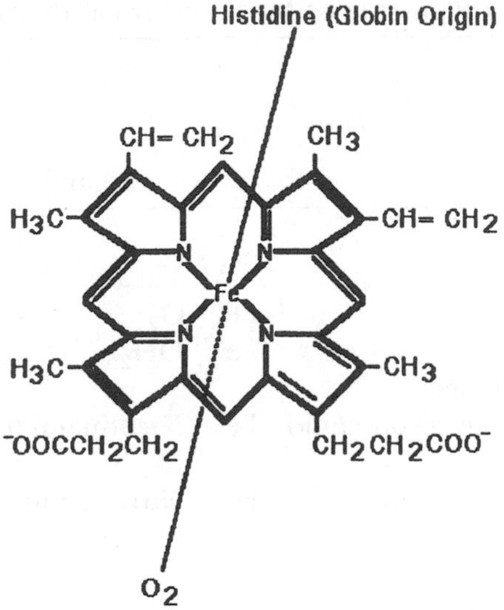

FIG. 12.9 The structure of oxymyoglobin. The central Fe atom has two nontetrapyrrole bonding positions. For oxymyoglobin, these two bondings are with histidine of the globin (fifth coordinate) and oxygen from the aqueous phase (sixth coordinate position).

on whether Fe exists in the ferrous (Fe^{2+}) or oxidized ferric (Fe^{3+}) state. Only native myoglobin, existing in an anaerobic environment, lacks a ligand in its sixth coordinate position. Characteristics of the major pigments of interest for fresh, cooked, and cured meat and meat products are given in Table 12.3, and their pathways of interchange or formation, with their respective visual color, are shown in Fig. 12.10. Readers are referred to Judge et al. (1989) and Price and Schweigert (1987) for other information on the importance of myoglobin in the processing and marketing of meat and meat products.

The globin of myoglobin is colorless yet contributes to the stability of the heme complex through steric and synergistic hisitidine residue interactions with the heme. Being embedded in a hydrophobic "cleft," the heme is generally restrictive to a small group of ligands unless the globin's native confirmation is altered by heat or acid treatment. Energy absorbed from photons of light of specific wavelengths incident to the myoglobin molecule is transferred to the electron clouds associated with bonding of the Fe atom and its ligand. The electronic configurations and

TABLE 12.3 Major Pigments of Meat and Meat Products

Major pigments	Fe	Ligand	Absorption property approximate λ_{max}(nm)
Fresh, Raw Meats			
Myoglobin	Fe^{2+}	Vacant	555–560 (broad)
Oxymyoglobin	Fe^{2+}	O_2	540 and 580
Metmyoglobin	Fe^{3+}	OH_2	500 and 630
Fresh, Cooked Meats			
Denatured globin nicotinamide hemichromes	Fe^{3+}	Nicotinamide	
Imidazole ferrihematins	Fe^{3+}	Nitrogenous base	
Cured, Raw Meat			
Nitrosylmyoglobin	Fe^{2+}	NO	540 and 580
Cured, Cooked Meats			
Nitrosylhemochrome	Fe^{2+}	NO	540 and 580
Dinitrosylhemochrome	Fe^{2+}	2NO	540 and 580
Hemichromes	Fe^{3+}		

bonding mechanisms of various ligands have been presented by Giddings (1977), Livingston and Brown (1981), and Wong (1989).

In comparisons among and within animal species and tissues, total meat pigment concentration significantly affects its color properties. Pigment concentration determines the amount of light energy that will be absorbed and, consequently, the intensity of color that is viewed and measured. Typical ranges of myoglobin concentrations found in the meat of species of fresh market age and in older, mature animals are given in Table 12.4. The intensity of coloration is reflected in both achromatic and chromatic descriptions used by meat scientists and some consumers such as the "light" (or "white") and "dark" meat of poultry and tuna and the "red," "intermediate," and "white" fiber types for beef, pork, and poultry. Other factors are involved in meat color and appearance in the marketplace and include light scattering at the lean tissue surface (Stewart et al., 1965; Synder and Armstrong, 1967; Hunt and Kropf, 1987), packaging and the physical environment in which the meat is stored and eventually displayed at retail (Kropf, 1980; Taylor,

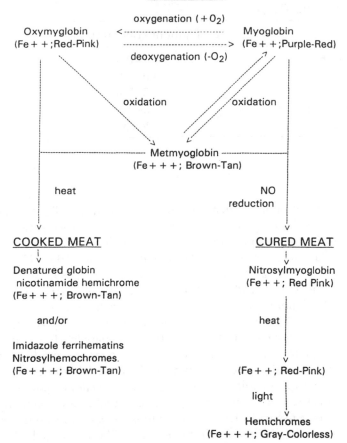

FIG. 12.10 Major pigments of importance in fresh, cooked, and cured meats and meat products. The ionic state of the Fe atom and the visual hue are stated for each pigment.

1982; Kropf, 1984; Acton and Dick, 1986), and biochemical characteristics of the tissue (Govindarajan, 1973; Livingston and Brown, 1981; Fox, 1987).

Pigment Removal. With various aqueous washing treatments applied to ground raw tissues of low economically valued fish during the processing of surimi, fat, heme pigments, and other water-soluble constit-

TABLE 12.4 Range of Myoglobin Concentration for
Various Muscle Tissues

Tissue	Pigment conc. (mg/g)
Poultry light	0.1–0.4
Poultry dark	0.6–2.0
Veal and pork	1.5–7.0
Beef, fresh market	2.5–10
Cow meat	10–18

uents are separated or leached and removed, creating a lighter-colored, higher-value protein product useful for further processing (Lanier, 1986). Washing techniques derived from surimi processes have also been applied to dark meat of poultry (Montejano and Ball, 1984; Hernandez et al., 1986; Elkhalifa et al., 1986; Dawson et al., 1989; Bowie et al., 1989; Lin and Chen, 1989) yielding washed tissue strips, flakes, and mechanically deboned tissue residue having a lighter, less red color. Generally, the effectiveness of decolorization treatments is dependent on the extent of tissue disruption such that adequate leaching can occur. Treatments involving oxidants such as hydrogen peroxide result in "lightening" and redness reduction of dark poultry thigh tissue yielding a meat material approaching or having nearly equivalent color characteristics found for raw breast tissue (Bowie et al., 1989). Removal of myoglobin from muscle tissues by leaching or extensive oxidation would account for a decrease in chromatic characteristics. The combined effects of reduced pigment content and increased light scatter occurring due to higher moisture contents within the remaining tissue structure would affect the increase in perceived and measured lightness. Although much research has been conducted in this area, the only product known to be commercially available as an ingredient is surimi.

Fat, Air, and Heating Effects in Products. When the fat content is reduced in cured meat products such as frankfurters (Hand et al., 1987) and bologna (Claus et al., 1989; Claus and Hunt, 1991), the products become darker and redder. The meat's darker appearance is due to a reduction in the overall light scattering associated with the scattering properties of the fat. However, since the myoglobin content was kept constant (by fat replacement with water in bologna), the redder appearance cannot be explained simply by a higher pigment content. The redder appearance is more likely due to a reduction of the yellow color component associated with the fat. If air bubbles are entrapped in the initial meat batter during its preparation, more light will be reflected from the product's surface due to scatter, and it will appear lighter and have a higher hue value (Palombo and Wijnaards, 1990). Palombo and Wijnaards (1990) reported two simultaneously occurring processes affecting lightness and hue during heating of meats. First, changes occur due to absorption characteristics by transitions of the meat pigments. Changes in the light scatter properties also occur due to thermal denaturation of the structural proteins of meat and the enlargement of any remaining incorporated air bubbles.

Flavoproteins and Cytochromes

Electron transport chain carrier proteins (flavoproteins and cytochromes) are chromophores whose color is determined by the protein's oxidation state (reduced or oxidized) (Fig. 12.11). The flavoproteins act as an electron donor and acceptor in the electron transport chain and in the catalyzation of a variety of reactions.

Flavin adenine dinucleotide (FAD) and flavin mononucleotide (FMN) exist in three spectrally different oxidation states (Fig. 12.12): the yellow oxidized form, the red or blue one-electron-reduced form, and the colorless two-electron reduced form. Nicotinamide adenine dinucleotide (NAD) is also a flavoprotein, which uses FMN as its prosthetic group and also contains an iron-sulfur protein.

The cytochromes are unique to aerobic cells and play a major role in the electron transfer from coenzyme Q to oxygen (Fig. 12.11). At least five different cytochromes are found in higher animals. Cytochromes b, c, and c_1 contain the identical heme configuration (iron-protoporphyrin IX) as is found in myoglobin and hemoglobin, while cytochromes a_1 and a_3 contain a modified heme called heme A differing from the other heme form only in the presence of a formyl group where the methyl group is found (Fig. 12.9). Cytochromes c and c_1 are the only cytochromes covalently bonded to a protein. While the chromatic property of the proteins of the electron transport chain has a minimal effect on food color, the application of this chromatic functional property to research has been significant.

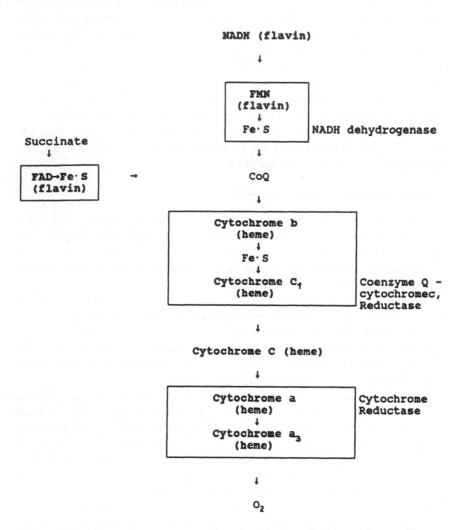

FIG. 12.11 Electron transport scheme showing flavinproteins and cyto-chromes involved in the production of ATP. Flavinproteins and cytochromes are both chromaphones.

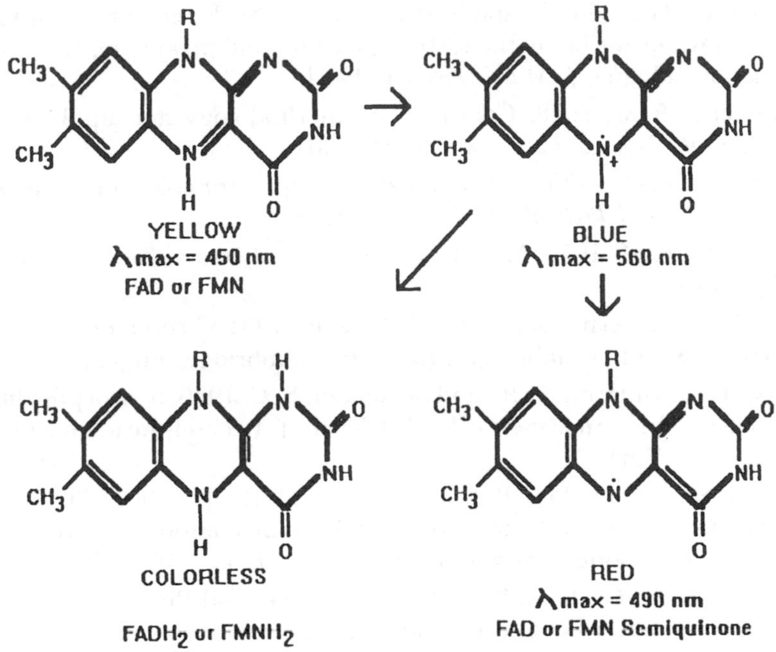

FIG. 12.12 Oxidation states of flavin coenzymes. R group is bound to coenzymes FAD, FMN, NAD, NADP. The flavin coenzymes are versatile one- and two-electron acceptors and donors.

REFERENCES

Acton, J. C. and Dick, R. L. 1986. Protecting color in fresh and processed meats. *Natl. Provisioner* 194(19): 12–17.

Bowie, B. M., Dick, R. L., and Acton, J. C. 1989. Composition and color alteration of small strips of chicken dark meat. *J. Muscle Foods* 1: 23–35.

Brockman, M. C. 1973. In *Food Dehydration*, Vol 2, W. B. Van Arsdel, M. J. Copley, and A. I. Morgan (Ed.), 2nd ed., p. 489. AVI Publishing Co. Westport, CT.

Chou, H. E. and Labuza, T. P. 1974. Antioxidant effectiveness in intermediate moisture content model systems. *J. Food Sci.* 39: 479–483.

Claus, J. R. and Hunt, M. C. 1991. Low-fat, high water-added bologna formulated with texture-modifying ingredients. *J. Food Sci.* 56: 643–647, 652.

Claus, J. R., Hunt, M. C., and Kastner, C. L. 1989. Effects of substituting added water for fat on the textural, sensory and processing characteristics of bologna. *J. Muscle Foods* 1: 1–21.

Clydesdale, F. M. 1978. Colorimetry—methodology and applications. *CRC Crit. Rev. Food Sci. Nutr.* 9: 243–301.

Clydesdale, F. M. 1976. Instrumental techniques for color measurement of foods. *Food Technol.* 30 (10): 52–55, 59.

Clydesdale, F. M. 1972. Measuring the color of foods. *Food Technol.* 26 (7): 45–51.

Commission Internationale de l'Eclairage, 1931. Proceedings of the Eighth Session. Cambridge University, Cambridge, England.

Cross, H. R., Durland, P. R., and Seideman, S. C. 1986. Sensory qualities of meat. Ch. 7. In *Muscle as Food.*, (Ed.) P. J. Bechtel, Academic Press, Inc., New York.

Dawson, P. L., Sheldon, B. W., and Ball, H. R., Jr. 1989. Pilot-plant washing procedure to remove fat and color components from mechanically deboned chicken meat. *Poultry Sci.* 68: 749–753.

Elkhalifa, E. A., Graham, P. P., Marriott, N. G., and Phelps, S. K. 1988. Color characteristics and functional properties of flaked turkey dark meat as influenced by washing treatments. *J. Food Sci.* 53: 1068–1071, 1080.

Fox, J. B., Jr. 1987. The pigments of meat. Ch. 5, In *The Science of Meat and Meat Products*, J. F. Price and B. S. Schweigert (Ed.). Food and Nutrition Press, Inc., Westport, CT.

Fox, J. B., Jr. 1966. The chemistry of meat pigments. *J. Agr. Food Chem.* 14: 207–210.

Francis, F. J. and Clydesdale, F. M. 1975. *Food Colorimetry: Theory and Applications*. AVI Publishing Co., Inc., Westport, CT.

Giddings, G. G. 1977. The basis of color in muscle foods. *J. Food Sci.* 42: 288–294.

Govindarajan, S. 1973. Fresh meat color. *CRC Crit. Rev. Food Technol.* 4: 117–140.

Hand, L. W., Hollingsworth, C. A., Calkins, C. R., and Mandigo, R. W. 1987. Effects of preblending, reduced fat and salt levels on frankfurter characteristics. *J. Food Sci.* 52: 1149–1151.

Hernandez, A., Baker, R. C., and Hotchkiss, J. H. 1986. Extraction of pigments from mechanically deboned turkey meat. *J. Food Sci.* 51: 865–867, 872.

Hodge, J. E. 1953. Dehydrated foods. Chemistry of browning reactions in model systems. *J. Agric. Food Chem.* 1: 928–943.

Hodge, J. E. 1967. In *Chemistry and Physiology of Flavors*. H. W. Schultz, E. D. Day, and L. M. Libbey (Ed.). pp. 465–491. AVI Publishing Co., Westport, CT.

Hunt, M. C. and Kropf, D. H. 1987. Color and appearance. *Adv. Meat Res.* 3: 125–159.

Hunt, M. C. 1980. Meat color measurements. Natl. Live Stock and Meat Bd., Chicago, IL. *Recip. Meat Conf. Proc.* 33: 41–46.

Hunter, R. S. 1975. *The Measurement of Appearance*. John Wiley and Sons, Inc., New York.

Judge, M. D., Aberle, E. D., Forrest, J. C., Hedrick, H. B., and Merkel, R. A. 1989. *Principles of Meat Science*. 2nd ed. Kendall/Hunt Publishing Co., Dubuque, IA.

Knightly, W. H. 1969. The role of ingredients in the formulation of coffee whiteners. *Food Technol.* 23(2): 37–42.

Kropf, D. H., Olson, D. G., and West, R. L. 1984. Objective measures of meat color. Natl. Live Stock and Meat Bd., Chicago, IL. *Recip. Meat Conf. Proc.* 37: 24–32.

Kropf, D. H. 1980. Effects of retail display conditions on meat color. Natl. Live Stock and Meat Bd., Chicago, IL. *Recip. Meat Conf. Proc.* 33: 15–32.

Kropf, D. H. 1984. Effect of display conditions on meat products. *Proc. Meat Industry Res. Conf.* pp. 158–176. Am. Meat Inst., Washington, DC.

Labuza, T. P., McNally, L., Gallagher, D., Hawkes, J., and Hurtado, F. 1972. Stability of intermediate moiture foods. I. Lipid oxidation. *J. Food Sci.* 37: 154–159.

Lanier, T. C. 1986. Functional properties of surimi. *Food Technol.* 40(3): 107–114, 124.

Lin, S. W. and Chen, T. C. 1989. Yields, color and composition of washed, kneaded and heated mechanically deboned poultry meat. *J. Food Sci.* 54: 561–563.

Little, A. C. and MacKinney, G. 1969. The sample as a problem. *Food Technol.* 23(1): 25–28.

Livingston, D. J. and Brown, W. D. 1981. The chemistry of myoglobin and its reactions. *Food Technol.* 35: 244–252.

Lundberg, W. D. (ed.) 1961. *Autoxidation and Antioxidants*, Vol. I and II. John Wiley and Sons, New York.

MacAdam, D. L. 1985. *Color Measurement*. 2nd ed. Springer-Verlag, New York.

Montejano, J. G. and Ball, H. R., Jr. 1984. Sensory, color and mechanical properties of washed broiler thigh meat. *Poultry Sci.* 63(Suppl. 1): 152.

Narayan, K. A. and Kummerow, F. A. 1963. Factors influencing the formation of complexes between oxidized lipids and proteins. *J. Am. Oil Chem Soc.* 40: 339–342.

Nassau, K. 1983. *The Physics and Chemistry of Color.* John Wiley and Sons, Inc., New York.

Obanu, Z. A., Ledward, D. A., and Lawrie. R. A. 1976a. The proteins of intermediate moisture meat stored at tropical temperatures. III. Differences between muscles. *J. Food Technol.* 11: 187–196.

Obanu, Z. A., Biggin, R. J., Neale, R. J., Ledward, D. A., and Lawrie. R. A. 1976b. The proteins of intermediate moisture meat stored at tropical temperatures. IV. Nutritional quality. *J. Food Technol.* 11: 575–581.

Palombo, R. and Wijnaards, G. 1990. Characterization of changes in psychometric colour attributes of comminuted porcine lean meat during processing. *Meat Sci.* 28: 61–76.

Pearson et al. 1962. The browning produced by heating fresh pork. I. The relation of browning intensity to chemical constituents and pH. *J. Food Sci.* 27: 177.

Pokorny, J., El-Zeany, B. A., and Janicek, G. 1973a. Reaction of oxidized lipids with protein. VII. Changes in oxidized butyl oleate in contact with albumin. *Nahrung* 17: 545–552.

Pokorny, J., El-Zeany, B. A., and Janicek, G. 1973b. Nonenzymic browning. III. Browning reactions during heating of fish oil fatty acid esters with protein. *Lebensunter u. Forsch.* 151: 31–35.

Price, J. F. and Schweigert, B. S. 1987. *The Science of Meat and Meat Products.* 3rd ed. Food and Nutrition Press, Inc., Westport, CT.

Schultz, H. W., Day, E. A., and Sinnhuber, R. O. (Ed.) 1962. *Lipids and Their Oxidation.* AVI Publishing Co. Westport, CT.

Snyder, H. E. and Armstrong, D. J. 1967. An analysis of reflectance spectrophotometry as applied to meat and model systems. *J. Food Sci.* 32: 241–245.

Stewart, M. R., Zipser, M. W., and Watts, B. M. 1965. The use of reflectance spectrophotometry for the assay of raw meat pigments. *J. Food Sci.* 30: 464–469.

Tappel, A. L. 1955. Studies of the mechanism of Vitamin E action. III. In vitro copolymerization of oxidized fats with protein. *Arch Biochem. Biophys.* 54: 266–280.

Taylor, A. A. 1982. Retail packaging systems for fresh meat. In *Proceedings: Intl. Symp. Meat Sci. and Technol.*, pp. 353–366. National Live Stock and Meat Board, Chicago, IL.

Wong, D. W. S. 1989. *Mechanism and Theory in Food Chemistry.* Van Nostrand Reinhold, New York.

Wyszecki, G. 1974. Proposal for study of color spaces and color-difference evaluations. *J. Opt. Soc. Am.* 64: 896–897.

Weast, R.C., ed., *Handbook of Chemistry and* ...

Wood, G.W., 1980, Rearrangement Theory, in Food Carbohydrates, Van Nostrand Reinhold, New York.

Young, F.G., 1994, Processes involved in taste perception, ...

13

Protein Gel Ultrastructure and Functionality

Shai Barbut

University of Guelph
Guelph, Ontario, Canada

INTRODUCTION

Production of acceptable protein gels is very important to the food industry in manufacturing products such as meat, dairy, and baked goods. The industry relies on the ability of various proteins to provide texture, mouthfeel, and water binding. The type of protein-protein interaction determines the way a gel is formed (i.e., during heating in whey proteins or during cooling in gelatin), the textural characteristics of the gel, and even the appearance of the gel (i.e., opaque vs. transparent gels).

The relationship between structure and function is of importance in studying the basic gelation phenomenon as well as practical aspects of food system stability. However, relationships between structure and functionality are not always so obvious, and one must be knowledgeable in both areas in order to correlate them.

Some food proteins can form gels, while others do not. The ability of a protein to form a gel, as well as the type of gel formed, depends on various factors such as pH, temperature, protein concentration, and the presence of ions. The term "gel" is not easy to define precisely because of the diversity and complexity of those structures. Different definitions can be found in the literature as reviewed by Clark (1992) and Ziegler and Foegeding (1990), who basically suggested the definition: a gel is a continuous network of macroscopic dimensions (the solid phase) immersed in a liquid medium (the liquid phase) and exhibiting no steady-state flow.

Protein gels can be generally divided into two major types. The first is an aggregated gel, typically formed by egg white proteins and dairy products made from casein. The aggregated gel comprises relatively large particles bound to one another to form a gel network. These gels are usually opaque (because of the relatively large aggregates), and their water-holding capacity is low. The second category is the fine-stranded gels, produced by association of small-diameter molecules to form an ordered network, e.g., a gel prepared from whey protein isolate (WPI) in the presence of low sodium concentrations. These gels are usually clear and have good water-holding capacity. Variation between these two groups exists, and sometimes mixed gels can also be formed, as will be discussed later.

Advances in high-resolution microscopy, computerized image analysis (Russ, 1991), and rheological testing (Hamann, 1987) have opened the way to more basic studies on relationships between structure and functionality and has initiated a change in some trial-and-error approaches to food product development. Currently it is possible to use more methodical ways to design/modify protein gels for a specific application by studying protein basic structure. In this review examples dealing with meat proteins, dairy proteins, and the use of proteins as a wall material for microencapsulation will be discussed. This chapter is designed to highlight some of the relationships between protein gel structure and functionality and demonstrate some techniques used in studying these relationships. Information regarding protein-protein interactions in gels can be found elsewhere in this book (see Chapter 7).

MEAT PROTEINS

Heat-Induced Gelation in Muscle Proteins

Actin and myosin are the two main proteins contributing to gel formation of meat products. Both are salt-soluble proteins that represent the major structural components of the muscle contractile unit—the sarcomere. The myosin molecule is a large, filamentous protein composed of two heavy and four light chains with a molecular weight ~480,000 daltons and represents 43% of the myofibrillar proteins (proteins building the muscle fiber). The actin molecule is smaller (molecular weight 42,000) and exists in either globular or fibrous form, depending on environmental conditions (Ashgar et al., 1985). It represents 22% of the myofibrillar proteins and is the major part of the thin filament structure. Myosin can form a heat-induced gel by itself, whereas actin cannot (Fig. 13.1). Yasui et al. (1980) used one of the early scanning rigidity monitors (developed in their laboratory) to follow the gelation of different mixtures of actin and myosin during heating (20–70°C). While actin alone could not form a gel, in the presence of myosin it exerts a synergistic effect, thereby considerably complementing the binding characteristics of myosin. It should be mentioned that studies by Brotschi et al. (1978) showed that actin filaments (in solution) can form three-dimensional networks of

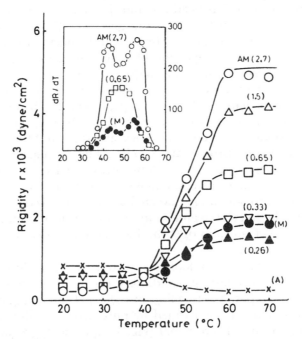

FIG. 13.1 Changes in heat-induced gel strength of actomyosin reconstituted at different myosin-to-actin ratios. Protein samples (5 mg/mL) were dissolved in 0.6 KCl solution containing 20 mM phosphate buffer, pH 6.0, and incubated for 20 min at temperature increments from 20 to 70°C. M, myosin alone; A, actin alone; AM actomyosin. Figures in parentheses indicate the mole ratio (corrected) for myosin to actin. Inset: derivative plot showing differential gel strength (dR/dT) as a function of temperature. (*From Yasni et al., 1980.*)

linear polymers, and in the presence of different cross-linking proteins the actin filament will produce gels. This can also be seen in Fig. 13.1. Yasui et al. (1980) showed that a specific myosin-to-actin ratio is essential in developing a stronger gel than is formed by myosin alone. The maximum strength was observed at a myosin:actin mole ratio of 2.7, which corresponds to the weight ratio of myosin to actin of about 15. An increase in the proportion of myosin beyond this point caused a decrease in gel strength. The state of actin (globular vs. filamentous) can also affect gel strength, since G-actin polymerizes very slowly at high salt concentrations. Other studies have indicated a remarkable enhancing effect of F-actin on heat-induced myosin gels with the formation of either reconstituted or natural actomyosin (Samejima et al., 1982). Ashgar et al. (1985) indicated that these findings suggest that a small amount of F-actomyosin complex formed in the system acts as a cross-linker of the free myosin molecule (on heating), and this may be a prerequisite for actin-induced improvement of myosin gels. Scanning electron microscopy studies demonstrated the difference in the microstructure of heat-induced gels when the proportion of actin and myosin varied. Figure 13.2a shows myosin forming an aggregated gel matrix consisting of assemblies of protein clumps. The addition of actin to the myosin in the ratio 1/0.43 (w/w) changed the type of gel; in Fig. 13.2b, a fine filament gel with some degree of cross-linking is seen. Overall, the variation in pore size distribution, shape of network structure, and the thickness of strands depended on protein aggregation. The aggregates range from assemblies of irregular protein clumps (myosin by itself) to filamentous strands with entangled clusters. When the ratio of myosin to F-actin was 4, a side-to-side cross-linking between filaments appeared to prevail, and cross-linkages between free and bound myosin seemed to dominate, but at a myosin to F-actin ratio of 0.43, the overall network structure of F-actin seemed to exist.

Hermansson et al. (1986) looked at myosin gel formation under various pH and ionic strength (μ) conditions (Fig. 13.3). They indicated the existence of two types of gels: a fine-stranded gel and an aggregated gel. The fine-stranded gel was formed by pure myosin under low μ (0.25 KCl) conditions at pH 5.5 or 6.0 (Fig. 13.4a); the aggregated gel was formed at high μ (0.6M KCl) conditions and at the same pH (Fig. 13.4b). The micrograph shows that all the globular aggregates building up the gel structure were approximately the same size. When the pH was raised to 6.0 the same type aggregates and fine-strand gels were produced, under high and low μ, respectively. However, one difference was that at pH 6.0 and high μ conditions, the structure was dominated by aggregates, and no fine strands could be seen. Like the aggregates formed at pH 5.5, aggregates building up the structure at pH 6.0 were uniform in

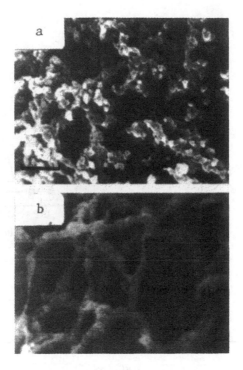

FIG. 13.2 SEM of heat-induced gel formed at various myosin-to-actin ratios in a system containing 0.6 M KCl (pH 6.0) at 65°C. (a) myosin alone; (b) myosin to F-actomyosin ratio = 0.43 (w/w). Bar = 1 μm. (*From Yasui et al., 1982.*)

size, but larger. The aggregated gel had lower rigidity (as measured by penetration test; Fig. 13.3) Overall, gel rigidity was higher at low μ conditions (pH 5.8–6.0) and decreased with an increase in salt concentration. Results obtained at pH 6.0 are in general agreement with studies of rabbit skeletal myosin (Ishioroshi et al., 1983).

Hermansson et al. (1986) also looked at the effect of heating temperature (55, 60, and 65°C) at high and low μ at pH 6.0. These heating temperatures had no effect on the structure formed, but some shrinkage of gel structure was observed, especially in the fine-strand gels. Overall, the diameter of the globular aggregates varied between 0.1 and 0.3 μm and that of the strands between 0.1 and 0.3 nm.

Hermansson et al. (1986) also observed that myosin preparation at low μ produced turbid solutions and at high μ (0.6 M) produced clear solutions. Turbidity can represent a rough estimate of the degree of

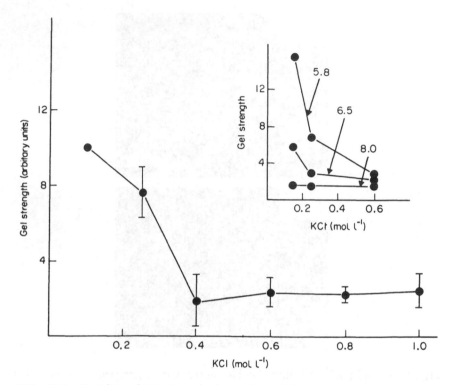

FIG. 13.3   Rigidity of myosin gels, heated to 65°C for 30 min, as a function of ionic strength at pH 6.0. Inset: graphs for KCl dependency of the gel rigidity at pH 5.8, 6.5, and 8.0. (*From Hermansson et al., 1986.*)

myosin aggregation. The authors produced a "phase diagram" showing the combined conditions of ionic strength and pH, which resulted in clear solutions (measured optical density at 340 nm) of the solutions (1.3 g/L) at 4°C. At pH 6.0 turbid solutions were produced at low μ (0–0.3 M), but an increase of μ resulted in clear solutions. Therefore, it was concluded that conditions required for formation of fine-strand–type myosin gel were present *prior* to heating. The strands were made up of myosin filaments formed under certain ionic strength and pH conditions and produced the turbid solution.

When the pH was lowered to 4.0 in the 0.6 M KCl, spontaneous formation of a strand-type gel structure occurred at room temperature (Hermansson et al., 1986) (Fig. 13.4c). Heating to 60°C (30 min) did not change the gel structure, indicating that the spontaneously formed structure was stable enough to withstand any major alterations that may have been produced by conformational changes during heating. It seems that at pH 4.0 the repulsive balance is such that the myosin filaments can interact spontaneously and form the gel without heating. A similar observation of gelation prior to cooking was presented by Ishioroshi et al. (1983) and Gordon and Barbut (1990a). It is well known that synthetic myosin filaments can be formed in solutions at pH 6.0–8.0 and μ of 0.1–0.2 by dialysis (Huxley, 1963). At higher pH (5.5–6.0) conformational changes taking place during heating are required for gelation. Overall, protein aggregates have previously been seen in gels of other systems such as casein proteins, whey proteins, and egg proteins (Aguilera and Stanley, 1990). This indicates that aggregation is a colloidal phenomenon, in which particle size is determined by reactions controlled by opposite forces (i.e., surface tension, electrostatic repulsion) and in which small changes in surface charges can significantly affect particle size.

In another study, Hermansson and Langton (1988) measured myosin filaments produced in a fine-stranded gel. The filaments produced at pH 5.5, under low μ conditions, had a backbone (25 nm diam) with myosin heads located close to the filament backbone. The total width of those filaments was 45 nm. The filaments were wider than the synthetic filaments of rabbit skeletal muscle observed at pH 7.0 and 0.1 M KCl by Pollard (1975) but were not as wide as the native thick filaments observed by Trinick and Elliott (1982). Hermansson and Langton (1988) indicated that heating the myosin solution (pH 5.5 and 0.25 M KCl) resulted in the formation of a parallel arrangement of filaments. A typical feature seen in these gels is the way the filaments interact to form "Y-junctions." The junctions and the parallel alignment give rise to a rather loose network structure (Fig. 13.5a). The "Y interactions" imply that the filaments have a tendency to interact in a very specific way. The authors also confirmed that the filaments seen prior to cooking are the ones forming the gel during cooking. They used three different microscopical preparation techniques and found no variation due to the preparation techniques. This is very important when studying microstructure, since some techniques can cause distortion of the sample and later result in erroneous interpretation.

The manner in which aggregated myosin gel network is formed was studied by Sharp and Offer (1992). Gel formation of rabbit skeletal myosin was studied at pH 6.5 and μ of 0.6 M KCl from 30 to 60°C. As indicated before, these conditions result in an aggregated gel. The beginning of microstructural changes was observed in a sample heated to 35°C (for 30 min). At that temperature the two heads of some myosin molecules seemed to coalesce and formed dimers. Heating to 40°C (for

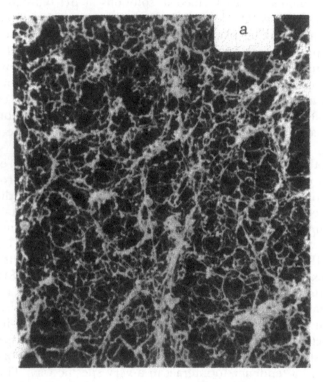

FIG. 13.4   SEM of myosin gels formed during heating: (a) pH 5.5, 0.25 M KCl; (b) pH 5.5, 0.6 M; and (c) gel formed spontaneously at 4°C at pH 4.0 and 0.6 M. (*From Hermansson et al., 1986*).

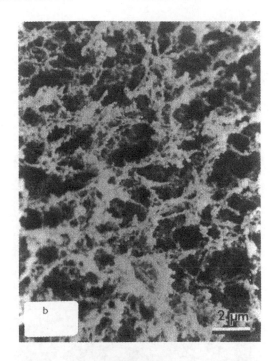

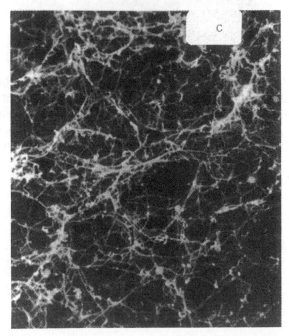

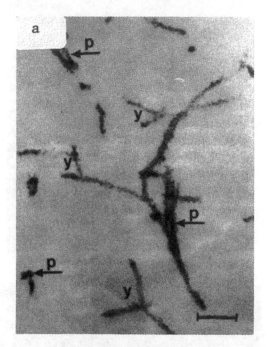

FIG. 13.5   (a) TEM of myosin gel at pH 5.5 and 0.25 M KCl heated to 60°C for 30 min showing the parallel alignment in pairs of filaments in the plane and perpendicular to the plane of the sections, marked with a ''p,'' and Y-junctions, ''y''. Bar = 200 nm. (b) Aggregates of myosin molecules observed by rotary shadowing of rabbit myosin heated to 35°C for 30 min; (c) heated to 45°C. Bar = 10 nm. [(a) *From Hermansson and Langton, 1988.* (b,c) *From Sharp and Offer, 1992.*]

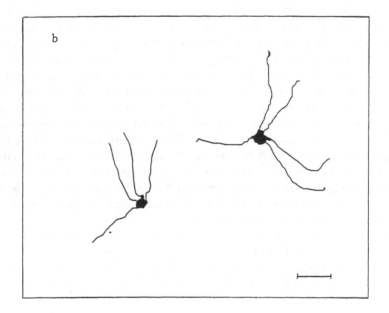

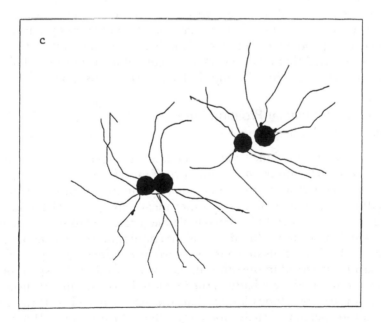

30 min) revealed that up to about a dozen myosin molecules aggregated through their heads to form a globular mass measuring up to 60 nm across, with the tails radiating outwards (Fig. 13.5b). As temperature increased, such head-linked oligomers further aggregated. At 50°C some oligomers coexisted with aggregates, which were formed by the coalescence of two or more oligomers. After heating to 50°C, the aggregates formed by coalescence of more oligomers and the tails were only seen indistinctly. At 60°C, the aggregated particles contained large numbers of globular masses and were typically 100–200 nm across, and occasionally up to 1 μm across (Fig. 13.5c). The authors suggested that those aggregates were formed by head-to-head interactions, but that tail-to-tail interactions may be important in providing the strength in cross-linking the gel network. They presented histograms of the size distribution of the oligomers at 44°C, and showed that the most frequently observed oligomer had four tails, but oligomers with 2 to 13 tails were also present. After heating to 48°C, the proportion of single oligomers was greatly reduced. They suggest that the oligomers were connected through the distal regions of the tails. Samejima et al. (1981) have indicated that the tails, as well as the heads, play a crucial role in myosin gel formation. They used different subunits of the myosin molecule, and have showed that myosin rods can form a gel in a 0.6 M KCl with a rigidity as high as 58% of that formed by the entire myosin molecule. On the other hand, they showed that a solution of subfragment 1 (S-1), obtained by proteolytic cleavage of the heads, formed only a weak gel. They concluded that myosin gelation occurred by head-to-head binding and network formation through linking of the myosin tails.

Role of the Interfacial Protein Film in Stabilizing Meat Batters

Finely comminuted meat batters (sometimes referred to as meat emulsions) are a fairly homogeneous mixture of muscle proteins, fat particles, water, salt, and other ingredients. Finely comminuted products include frankfurters, bologna, and some other meat loaves. The different components are combined to form what has been referred to either as a meat emulsion (Theno and Schmidt, 1978) or a nonemulsion meat batter (Lee, 1985). A classical emulsion consists of two immiscible liquid phases, one of which is dispersed in the other in the form of colloidal suspension. In finely comminuted meat batters, the fat globules constitute the dispersed phase but are sometimes larger than the size required to form a true emulsion-dispersed particle not larger than 20 μm (Lee, 1985). The mechanisms by which meat batters are stabilized has practical applications to further meat processing. In the past, the most widely accepted theory viewed finely comminuted products as a "classic oil-in-water" emulsion. However, some more recent research indicates that the gel-forming ability of the meat proteins can also be a major factor in stabilizing the fat (Lee, 1985; Gordon and Barbut, 1992a). According to emulsion theory, the fat globules are stabilized by the formation of an interfacial protein film around the fat globule, whereas the nonemulsion theory claims that the fat globules are physically entrapped within the protein matrix. So far, the mechanism most important in determining stability has not been precisely determined (Lee, 1985; Gordon and Barbut, 1992a). Overall, understanding the mechanisms involved in stabilizing meat batters is very important in optimizing processing conditions (chopping time, final chopping temperature, and equipment used) to achieve high yields and prevent "emulsion breakdown" (a condition describing the separation of the fat during cooking). Furthermore, new ingredients available to the meat industry and the demand for low-fat meat products also require a better understanding of the mechanisms involved. Studying the microstructure of different meat batters, and relating it to textural and physical properties (gel strength, moisture and fat losses during cooking), has already assisted the industry in improving process control and optimizing product formation. The following discussion presents research data published in support of both the emulsion and the nonemulsion theories.

Early studies (Hansen, 1960) suggested the involvement of myofibrillar proteins in the formation of interfacial protein film around fat globules. Hansen suggested that the salt-soluble proteins are drawn to and concentrated at the fat globule surface to form a stabilizing membrane. Borchert et al. (1967) showed some transmission electron microscope (TEM) pictures indicating the existence of this film around the fat globules, both in the cooked and raw meat batters. They also discovered the existence of small holes (or pores) in the protein film. Jones and Mandigo (1982) showed the existence of a large number of pores in the protein envelope surrounding large fat globules, and the presence of several small fat globules in the vicinity of these pores. They suggested that the pores acted as a "pressure release valve" allowing some of the thermally expended fat to come out during cooking. The myofibrillar proteins involved in the film formation begin to denature at around 45°C; however, meat batters are cooked to 70°C, and at this temperature the encapsulated fat would expand. The creation of weak spots in the protein film would allow release of small amounts of fat while maintaining the integrity of the film. Later Gordon and Barbut (1990b) also showed the existence of the pores in various stable and unstable meat batters (Fig. 13.6). We used five different chloride salt treatments to study the effects of different ions on meat batter stabilization (Table 13.1). We have indi-

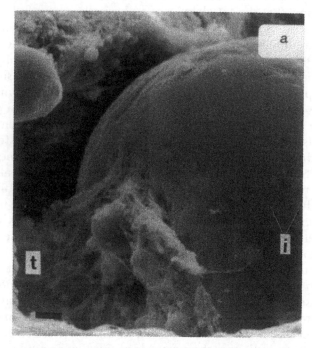

FIG. 13.6   TEM micrographs of meat batters: (a) fat globules in a 2.5% NaCl treatment showing holes in the interfacial protein film (i = poers, t = protein threads; bar = 2 μm); (b) the interfacial protein film in a stable batter (bar-1 μm); (c) unstable fat in a CaCl₂ treatment (bar = 1 μm). (*From Gordon and Barbut, 1990b.*)

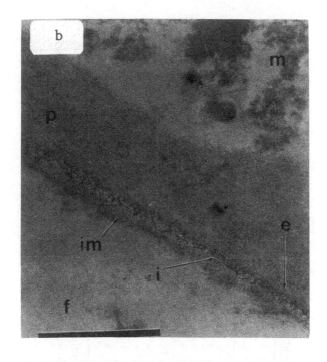

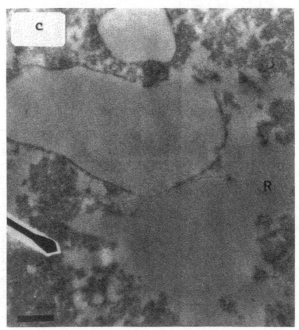

TABLE 13.1 Effect of Chloride Salts on Meat Batter Stability during Cooking and Protein Extraction

Treatment[g]	Batter stability[h] (% loss)			Protein extraction[i] (mg/g)		
	Liquid (cooked)	Fat (cooked)	Fat (raw)	Total	Actomyosin	Myosin
1.5% NaCl	4.62[c]	0.15[bc]	6.1[b]	13.5[d]	2.0[c]	7.0[c]
2.5% NaCl	3.09[cd]	0.08[bc]	5.3[b]	26.8[a]	6.4[a]	9.0[b]
1.35% MgCl$_2$	21.06[b]	0.49[a]	7.9[a]	10.3[e]	0.0[d]	4.5[d]
1.58% CaCl$_2$	23.9[a]	0.55[a]	6.0[b]	4.5[f]	0.0[b]	0.0[e]
3.19% KCl	4.09[c]	0.21[d]	6.0[b]	19.3[e]	5.3[b]	6.9[c]
1.81% LiCl	1.94[d]	0.06[c]	5.7[b]	25.7[a]	6.5[a]	9.9[a]

[a-f]Values followed by a different superscript are significantly different ($P<0.05$).

[g]All treatments were formulated with the equivalent ionic strength of 0.42 ($=2.5\%$ NaCl), except the first treatment where $\mu = 0.25$.

[h](From Gordon and Barbut, 1989.)

[i](From Gordon and Barbut, 1992c.)

cated that monovalent cations positively contributed to meat batter stabilization, while divalent cations (Mg and Ca) destabilized the meat batters. The use of electron microscopy helped to determine the differences in the mechanism by which the two divalent cations destabilized the meat batters (Barbut and Findlay, 1989). In the past it was believed that Mg^{2+} and Ca^{2+} act by the same mechanism (i.e., forming cross-bridges between protein molecules). In order to understand this, it is first important to see how stable meat batters are formed. Figure 13.6a shows a stable product where the fat globules are kept in spherical globules. This batter was formed by the use of 2.5% NaCl, and as seen in Table 13.1 was very stable (i.e., releasing very small amounts of fat and water). The fat globules also looked very stable, and no major fat separation was observed (Fig. 13.6b). The protein film in Fig. 13.6b shows three distinct layers around the fat globules. It should be emphasized that protein film thickness is not the same in all treatments and also not within the same treatment. Sometimes thinner coats are seen, and in some cases an incomplete cover of fat globules is evident (Fig. 13.6a). Similar findings were indicated by Galluzzo and Regenstein (1978). Overall, the formation of the protein film is believed to be an energy-driven process that reduces the free energy of the system. Galluzzo and Regenstein used a model system to show that myosin is absorbed to form a film around the fat globules during emulsification. Consequently, myosin appears to act as an emulsifier even in its native state and formed a film of defined viscoelastic and mechanical properties at the oil-and-water interface, thereby assisting the stabilization of fat in uncooked meat batters. Table 13.1 indicates that the 2.5% NaCl treatment extracted high levels of myosin, and this level was dependent on the concentration of NaCl used. The $CaCl_2$ extracted much less protein from the meat (when used at the same ionic concentration as the 2.5% NaCl) and resulted in a very unstable batter (Fig. 13.6c). This treatment released a lot of fat during cooking and during mild centrifugation in the raw state. The microstructure of the cooked batter showed that the fat was not confined to globules and was basically spread throughout the meat batter, resulting in a very unstable batter or an "emulsion breakdown." The $MgCl_2$, also resulted in a very unstable meat batter, but, as can be seen in Table 13.1, the amount of protein extracted was about three times as high as from the $CaCl_2$ (but still less than half of the NaCl treatment). However, the mechanisms of destabilization were very different, since in this treatment not as much fat was released during mild centrifugation in the raw state. Overall, it was suggested that Mg^{2+} ions destabilized the batter mainly by causing extensive precooking protein matrix aggregation and poor fat stabilization because of insufficient protein film formation. Calcium ions destabilized the batters by causing widespread protein aggregation during cooking, which led to more extensive fat and water losses than caused by Mg^{2+}.

The physical entrapment theory suggests that the fat is stabilized by physical entrapment within the protein gel matrix. According to the theory, proteins in uncooked batter exist in a sol form and later coagulate during cooking to form a gel, which physically traps the fat particles. The fact that some meat proteins can form a sort of gel structure prior to cooking was introduced earlier in this review (Hermansson et al., 1986; Gordon and Barbut, 1990a). Consequently, this means that some kind of a structure that is viscous enough is formed (in the raw state), and it can hold the fat globules in place (i.e., prevent their coalescence) prior to cooking. Theno and Schmidt (1978) and Gordon and Barbut (1990b), have shown the physical binding of fat globules to the protein matrix in meat batters (Fig. 13.6a). The micrograph shows that thread-like protein strands are bound to the fat globule and also entrap a small fat globule on the side. It is therefore logical to assume that protein coagulation during cooking increases the immobilization of the fat glob-ules by binding them to the protein matrix and prevents their coales-cence. Some of the support given to this theory is the fact that various synthetic emulsifiers are detrimental rather than beneficial to meat batter stability (Meyer et al., 1964). Those observations have been used as evi-dence that interfacial activity is of less importance than the structure of the protein matrix in determining fat stabilization. The fact that those synthetic emulsifiers (e.g., Tween 80) can destabilize the emulsion has been shown by various researchers (Gordon and Barbut, 1992a). How-ever, more recent studies (Whiting, 1987; Gordon and Barbut, 1992b) suggest that the destabilization is due to the disruption of the protein matrix.

Overall, the physical and chemical properties of the fat have been shown to be important in determining the stability of the meat batter. The fat-dispersion patterns and fat particle size (determined by chop-ping time and temperature) were indicated to vary with the hardness, density, and melting point of the fat, which therefore affect batter stabil-ity (Lee, 1985). The question of which of the mechanisms is prevalent in meat batter stabilization is still far from being resolved. However, the formation of a cohesive protein matrix is of unquestionable importance in helping stabilize the fat. It assists in entrapping and localizing the larger fat particles (mainly existing as fat cells or cell clumps). However, formation of the interfacial protein film was also shown to take place and affect fat stability. Furthermore, this film has been shown to play a role in facilitating direct fat binding by the protein matrix. Hence it appears that fat stabilization is a combination of the effectiveness of the protein film in localizing fat and the physical restriction and binding provided by a cohesive protein matrix.

Effects of Chemical Modifications

Chemical modifications can be used to study the mechanisms involved in structure/function relationship in protein gels. In this section, examples of using specific chemicals that affect the microstructure of meat batters will be provided. The functionality of proteins is directly dependent on their conformation within a given environment. Even a small change in the environment can lead to a large change in protein conformation. These changes can influence the native structure of the protein, which later on determines the type of interactions the protein will be involved in. Therefore, investigating the effects of different modifications can assist in clarifying the role of meat proteins in the formation and stabilization of a meat batter. In general, functionality of the proteins is examined before and after modification, and this information can be used to determine the effects of a specific modification. Chemical agents such as hydrogen peroxide (a powerful oxidizing agent that oxidizes mainly the free sulfhydryl groups of exposed cysteine side chains) can be used for oxidation (Stark, 1970). Other chemicals that have been used in studying meat proteins are β-mercaptoethanol (β-ME), which effectively reduces exposed disulfide bonds in protein; urea, which disrupts noncovalent bonds (hydrogen bonds and electrostatic interaction) and increases the solubility of hydrophobic groups; EDTA, which participates in electrostatic and H-bonds and therefore can increase protein cross-linking; and Tween 80, which is a very powerful nonionic detergent and an effective emulsifier in food systems (Stark, 1970; Tanford, 1970).

Most of the published information on the functionality of chemically modified meat proteins was derived from pure isolated proteins or subfragments (Jiang et al., 1988). The effect of chemical modification on a commercial type meat batter is also of great interest to the industry. Whiting (1987) was among the first to examine the effects of chemical modification on the functionality of commercial-type meat batters. He reported that H_2O_2 and β-ME had no effect on cooked batter stability and that urea led to a significant increase in water binding in the cooked meat batter (Table 13.2). However, no relationship to the microstructure was reported by Whiting. We (Gordon and Barbut, 1992b) looked at the effect of some of these reagents and related their effects to the microstructure obtained. The meat batters with H_2O_2 and β-ME showed no significant differences in the raw or cooked batter stability compared

TABLE 13.2 Comparing the Effects of Various Chemicals on the Functionality of Meat Batters in Two Different Studies

Treatment	Water loss[e] (%)	Fat loss[e] (%)	Gel strength[e] (kg)	Water loss[f] (%)	Fat loss[f] (%)	Hardness[f] (N)
Control	10.3	0.5	0.45	4.82[b]	0.18[b]	13.06[b]
β-ME	11.4	0.5	0.42	6.94[b]	0.32[b]	11.77[c]
H_2O_2 (0.3%)	10.7	0.7	0.42	5.24[b]	0.18[b]	14.53[a]
(0.6%)						
EDTA (0.20%)	20.0[a]	0.6	0.40	10.32[a]	0.65[b]	6.70[a]
Urea (4.5%)	0.0	0.0	0.57	0.44[c]	0.03[b]	13.41[b]
Tween 80 (0.66%)	30.5[a]	27[a]	0.55	11.74[a]	4.77[a]	9.23[d]

[a–d]Values marked with superscript a are significantly different from the control. Values followed by a different superscript (b–d) are significantly different.
[e]*Results from Whiting (1987). Meat batter (pork/beef) with approximately 10% protein, 27% fat, and 2% salt.*
[f]*Results from Gordon and Barbut (1992b). Emulsified poultry meat batters (18% fat, 15% protein, 2.5% salt).*

to the control (Table 13.2), and the microstructure of these treatments was basically similar to that of the control with 2.5% NaCl (Fig. 13.7a). Based on this observation, it was concluded that disulfide bond formation within the gel structure during cooking does not appear to strongly affect the stability of the system and consequently plays no major role in water and fat binding. If disulfide bonds were essential for water and fat binding, the changes in the amount of these bonds would significantly increase/decrease the results (Table 13.2). Furthermore, the microstructure of the cooked batter as well as the raw batters (micrographs not presented here) showed that all the treatments were stable, and no major water or fat separation occurred. However, the H_2O_2 treatment resulted in a slightly more cohesive matrix. On the other hand, the addition of β-ME caused a slightly more aggregated matrix, which was probably due to more disulfide bonds formation during cooking. The urea-treated batter showed a cohesive and uniform structure with a very fine, highly interconnected network (which had numerous small pores) compared to the control (Fig. 13.7b). This microstructure provided excellent fat- and water-binding properties (Table 13.2). The urea increases the effective hydrophobicity of the meat proteins by solubilizing hydrophobic residues (Nakai, 1983; Whiting, 1988). The micrograph of this treatment (Fig. 13.7b) revealed that this resulted in a much more cohesive structure of the protein matrix.

Using EDTA resulted in an unacceptable texture and high moisture and fat losses. EDTA caused a significant reduction in the gel strength (Table 13.2). The structure of this treatment revealed an excessive cross-linking among the proteins, which EDTA is known to cause (Whiting, 1988). From the microstructural study this appeared to be because EDTA acted mainly on the proteins forming the matrix, and not so much because it affected the interfacial protein film. In fact, EDTA also caused extensive protein matrix aggregation in the raw state (Gordon and Barbut, 1991), which was compounded by cooking. The microscopy evaluation showed a meat batter with large, well-interconnected channels spread throughout the batter. These channels probably formed the route by which water and fat were able to leave the structure. Therefore, it was suggested that the increase in the number of electrostatic and H-bonds in the system is very important in providing good structure and high batter stability.

The treatment of the meat batters with Tween 80 resulted in low gel firmness, and the resulting batter lost a lot of water and fat (Table 13.2). Whiting (1987) also reported a significant deterioration effect due to Tween 80 addition. The structure of the Tween 80 treatment (which lost a lot of liquid) also showed small pores within the gel (Fig. 13.7c).

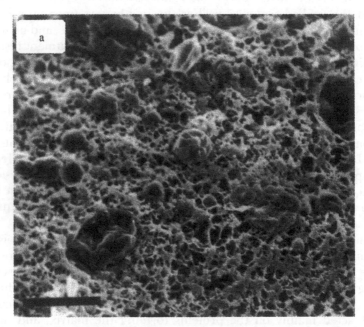

FIG. 13.7   Cryo-SEM of a cooked meat batter prepared with (a) 2.5% NaCl; (b) 2.5% NaCl + 4.5% urea; and (c) 2.5% NaCl + 0.66% Tween 80. Bar = 20μm. (*From Gordon and Barbut, 1992b.*)

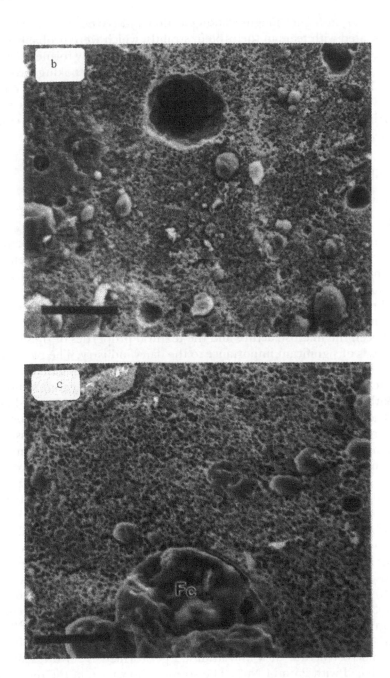

However, at higher magnification Gordon and Barbut noted that it was associated with tears in the protein strands, which were probably due to excessive shrinkage of the structure after it was initially formed. Overall, this treatment lost about 20% of its weight during cooking processes, resulting in a much softer texture compared to the control. Whiting (1987) reported that the Tween 80 treatment lost 50% of its weight during cooking, which resulted in a tougher texture than the control. This was basically because the resulting gel was so dry.

DAIRY PROTEINS

Structural Differences Between Opaque and Transparent Gels

In this section whey proteins will be used as an example because of their ability to form two different types of gels depending on environmental factors (e.g., pH), ions present (e.g., Na vs. Ca salt), and pre-heating conditions. The two types of gels are a transparent gel, which is defined as a fine-stranded gel, and an opaque gel, which is defined as an aggregate gel (Clarke and Lee-Tuffnell, 1986). Studying the relationship between the microstructure and the textural properties of these two gel types is of significant importance to the dairy industry. The two types of gels differ not only in their microstructure (Fig. 13.8a,b) but also in their water-holding capacity, resistance to shear, and critical gel concentration (Barbut, 1994). The aggregated gels are known to have poor water-holding capacity, and water can be expressed by a very low compression force. An interesting observation is that they will reabsorb the moisture (just like a sponge) after the release of pressure (Langton and Hermansson, 1992; Barbut and Foegeding, 1993). On the other hand, the transparent gels have very good water-holding capacity and form elastic structures.

Microscopy has been used very successfully to study the differences between the two gel systems. Methods such as x-ray scattering are more problematic for studying these gels (Clarke and Lee-Tuffnell, 1986), since the opaque gels produce little scattered intensity in the accessible angular region, while the transparent gels produce a strong signal. Microscopy (light, scanning, and transmission) studies have shown large differences in the magnitude of the gel network structure (Kalab, 1985; 1993). Figure 13.8a shows a TEM micrograph of the white coarsely aggregated gel produced from WPI (pH 7.0) in the presence of 25 mM Ca^{2+}. The aggregates seen are in the μm range. Figure 13.8b shows the microstructure of a fine-stranded gel formed from the same WPI produced with 25 mM Na^+. The strands shown are in the nm range.

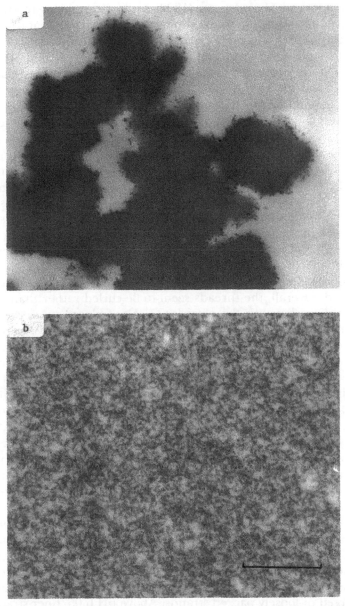

FIG. 13.8   TEM micrographs of heated whey protein isolate gels (10%) produced at pH 70: (a) aggregated white gel produced in the presence of 25 mM Ca and (b) fine-stranded clear gel produced in the presence of 25 mM Na. Bar = 0.5 μm. (*From Barbut, 1994.*)

Langton and Hermansson (1992) studied the microstructure of β-lactoglobulin gels produced at different pH, reporting that white coagulated gels were produced in the pH range 4.0–6.0 and transparent (fine-stranded) gels were produced above and below this region. The aggregated gels are composed of spherical aggregates linked together to form threads, and the size of the particles was found to be pH dependent. The gels could be divided into two groups. The first occurred between pH 5 and 6, where regular network structures could be observed. They were composed of small particles in the pH 5.5–6.0 range and larger particles in the range 5.0–5.5. The networks had uniform distribution of particles and pores throughout the gel. Size distribution was observed to be fairly narrow, and nearly all particles were within the same size in each of the pH treatments. At pH 6.0 the diameter of the spherical particle was about 300 nm. In the pH range 5.5–5.0, the large spherical particles had a diameter of about 1–2 μm. The micrographs also showed that the threads were thicker than the diameter of each individual particle. Cross sections of the gel showed (using TEM) that the particles did not associate in a perfect linear arrangement but were arranged in a zig-zag band. Overall, the threads seem to be curled rather than stiff, and the junctions between the particles were not placed precisely opposite one another. Langton and Hermansson (1992) reported that those gels were elastic and lost moisture during moderate compression. At the lower end of the particulate gel formation (i.e., pH 4.5), an irregular network of coarse and uneven aggregates with subunits at different sizes was observed. Fine-stranded gels (transparent) were formed at pH lower than 4.0 and higher than 6.0; however, the microstructure was not identical at the low and high pH. At the low pH (3.5) very fine, regular networks were seen (fine strands with <5 nm diameter). The strands themselves were about one monomer thick, which supports the idea that the monomers associated in a linear arrangement are forming the strands at this pH (Clark et al., 1981). As the pH was raised toward that of the particulate gels (pH 4.0), the network showed more density fluctuations, resulting from the beginning of particulate area formation. At pH 4.0 the thickness of the strands was estimated to be 4–6 nm, which was slightly thicker than at pH 3.5. Overall, the shift between aggregated and fine-stranded gels was fairly gradual and less sharp than seen at pH 6. Fine structures were formed just about pH 6 (i.e., pH 6.05) and were composed of loosely packed strands. Above pH 6.05, finer strands were formed, and at pH 9.0 the finest strands were produced. The microstructural observations were related to some of the rheological properties by Stading and Hermansson (1991). They reported that gels formed at pH 3.0 were brittle, whereas gels formed at pH 7.5 could be extended to large deformation (Fig. 13.9). The stress/strain curves indicate that the gels formed at pH 4.5, 5.5, and 6.0 (aggregated gels) had similar strain at fracture but that their stresses at fracture differed. The strain for the transparent gels differed depending on the pH. At low pH, the gels were brittle with low strain, while at high pH they had large strains, but they all showed small stress. The authors indicated that small deformation viscoelastic measurement did not distinguish between the different types of gels as well as the large deformation tests. They indicated that small deformation could differentiate between transparent and particulate gels but not within the two types. Other researchers have also indicated that it is difficult to differentiate between various particulate gels using small deformation viscoelastic measurements (Roefs et al., 1990). The reason given was that in large deformation measurements the structure is broken and, basically, the weaker areas are broken first, whereas with the small deformation test, the whole structure contributes equally and no part of the structure is broken. In the study by Stading and Hermansson (1991), the large deformation measurements of the coagulated gels showed differences between the regular and irregular particulate networks. At pH 5.5–6.0, higher stress at fracture was obtained than at pH 4.0–5.2. Those intervals match well with the microstructural evaluation (pH 5.0–6.0, a regular network; pH 4–5.2, an irregular microstructure). At pH 7.0 the fine strands were more curled and had longer contour length between junctions. This might have contributed to the rubberlike behavior measured at pH 7.5, whereas the stiff strands at pH 3 could be responsible for the formation of a fragile gel. These observations suggest that the length and flexibility of the strands, rather than the number of junctions, correlate with the stress and strain at fracture.

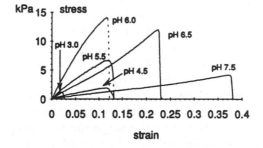

FIG. 13.9 Stress-strain curves in tension for 12% β-lactoglobulin gels. (*From Stading and Hermansson, 1991.*)

A recent study showed the ability of WPI to form a gel at room temperature after the WPI has received a preheat treatment (Barbut and Fogoeding, 1993). In this experiment, WPI (10%) was heated to 80°C and kept at that temperature for different lengths of time (10, 30 min). After cooling, Ca^{2+} was added to the solutions, and gels were formed (referred to as Ca-induced gels). An interesting observation was that clear-type gels were produced when Ca^{2+} was added after preheating. This is opposite to the traditional aggregated gels formed by WPI heated in the presence of Ca^{2+} (referred to as heat-induced gels). The gels resembled the characteristics of a WPI gel prepared with low levels of Na^+ (during heating), which produces a transparent gel (Fig. 13.8b). This is also a Na^+-dependent phenomenon: at low Na^+ level a clear gel is formed, and as Na^+ level increases a more opaque gel is formed. However, the Na^+- and Ca^{2+}-induced gels (regardless of their clarity) have much better water-holding capacity than gels prepared during heating in the presence of Ca^{2+} (which release moisture under moderate compression). The typical heat-induced gels formed in the presence of Ca^{2+} are aggregated gels (Fig. 13.10a) as has also been shown by other researchers (e.g., Foegeding et al., 1992). The microstructure of the gels studied by SEM revealed that a fine-stranded gel matrix was formed in the Ca^{2+}-induced WPI gels (Fig. 13.10b,c). It was found that the textural properties of the Ca^{2+}-induced gels depended on the preheating time (Fig. 13.11). The figure shows different shear stress and shear strain values for the gels after the protein solution was held at 80°C for either 10 or 30 min. In addition, the Ca^{2+} level added also affected the results. It is clear that heating time is related to the amount of protein that can be converted to be "Ca sensitive." This was seen in the fact that the shear stress values of the samples preheated for 30 min were higher than those undergoing the 10-min heating treatment. This and other tests (penetration tests, results not shown here) indicated that heating time had a strong effect on the amount of proteins being "modified" which later on can form a gel. Samples prepared for the penetration test were heated to 70, 80, and 90°C and held for 0–80 min. The results showed that at 70°C, it was necessary to preheat the samples for at least 20 min in order to form a gel. At 80 and 90°C a period of about 5 min was required. The gels produced from the 90°C preheated treatment showed higher penetration force values compared to the treatments preheated to 80 and 70°C. However, all these gels showed lower penetration force values compared to the same samples prepared with Ca^{2+} during heating. This indicated that the type of gels formed in the heat-induced gels and the Ca^{2+}-induced gels were very different. The shear stress values of the traditional heat-induced gels were around 16–18

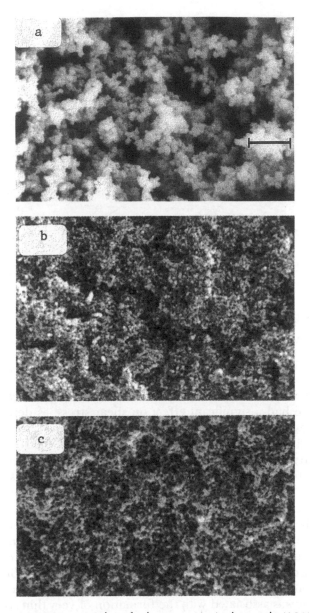

FIG. 13.10   SEM micrographs of whey protein isolate gels (10%): (a) heat-induced gel prepared with 10 mM Ca^{2+}; (b) Ca^{2+}-induced gel prepared from sample heated for 30 min, cooled and 10 mM Ca^{2+} was added; (c) same as (b) but 150 mM Ca^{2+} added. Bar = 2 µm. (*From Barbut and Foegeding, 1993.*)

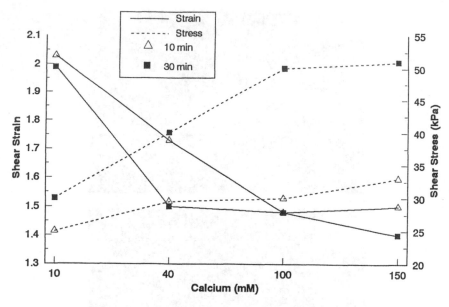

FIG. 13.11   Effects of preheating time and Ca^{2+} level on Ca^{2+}-induced gel made from whey protein isolate. (*From Barbut and Foegeding, 1993.*)

kPa, much lower than values obtained for the Ca^{2+}-induced gels. Overall, heating time had a significant effect on shear stress, as determined by linear regression of the slopes of the two lines (representing shear stress in Fig. 13.11) which were significantly different ($p < 0.001$). The strain values decreased as Ca^{2+} levels increased; the regression analysis comparing the two heating times indicated that they were not significantly different ($p > 0.05$). There was an overall trend of the stress to increase as strain decreased when Ca^{2+} concentrations were raised. The micrographs indicated that the heat-induced gels were aggregated gels composed of beadlike particles attached to one another. Such a structure has been previously reported in thermally induced whey protein concentrate gels (Beveridge et al., 1983) and in β-lactoglobulin gels (Stading and Hermansson, 1991). The Ca^{2+}-induced gels had a fine-stranded network, which was associated with the formation of very thin strands (Fig. 13.10b). The microstructure was mainly affected by the amount of Ca^{2+} used. Figures 13.10b and c show the samples prepared from preheated solutions (80°C, 30 min) to which 10 and 150 mM Ca^{2+} was added, respectively. The treatment with the 150 mM solution had larger pores within the structure compared to the 10 mM treatment. Those differences were probably the cause for the differences in shear strain and shear stress results previously discussed. Overall, the combination of the microstructural study and textural measurements helped to reveal the formation of these Ca-induced gels.

Effect of Fat on Dairy Protein Gel Formation

Dairy gels can be obtained by different mechanisms, from either casein or the whey protein fraction. Yogurt-type gels derived from casein are produced by acid solubilization (citrate) followed by aggregation and fusion of casein micelles in the form of chains and clusters (Kalab, 1985, 1993). Gelling of whey proteins involves aggregation of both the α-lactalbumin and β-lactoglobulin in the denatured state into a structure formed by polypeptide chains cross-linked by disulfide linkages. The microstructure of whey proteins is highly pH dependent, as was previously discussed. Dairy proteins can form simple, mixed, filled, and/or filled-mixed gels (Fig. 13.12). Simple protein or carbohydrate gels are formed from a single repeating component, whereas mixed gels are formed by two or more types of structures (Aguilera and Kessler, 1989; Ziegler and Foegeding, 1990). In the case of dairy proteins, whey proteins (about 60% β-lactoglobulin) will form a simple gel at pH 7.5 (Fig. 13.13B). Skim milk powder will form a mixed gel since it usually contains about 75% casein and 25% whey proteins (Aguilera and Kinsella, 1991). The casein gels are formed by interactions of casein micelles producing a "string-of-beads" arrangement (Kalab, 1985). The types of gels presented in Fig. 13.12 show that in a mixed gel three different forms can exist: incompatible, compatible, and synergistic. The skim milk and WP gel presented in Fig. 13.13C shows a gel composed of two components: (1) the casein micelles, and (2) the whey protein strands, which produced a compatible gel (the influence of fat globules on the structure is a different issue and will be discussed later). The authors (Aguilera and Kessler) concluded that this is a compatible gel after they had studied the firmness of different mixtures of skim milk powder (SMP) and whey protein concentrate (WPC). They compared the firmness of the combinations to a pure SMP gel and showed that a WPC gel was about 1.4 times firmer than SMP gel. No mixture showed a gel firmness lower than that of the pure SMP gel, so compatibility between those proteins was assumed. Maximum firmness was observed in blends having a ratio of 1:1 SMP:WPC. In a later study (Aguilera and Kinsella, 1991), it was indicated that the addition of WP to predominantly SMP gels strengthened the structure of the SMP gels and allowed gelation to occur at lower SMP concentrations. However, this was not true when small quantities of

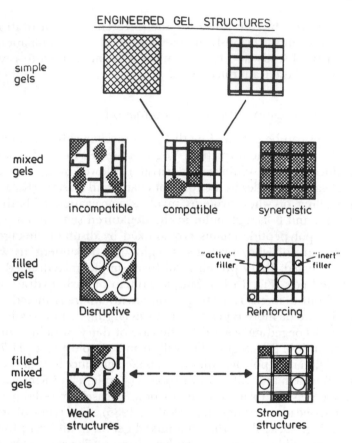

FIG. 13.12   Possibilities for engineered structures of simple, mixed, filled, and filled mixed gels. (*From Aguilera and Kessler, 1989.*)

SMP were added to a predominantly WPI gel, which showed a weakening effect. The weakening effect could only be overcome when SMP concentrations increased above 4–5%. The authors hypothesized that the undeveloped casein gel behaved as a filler (at low concentrations) that interfered and weakened the basic structure of WP gels (see Fig. 13.12 for an illustration of a filler disrupting the gel structure). They confirmed that casein gels are forming the "string-of-beads" structure (Kalab, 1985), and therefore their gelation requires a minimum number of casein units to form chains long enough to develop a continuous network. Their hypothesis is consistent with the report of Langley et al. (1988), who observed that the strength of whey protein gels filled with small glass beads, interacting with the matrix, was reduced at a low volume fraction.

When emulsified fat globules were added to the mixed gel (Fig. 13.13C), an increase in gel strength was observed. The fat globules were preemulsified with sodium caseinate. This was a very important factor in allowing their compatibility with the gel structure, since the protein coat (around the fat globules) allowed their easy interaction with the other proteins. In both the SMP or WPC gels, the addition of the emulsified fat globules showed only a slight increase in firmness. However, substitution of one protein source by the other with simultaneous addition of fat resulted in firming of the gel, which reached a distinct maximum at the ratio of 1:1 SMP/WPC. The reinforcement and the production of filled gels can be interpreted in two ways. First, the fat globule membrane interacted positively with the protein matrix, as was suggested by van Vliet and Denterner-Kikkert (1982). The second interpretation is that there was a steric effect caused by the exact fitting of the fat droplets in the porous structure (of the gel) without disturbing the reticulation (Jost et al., 1989). In Fig. 13.13A (i.e., casein micelles and 7% fat), the relative firmness (expressed as 1 for the SMP gel) increased $\times$ 1.1, whereas the relative firmness of the gel shown in Fig. 13.13C was $\times$ 2.2 (Aguilera and Kessler, 1989). When 20% fat was added to this mixture, the relative firmness increased $\times$ 2.4.

Understanding the effect of particles filling spaces is important in the production of dairy products such as cheese and yogurt (which represent composite gels) and in future product development. In dairy gels denatured serum proteins can act as a filler or binder (within the casein matrix) and fat globules can be incorporated into the structure (Kalab, 1985). Comparing this information to data obtained in nonfood materials can be very helpful. Some fillers are known to result in reinforcement of a polymeric matrix due to volumetric and/or surface phenomena (Jastrzebski, 1976). A well-documented phenomenon is the addition of carbon black to rubber, which can result in an enormous increase in the mechanical moduli of pure rubber. In this case, not only the actual size of the carbon black particles, but also their strong surface adsorption properties play a critical role. Aguilera and Kessler (1989) indicated that the fat globules in their experiment were uniformly distributed within the matrix. This may have contributed to good fitting of the fat within the pores of the protein matrix (Fig. 13.13C). An adsorption of the caseinate at the fat globule interface was also seen; however, no conclusions could be drawn as to the exact physico chemical basis of this interaction. Overall, based on the moderate increase in relative firmness of the filled dairy gels, it is possible that the globules assisted by hydrostatically

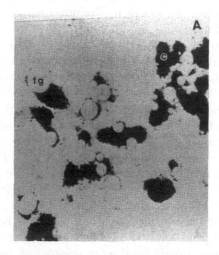

FIG. 13.13   TEM of filled gels from skim milk powder (SMP), whey protein concentrate (WPC) and fat: (A) SMP/fat; (B) WPC/fat; and (C) WPC/SMP fat. Arrows point to casein chains (c) and whey protein domains (W); fg = fat globule. (*From Aguilera and Kessler, 1989.*)

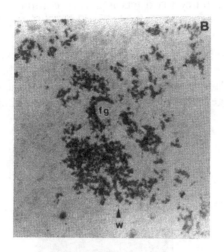

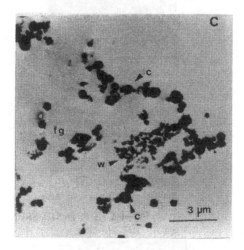

restraining the movement of the matrix close to them. This is suggested since reinforcement based on surface phenomena would induce a much higher increase in the mechanical properties of the gel.

Fat addition to dairy products is a very common practice. Xu et al. (1992) studied the effect of fat addition (2.0, 4.0%) on the microstructure of milk/hydrocolloid (10.5% skim milk powder/0.5% carrageenan) gels. The gel without fat (Fig. 13.14a) shows a typical honeycomb gel structure (referred to as a "yogurt-type gel") of milk proteins. When 2.0% preemulsified fat (with WPI) was added (Fig. 13.14b), most of the globules fitted into the pores of the gel matrix, while some were situated at the surface.

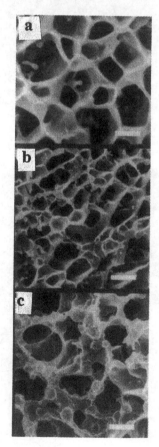

FIG. 13.14   Effect of fat on gel structure (TS = 10.5%, pH 7.0) as determined by cryo-SEM. (a) 0% fat; (b) 2% fat; and (c) 4% fat. Bar = 30 μm. (*From Xu et al., 1992.*)

Overall, a much denser gel network was produced when 2.0% fat was added. The gel strength obtained for this microstructure was higher than the gel without fat (2.71 ± 0.1 N vs. 2.38 ± 0.08, respectively). When 4.0% fat was added (Fig. 13.14c), a more open structure was obtained than with the 2.0% fat addition. The fat globules were adsorbed on the surface of the gel network, while some were observed situated in the pores. This gel showed a lower gel strength (2.51 ± 0.09 N) compared to the 2% fat treatment. The authors (Xu et al., 1992) indicated that this was expected since in another experiment they observed that an increase in gel density (by increasing the amount of total solids) resulted in a higher gel strength. They pointed out that the gel density of the 4.0% added fat and no-fat treatments were very similar, which might have been related to the very close gel strength results obtained. However, they have also indicated an alternative explanation which emphasizes the presence of higher volume of the filler (fat), resulting in more fat globules and less separation among them. This can disrupt the gel conformation and lead to a weaker structure. Even though it is not easy to fully explain the mechanical properties at the microstructure level (due to the complexity of the interactions), this study is useful in studying the effect of fat on gels, where fat was observed to be a filler as well as contributing to the texture. Aguilera and Kessler (1989) discussed the properties of fat modification by homogenation and the role of protein membranes as gel ingredients. They indicated that fat globules can sometimes behave as "active" large protein units and act as polymerizing nucleide to reinforce the gel by forming a co-polymer network. Thus, fat coated with protein can bind additional protein molecules and assist in gel matrix formation. However, a very critical factor in this process is the homogenation of the fat, which should result in small size fat globules coated with protein. Obviously, large fat droplets can disrupt the protein matrix.

Effects of Freezing Rate on Ice Cream

Ice cream has a unique composition of both an oil-in-water emulsion and a fat-rich foam. Proteins in an ice-mix (1) assist in emulsifying the fat, (2) contribute to foam formation by adsorbing to the lamellae, and (3) increase the viscosity by binding water. The texture of ice cream is initially produced in the scraped-surface freezer, where nucleation of ice crystals is initiated and air is incorporated. Many ice crystal nuclei are scraped into the bulk of the ice cream mix and crystal growth occurs later during the hardening process. Once the product has left the scraped-surface freezer, no additional nucleation will occur. However, the number of nuclei initially produced and their final size will significantly affect the quality of the ice cream. It is accepted that if the average ice crystal size is >55 μm, a coarse texture (sandy/icy) will occur. The effect of processing conditions, mainly hardening rate, will be used here to illustrate the relationship between texture and microstructure.

Caldwell et al. (1992) used cryo-SEM to examine the effect of hardening rate on the relative proportion of different structures in ice cream. The use of cryo-SEM permitted studying the ice cream in its natural state (after sublimating the water from the surface of the frozen sample) thus eliminating chemical fixation. Three hardening rates were used: (1) a very rapid freezing by liquid nitrogen, (2) plate hardening at −40°C, and (3) slow freezing at −25°C. A typical micrograph (Fig. 13.15) shows the four-phase structure of batch-frozen ice cream. The ice crystal area has the typical appearance of a freeze-etched area of a solute solution containing salt. It shows a network of a reticulate structure with small spheres (approximately 0.2–0.4 μm), which are thought to be the con-

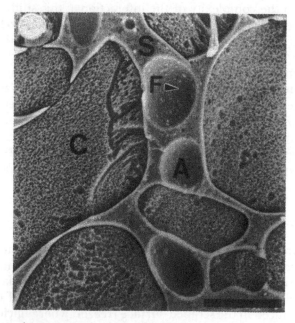

13.15 A typical micrograph showing the four phase structure of batch-frozen ice cream: partially freeze-etched ice crystal socket (c), air bubble (A), fat globules (F), and serum phase (S). Bar = 25 μm. (*From Caldwell et al., 1992.*)

centrates of the salts (and/or solids) dissolved in the aqueous phase. The fast freezing by liquid nitrogen (treatment 1) represents the structure formed in the continuous freezer. This was captured by dipping the product coming out of the scraped-surface freezer in liquid nitrogen. At this point the ice cream is in a nonequilibrium state and the sample contains a large number of small air bubbles and small ice crystals with poorly defined borders. The area occupied by air (as measured on a fractured surface) represented almost 50% of the area (Table 13.3). The open serum structure and the poorly defined ice crystal borders demonstrated that at this stage all the freezable water was frozen and the freeze concentration process was incomplete. The water did not have time to migrate to the freezing front. In plate hardening (treatment 2), which represents industrial conditions (freezing to $-25°C$ takes about 1 hr), a different microstructure was obtained. The ice crystals were relatively small (approximately 25 μm diameter), yet larger than in the nitrogen freezing. The ice crystal borders were well defined, indicating that the solutes had more time to migrate from the freezing front. The air bubbles had become larger and many assumed a more elongated shape. The slowest freezing rate (i.e., at $-25°C$), which took 5 hr to harden, showed a different structure. The ice crystal size was larger and their average size was 35 μm. Overall, ice cream—hardening rate should be fast enough to allow sufficient ice crystal formation without causing a large degree of foam collapse. As the hardening rate becomes slower, the average area of the ice crystals became larger and the average area of air decreased (Table 13.3). Although the total amount of air in the samples was consistent (after drawing the product from the scraped-surface freezer), it appeared that the smaller air bubbles seen in the nitrogen-freezing treatment had collapsed and formed larger air bub-

TABLE 13.3   Percentage of Total Area of Micrograph for Three Ice Cream Phases as a Function of Hardening Rate and Storage Time

| Phase | Hardening rate[a] | | | |
	Fast N_2	Plate $(-40°C)$	Freezer $(-25°C)$	Fast N_2 (2 weeks)
Ice	17.3	29.1	34.6	33.9
Air	49.4	41.3	25.7	34.0
Serum	33.3	29.6	39.7	32.1

[a]For first three columns, t = 0 wk; fourth, t = 2 wk.
Source: Caldwell et al., 1992.

bles, thus giving the impression of air reduction in the total surface area occupied by the air bubbles. This observation coincides with the theory that ice cream is at a nonequilibrium state as it leaves the continuous freezer.

Another observation was made after 2 weeks of storage (at $-25°C$) of the initially fast-frozen ice cream. In this treatment, ice crystal surface area significantly ($p < 0.05$) increased (during storage) and air bubble surface area significantly decreased (Table 13.3). The initial fast freezing resulted in water solidifying into a glass state, because there was not enough time for crystallization. However, when the ice cream was stored at a temperature above its glass transition point (approximately $-35°C$), the mobility of the constituents increased. The serum changed from a glassy solid to a viscoelastic liquid, which allowed recrystallization. This kind of study, using modern cryo-fixation techniques, opens the door for examining samples in their natural state (without the need for chemical fixation) and relating it to textural properties.

MICRO-ENCAPSULATION

Use of Whey Proteins

Micro-encapsulation is micro-packaging of solid particles or liquid droplets coated by a thin film (usually called the "wall"). The wall, or protective coat, encapsulates the material inside from deterioration such as oxidation, interaction with water, or changes in aroma (Reineccius, 1988). Another reason for encapsulation is to enable controlled release of various chemicals under predetermined conditions (such as salts/acids requiring slow release during processing of dairy/meat products). In the food industry, typical materials that are coated consist of flavor and aroma compounds, fats, essential oils, minerals, vitamins, and oleoresins. The most common method for micro-encapsulation in the food industry consists of spray drying the core material with the coating ingredient. The film materials used by the food industry mainly include natural gums, carbohydrates, waxes, gelatin, and some chemically modified polymers. So far, natural gums and carbohydrates (i.e., starches) are the most commonly used ingredients for micro-encapsulation. The requirements from these materials are high solubility, film forming and drying characteristics, low viscosity at high concentrations, and in some cases emulsification (Reineccius, 1988). Besides protection from chemical deterioration, the film also determines the flowability and wettability of the final product, which is essential to the quality of the product. Over the years, microscopy has been used successfully for quality control and evaluating the potential of using new "wall" materials. In this section an example of the potential use of whey protein (WP)–based microcapsules will be provided in order to demonstrate the beneficial effect of microscopy in determining the feasibility of new microencapsulation agents.

Rosenberg and Young (1993) reported the potential use of WP (two types of concentrates and one type of isolate) as a micro-encapsulating agent for anhydrous milk fat. The WP solutions were prepared (10–20% w/w) and then homogenized with the fat followed by spray drying at different inlet air temperatures (105–210°C) and outlet air temperatures (50–95°C). The micrographs shown in Fig. 13.16 deal with the 20% whey protein concentrate loaded with 75% milk fat produced at (1) 105°C inlet and 50°C outlet, (2) 160°C and 80°C, and (3) 210°C and 90°C, respectively. The capsules dried at 105°C were characterized by deep surface dents (Fig. 13.16a). The proportion of capsules exhibiting surface dents was significantly reduced when drying temperature was increased to 160°C, and this phenomenon was almost completely eliminated at an inlet air temperature of 210°C. Thus, the surface topography of the final product depended on drying temperatures or drying rates. At high drying rates, the thermal expansion of either the air or water vapors inside the drying particles could "erase" the dents. The structure of the whey protein microcapsules reported here differs from that reported when gum arabic or dextrins served as "wall" materials during spray drying (Chang et al., 1988; Rosenberg et al., 1985). An example of capsules with high proportions of dents is shown in Fig. 13.17a. It has been reported that the formation of deep surface dents in micro-encapsulation, by spray drying, could be eliminated by introducing a fast drying rate (Rosenberg et al., 1985). In such cases, the expansion of the capsules must take place after the dents are formed, but at a stage when the "wall" system is still elastic enough to allow "ballooning." In extreme cases, excessive expansion of the capsules, during the late stages of drying, may result in rupturing. Such fragmentation has been reported by Rosenberg and Young (1993) when 20% WPI was used by itself. The use of microscopy to examine the morphology and/or fragmentation of the capsules provides a fast and convenient method for evaluating the quality of the process. Otherwise, indirect analysis, which is more time consuming, should be used (the presence of cracks and pores in the wall can be identified by long-time incubation of the powders and continuous detection of the retention of the volatile material by checking their strength).

Results obtained for the WP concentrate–based capsules are in agreement with reported effects of drying temperature on the surface morphology of spray-dried particles (Greenwald and King, 1982). Over-

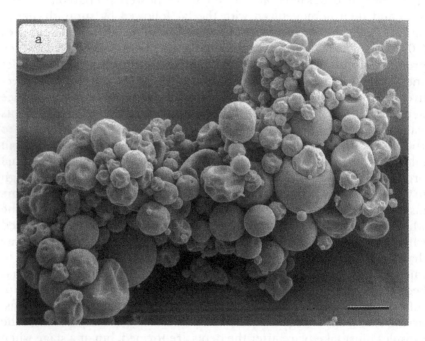

FIG. 13.16 SEM of spray-dried WP microcapsules containing milk fat. Three different drying conditions are presented: (a) inlet = 105°C and outlet = 50°C; (b) 160 and 80°C; (c) 210 and 90°C, respectively. Bar = 20 μm. (*From Rosenberg and Young, 1993.*)

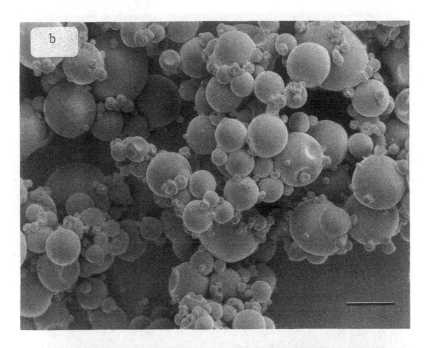

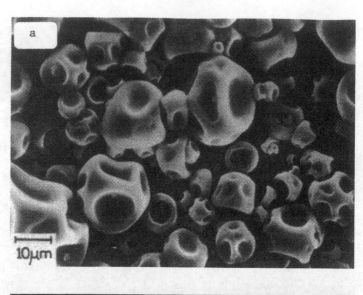

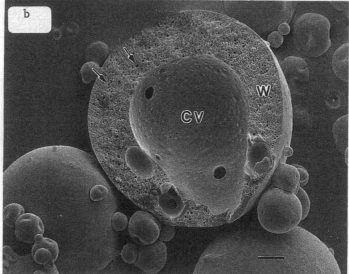

FIG. 13.17   SEM of spray-dried microcapsules. (a) Gum arabic used as a wall material. (*From Rosenberg et al., 1985*); (b) SEM of typical inner structure of a WPI-based microcapsule. CV = central void; W = wall; small fat droplets can be seen inside the wall material. Bar = 5 μm. (*From Young and Rosenberg, 1993.*)

all, dent formation during spray drying is related to an uneven shrinkage of the atomized droplets at the early stages of spray drying. An attempt to develop mathematical models correlating surface topography (studied by SEM) with the effects of compositional or process parameters has been reported (Keith, 1983). It has been established that the tendency to develop faults depends on the ratio of low-to-high molecular weight solutes in the feed. A mechanism based on a surface tension–driven process and the viscous flow of surface folds (formed by uneven shrinkage of the drying droplets) was hypothesized. Based on this theory it was speculated that a critical viscosity for folding exists, below which the surface tension forces are sufficient to smooth out surface irregularities.

The microcapsules were also fractured in order to examine their inner structure. Three different procedures were used. The first included embedding the capsules in a Lowicryl HM-20 resin. This resulted in a significant shape distortion, which was attributed to softening of the capsule walls by the embedding medium. Furthermore, many of the capsules were cracked during the resin polymerization (under UV light). The second method included gluteraldehyde fixation prior to embedding. This also did not eliminate the adverse effects mentioned above. The third method, consisting of fracturing the capsules with a razor blade (after gluing them to a specimen holder), was the most successful. This again illustrates the importance of using different techniques to study the microstructure of various fragile protein structures. Figure 13.17b shows the inner structure of the spray-dried capsule. Small particles of the core material (anhydrous milk fat) were evenly distributed throughout the protein matrices of the wall. These structural features are similar to those reported for spray-dried gum arabic–based microcapsules (Rosenberg et al., 1985). A central void was seen in all the WP-based capsules regardless of drying conditions or composition. This is in contrast to results reported for the gum arabic–based microcapsules where low drying temperature resulted in limited formation of central voids. In all cases, the milk fat was distributed as small droplets (50–600 nm) embedded in the capsule wall. Similar structural details were reported in spray-dried whole milk powder (Buma, 1971). The diameter range of the encapsulated milk fat (Fig. 13.17b) was similar to that of the milk fat droplets at the emulsion stage as determined prior to drying. This indicates that no coalescence of the milk fat droplets occurred during the drying process. Figure 13.17b also shows no visible cracks or channels exposing the encapsulated milk fat droplets to the environment. Some voids were found in the wall, and they were also found in the milk fat–free whey protein capsules used as a control. This finding suggests that either air bubbles were trapped in the wall solution or small capsules were trapped in the walls of larger capsules.

In studies relating to the structure of spray-dried whole and skim milk powders, it has been suggested that the observed surface faults, dents, and pores are the effect of mechanical stresses, including uneven drying at different parts of the droplets and shrinkage of casein (Buma, 1971). The author concluded that casein, rather than lactose, was probably responsible for the surface dents. In whole milk, it was suggested that the indentations are the result of a large number of fat globules that might hinder the shrinkage of casein. On the other hand, in ultrafiltered and diafiltered skim milk (Mistry et al., 1992), surface dents and wrinkles were attributed to lactose content. Overall, the presence of surface dents adversely affects the flowability, as well as the wettability of powders because it leads to the formation of clumps or aggregates. Research by Rosenberg and Young (1993) showed that whey protein–based spray-dried microcapsules have a different structure than other milk-derived powders. The shallow surface dents found at low drying temperatures (Fig. 13.16a) were eliminated by increasing the drying temperature. Regardless of lactose or fat content, surface wrinkles were not detected. The results indicated that whey proteins can be considered an effective micro-encapsulating agent for milk fat in preparing an "all-milk-derived" powder with high fat loads. The microstructure of the inner and outer structural features of the capsules indicated that good physical protection is provided. However, the stability of the milk fat remains to be determined.

CONCLUSIONS

Studies relating the microstructure of food protein gels to their textural characteristics and the type of gel they can form (i.e., aggregated or fine-stranded) have provided much practical information to the food industry. In addition, some of the basic mechanisms involved in gel formation have been described. However, as discussed in this chapter, the relationships between structure and functionality are not always so obvious, and there is still much to be learned. The rapid progress in computer software/hardware during recent years has made it possible to monitor processes and acquire data (by dynamic monitoring of gelation, image analysis, etc.) at a rate and precision not previously available. It is hoped that this progress will assist us in obtaining better insight into the relationships between structure and functionality, which would be useful in improving existing products and in designing new food products.

REFERENCES

Aguilera, J. M. and Kessler, H. G. 1989. Properties of mixed and filled-type dairy gels. *J. Food Sci.* 54: 1213.

Aguilera, J. M. and Kinsella, J. E. 1991. Compression strength of dairy gels and microstructural interpretation. *J. Food Sci.* 56: 1224.

Aguilera, J. M. and Stanley, D. W. 1990. *Microstructural Principles of Food Processing and Engineering.* Elsevier Applied Science, New York.

Ashgar, A., Samajima, K., and Yasui, T. 1985. Functionality of muscle protein in gelation mechanisms of structured meat products. *CRC Crit. Rev. Food Sci. Nutr.* 22: 27–107.

Barbut, S. 1994. Effect of sodium levels on the microscructure and texture of whey proteins isolate gels. (Submitted for publication.)

Barbut, S. and Findlay, C. J. 1989. Sodium reduction in poultry products—a review. *CRC Crit. Rev. Poultry Biol.* 2: 59.

Barbut, S. and Foegeding, A. 1993. Ca^{2+}-induced gelation of preheated whey protein isolate. *J. Food Sci.* 58: 867–871.

Beveridge, T., Jones, L., and Tung, M. A. 1983. Progel and gel formation and reversibility of gelation of whey, soybean and albumen protein gels. *Food Microstruct.* 2: 161.

Borchert, L. L., Greaser, M. L., Bard, J. C., Cassens, R. G., and Briskey, E. J. 1967. Electron microscopy of a meat emulsion. *J. Food Sci.* 32: 419.

Brotschi, D. A., Hartwig, J. H., and Stossel, T. P. 1978. The gelation of actin by actin-binding protein. *J. Biol. Chem.* 253: 8988.

Buma, T. J. and Henstra, S. 1971. Particle structure of spray-dried milk products as observed by scanning electron microscope. *Neth. Milk Dairy J.* 25: 75–80.

Caldwell, K. B., Goff, H. D., and Stanley, D. W. 1992. A low-temperature scanning electron microscopy study of ice cream. II. Influence of selected ingredients and processes. *Food Struct.* 11: 11.

Chang, Y. I., Scire, J., and Jacobs, B. 1988. In *Flavor Encapsulation*, G. A. Reineccius and S. J. Risch (Ed.). American Chemical Society Symposium No. 370, Washington, DC.

Clark, A. H. 1992. Gels and gelling. In *Physical Chemistry of Foods*, H. G. Schwartzberg and R. W. Hartel (Ed.), pp. 263–305. Marcel Dekker, New York.

Clark, A. H. and Lee-Tuffnell, C. D. 1986. Gelation of globular proteins. In *Functional Properties of Food Macromolecules*, J. R. Mitchell and D. A. Ledward (Ed.), pp. 203–272. Elsevier Applied Science, London.

Clark, A. H., Judge, F. J., Stubbs, J. M., and Suggett, A. 1981. Electron microscopy of network structures in thermally induced globular protein gels. *Inter. J. Peptide Prot. Res.* 17: 380.

Foegeding, E. A., Kuhn, P. R., and Hardin, C. C. 1992. Specific divalent cation-induced changes during gelation of β-lactoglobulin. *J. Agric. Food Chem.* 40: 2092–2097.

Galluzzo, S. J. and Regenstein, J. M. 1978. Role of chicken breast muscle proteins in meat emulsion formation: Myosin, actin and synthetic actomyosin. *J. Food Sci.* 43: 1761.

Gordon, A. and Barbut, S. 1989. The effect of chloride salts on the texture, microstructure and stability of meat batters. *Food Microstruc.* 8:271.

Gordon, A. and Barbut, S. 1992a. Mechanisms of meat batter stabilization—a review. *CRC Crit. Rev. Food Sci. Nutr.* 32(4): 299.

Gordon A. and Barbut. S. 1992b. Use of carrageenans and xanthan gums in reduced fat breakfast sausage. *Food Struct.* 11: 133.

Gordon A. and Barbut, S. 1992c. The effect of chloride salt on protein extraction and interfacial protein film formation in meat batters. *J. Sci. Food Agric.* 58: 227.

Gordon, A. and Barbut, S. 1991. Effect of chemical modification on the microstructure of rat meat batters. *Food Struct.* 10: 241.

Gordon, A. and Barbut, S. 1990a. The microstructure of raw meat batters prepared with mono and divalent chloride salts. *Food Struct.* 9: 279.

Gordon, A. and Barbut, S. 1990b. The role of the interfacial protein film in meat batter stabilization. *Food Struct.* 9: 77.

Greenwald, C. G. and King, C. G. 1982. The mechanism of particle expansion in spray drying of foods. *AIChE Symp. Ser.* 218: 101–108.

Hamann, D. D. 1987. Methods for measurement of rheological changes during thermally induced gelation of proteins. *Food Technol.* 41(1): 100–108.

Hansen, L. J. 1960. Emulsion formation in finely comminuted sausage. *Food Technol.* 14: 565.

Hermansson, A. -M., Harbitz, O., and Langton, M. 1986. Formation of two types of gels from bovine myosin. *J. Sci. Food Agr.* 37: 69.

Hermansson, A. -M. and Langton, M. 1988. Filamentous structure of bovine myosin in diluted suspensions and gels. *J. Sci. Food Agr.* 42: 335.

Huxley, H. E. 1963. Electron microscope studies on the structure of natural and synthetic protein filaments from striated muscle. *J. Mol. Biol.* 7: 281.

Ishioroshi, M., Samajima, K., and Yasui, T. 1983. Heat-induced gelation of myosin filaments at a low salt concentration. *Agric. Biol. Chem.* 47: 2809.

Jastrzebski, Z. D. 1976. *The Nature and Properties of Engineering Materials.* John Wiley, New York.

Jiang, S. T., Hwang, D. C., and Chen, C. S. 1988. Effect of storage temperatures on the formation of disulphides and denaturation of milkfish actomyosin (*Chanos Chanos*). *J. Food Sci.* 53: 1333.

Jones, K. W. and Mandigo, R. W. 1982. Effect of chopping temperature on the microstructure of meat emulsions. *J. Food Sci.* 47: 1930.

Jost, R., Dannenberg, F., and Rosset, J. 1989. Heat-set gels based on oil/water emulsions: An application of whey functionality. *Food Microstruct.* 8: 23.

Kalab, M. 1985. Microstructure of dairy foods. 2. Milk products based on fat. *J. Dairy Sci.* 68: 3234.

Kalab, M. 1993. Dairy foods microstructure. *Food Struct.* 12: 95.

Keith, A. 1983. Factors governing surface morphology in the spray-drying of foods. Ph.D. thesis, University of California-Berkeley. (Microfilm #8413299, University Microfilms International, Ann Arbor, MI.)

Langley, K. R., Green, M. L., Brooker, B. E., and Smith, A. C. 1988. Mechanical properties of whey protein gels in relation to composition and microstructure. In *Gums and Stabilizers for the Food Industry.* Elsevier Applied Science, New York.

Langton, M. and Hermansson, A. -M. 1992. Fine-stranded and particulate gels of β-lactoglobulin and whey protein at varying pH. *Food Hydrocolloids* 5(6): 523–539.

Lee, C. M. 1985. Microstructure of meat emulsions in relation to fat stabilization. *Food Microstruct.* 4: 63.

Meyer, J. A., Brown, W. L., Giltner, M. E., and Grinn, J. R. 1964. Effect of emulsifiers on the stability of sausage emulsions. *Food Technol.* 18: 1796.

Mistry, V. V., Hassan, H. N., and Robinson, D. J. 1992. Effect of lactose and protein on the microstructure of dried milk. *Food Microstruct.* 11: 73–82.

Nakai, S. 1983. Structure-function relationships of food proteins with an emphasis on the importance of protein hydrophobicity. *J. Agric. Food Chem.* 31: 676.

Pollard, T. D. 1975. Electron microscopy of synthetic myosin filaments. *J. Cell Biol.* 67: 93.

Reineccius, G. A. 1988. Spray drying in food flavors. In *Flavor Encapsulation*, G. A. Reineccius and S. J. Risch (Ed.), pp. 55–66. American Chemical Society Symposium No. 370, Washington, DC.

Roefs, S. D., DeGroot-Mostert, A. E., and van Vliet, T. 1990. Measurements of particulate gels. *Coll. Surf.* 50: 141.

Rosenberg, M., Kopelman, I. J., and Talmon, Y. 1985. A scanning electron microscopy study of microencapsulation. *J. Food Sci.* 50: 139.

Rosenberg, M. and Young, S. L. 1993. Whey proteins as microencapsulating agents. Microencapsulation of anhydrous milkfat—structure evaluation. *Food Struct.* 12: 31.

Russ, J. 1991. In *Computer Assisted Microscopy*, p. 8. Plenum, New York.

Samajima, K., Ishioroshi, M., and Yasui, T. 1981. Relative role of the head and tail portions of the molecule in heat-induced gelation of myosin. *J. Food Sci.* 46: 1412–1418.

Samajima, K., Ishioroshi, M., and Yasui, T. 1982. Heat-induced gelling properties of actomyosin: Effect of tropomyosin and troponin. *Agric. Biol. Chem.* 46: 535.

Sharp, A. and Offer, G. 1992. The mechanism of formation of gels from myosin molecules. *J. Sci. Food Agr.* 58: 63.

Stading, M. and Hermansson, A. -M. 1991. Large deformation properties of β-lactoglobulin gel structures. *Food Hydrocolloids* 5: 339–352.

Stark, G. R. 1970. Recent developments in chemical modifications and sequential degradation of proteins. *Adv. Prot. Chem.* 24: 261.

Tanford, C. 1970. Protein denaturation, Part C. Theoretical models for the mechanism of denaturation. *Adv. Prot. Chem.* 24: 2.

Theno, D. M. and Schmidt, G. R. 1978. Microstructural comparison of three commercial frankfurters. *J. Food Sci.* 43: 845.

Trinick, J. and Elliott, A. 1982. Effect of substrate on freeze-dried and shadowed protein structures. *J. Microsc.* 126: 151.

van Vliet, T. and Denterner-Kikkert, A. 1982. Influence of the composition of the milk fat globule membrane on the rheological properties of acid milk gels. *Neth. Milk Dairy J.* 36: 261.

Whiting, R. C. 1987. Influence of lipid composition on the water and fat exudation and gel strength of meat batters. *J. Food Sci.* 52: 1130.

Whiting R. C. 1988. Solute-protein interactions in a meat batter. In *Reciprocal Meat Conference Proceedings*, Amer. Meat Sci. Assoc., p. 53.

Xu, S. Y., Stanley, D. W., Goff, H. D., Davidson, V. J., and LeMaguer, M. 1992. Hydrocolloid/milk gel formation and properties. *J. Food Sci.* 57: 96–102.

Yasui, T., Ishioroshi, M., and Samajima, K. 1982. Effect of actomyosin on heat-induced gelation of myosin. *Agric. Biol. Chem.* 46: 1049.

Yasui, T., Ishioroshi, M., and Samajima, K. 1980. Heat-induced gelation of myosin in the presence of actin. *J. Food Biochem.* 4: 61.

Ziegler, G. R. and Foegeding, E. A. 1990. The gelation of proteins. *Adv. Food Nutr. Res.* 34: 203–298.

14

Proteins as Fat Substitutes

Mark S. Miller

Kraft General Foods, Inc.
Glenview, Illinois

INTRODUCTION

The introduction of protein-based fat mimetic ingredients in the late 1980s is considered by some to be a breakthrough in food-processing technology which revolutionized the category of reduced-fat and fat-free foods. Prior to that time, the food industry took an incremental approach to fat reduction, and the marketplace saw the introduction of many *light* products, which were one quarter to one half reduced in fat. Very few fat-free versions were introduced or even in development because of a belief that lipid-based fat substitutes were needed for these products. Protein-based fat mimetics broke that paradigm. The creative breakthrough brought about by protein-based fat mimetics opened up lower-fat food product development to a variety of water-based fat-replacement technologies. The flood gates were open for fat-free products to pour into the marketplace.

Two years ago, I was approached about participating in this IFT-

IUFoST Basic Symposium. I was overjoyed and eager to review the great progress that I was sure would have come about in understanding the structure and function of protein-based fat replacers. At that time, protein-based fat mimetics were regarded as magic bullet ingredients, which could be used in almost any application to create fat-free products indistinguishable in taste or texture from their full-fat counterparts. With all the emphasis on these ingredients in both the food trade journals and the public press, there was sure to be a flood of basic studies on their functionality in lower-fat foods.

Well, this has not happened. While there have been dozens of fat mimetics introduced into the market in the last 3–4 years, very few of them are based on protein technology. Many use carbohydrate polymers as their structural units. A variety of factors have contributed to this shift from protein to carbohydrate. First of all, many of the functional roles provided by proteins in lower-fat products are performed as well or better by carbohydrates. These roles include water binding, viscosity, lubricity, and structure forming. Other factors include cost, microbial instability, and flavor and mouthfeel issues. On a dry solids basis, protein products are generally more expensive than those based on carbohydrates, especially those based on starch. Proteins can encourage microbial spoilage, acting as a major source of nitrogen in products that would normally contain very little, such as fat-free salad dressings. And while many fat mimetics affect flavor balance, protein-based mimetics were shown to specifically bind aldehyde-containing flavor compounds (Schirle-Keller et al., 1992).

On the other hand, there have been numerous recent patents for various protein-based fat replacers, and the products that are currently on the market are constantly being improved. The potential for proteins as fat-replacement materials remains high.

APPROACHES TO FAT REPLACEMENT IN FOOD PRODUCTS

When fat is removed from a product, it needs to be replaced by something, regardless of whether fat mimetics are used. Several approaches to fat replacement are possible. The most direct approach is a passive one in which the remaining components are allowed to increase proportionately. This option is most successful when the product is only partially reduced in fat. Eventually the texture or stability of the product becomes unacceptable as more fat is removed. Typically, moisture is adjusted to control these factors. The carbohydrate and protein components must then be rebalanced to maintain the desired rheological properties, a task that becomes difficult as products approach fat-free status. The active approach to fat replacement is to add fat-mimetic ingredients. These ingredients either physically replace fat or modify the interactions of the remaining components. Control of moisture is one of the most important roles for fat-mimetic ingredients.

Currently many ingredient suppliers are promoting a "systems approach" to fat replacement. It is generally agreed that no single nonfat ingredient can provide all functionality of fat. Using the systems approach, multiple fat-mimetic ingredients are combined to replace the numerous and varied roles that fat plays in food products. In many fat-free product formulations, moisture is the main component of the fat-replacement system.

Another "systems approach" is to consider the entire food matrix as the system to be manipulated. Proteins are the major structural elements of several types of food, including dairy and processed meat products, which are essentially protein gels. The protein matrix in these products is drastically altered by the removal of fat. In protein-based systems such as these, fat replacement not only requires building in positive fatlike attributes by addition of fat mimetics, but also minimizing the detrimental impact of protein interactions in their new fat-free environment. Since the topic of this chapter is protein as an ingredient, only the former option will be covered. But it should be recognized that by manipulating the process for production of lower-fat food products, it may be possible to alter the functionality of the native proteins and achieve a fat-mimetic effect without using any fat mimetics per se.

PROTEIN-BASED FAT-REPLACEMENT INGREDIENTS

The use of protein as a fat replacer is not a new concept. A Kraft patent (Flick and Pisani, 1970) describes a cottage cheese creaming mixture made from homogenized cottage cheese curd and/or fines. This product was used to produce low-fat cottage cheese products. A later Kraft patent (Hynes and Vakaleris, 1975) used homogenized cottage cheese curd as a fat-mimetic ingredient in low-fat cream cheese products. The microparticulate structure of homogenized cottage cheese curd allows it to physically replace fat droplets in lower-fat food products, ideally occupying the same space left vacant after fat removal. The space also contains a significant quantity of water, since the protein aggregates are substantially hydrated. In fact, moisture is the main fat replacement ingredient, with proteins functioning to hold moisture in a fat droplet morphology.

Protein-based fat mimetics cannot replace all the functional roles of fat in fat-free products. They have proven to be most valuable for fat replacement in oil-in-water emulsion products. They can be structured by many techniques into microparticles that mimic the morphological properties of emulsified fat droplets. They are less useful for replacing fat in bulk fat applications such as frying, liquid oil, or solid fat products. The physical requirements of these applications are not a close match for properties of proteins. Frying requires a nonaqueous liquid to act primarily as a heat transfer medium. There are no aqueous solutions or dispersions of protein that remain stable at frying temperatures (>150°C). Liquid oil products, such as salad oils, require highly soluble mimetics to provide thickness, lubricity, and mouthcoating. Certainly proteins can provide this functionality, but these roles are traditionally carried out by polysaccharide hydrocolloids. Solid fat products, such as tablespread or shortening, require thermal-reversible gelation. With the exception of gelatin, this role is more suitable for carbohydrate-based ingredients.

Proteins are ideal materials for forming structural analogs of fat globules. The diversity of proteins and their properties, resulting from their

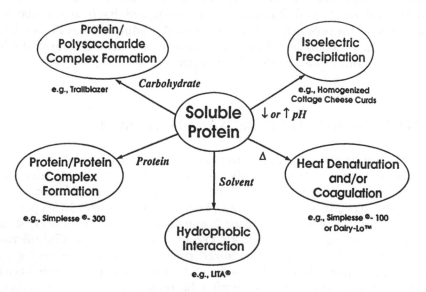

FIG. 14.1 Mechanisms used to form protein-based fat mimetics. The various mechanisms and examples are discussed in the text.

amino acid sequence, structure, and side chain modifications, leads to numerous possible reactions and interactions which can be utilized to build microparticles. Such mechanisms include isoelectric precipitation, heat denaturation and coagulation, hydrophobic effects, as well as protein-protein and protein-carbohydrate complex formation. In the following sections we will discuss the examples of fat mimetics made using each of these mechanisms, as illustrated in Fig. 14.1. Protein microparticles are characterized by their size, shape, surface properties, and degree of hydration, all of which impact their ability to act as fat replacers.

Milk Proteins and Isoelectric Precipitation

Casein and other milk proteins have been used as fat mimetics for many years (see, e.g., Riisom, 1991). It is common practice in formulating lower-fat dairy products, such as frozen desserts and yogurt, to fortify them with milk proteins to provide the proper mouthfeel and control moisture. Both native milk proteins and those that have been functionally modified to increase water binding or emulsification have been used to provide fatlike attributes and/or to correct defects resulting from fat reduction. A commercial example of protein fortification is "Extra Light Milk," 1% fat milk in which nonfat milk solids have been increased from 9 to 11% to give it the taste and texture qualities of 2% fat milk (Keck, 1990). Increasing the milk protein content replaces some of the creaminess lost upon removing fat, and casein micelles acting as microparticles are major contributors to the mouthfeel.

Casein micelles are obvious examples of naturally occurring protein microparticles. These round particles of protein and colloidal calcium phosphate average about 0.14 μm in diameter and are held together by a variety of forces including hydrophobic and electrostatic interactions (see, e.g., Farrell, 1988). A recent patent describes concentrated, substantially nonaggregated casein micelles as a fat or cream substitute (Podolski and Habib, 1992). The casein micelles are obtained by ultrafiltration of skim milk and are used at a concentration of at least 6% in fat-free frozen dessert formulations. They can also be used in fortified skim milk and in a variety of other products, provided that the particles are not allowed to aggregate.

It is important for microparticulate protein-based fat mimetics to remain as discrete particles in the food product. Later in this chapter we will discuss how this is achieved using various mimetics as examples. In the case of native casein, this property is integral to the micellar structure. The glycomacropeptide (GMP) portion of κ-casein provides an anionic surface charge, which keeps the micelles apart by electrostatic repulsion (see, e.g., Farrell, 1988). Once the GMP is removed by peptide bond cleavage with chymosin or the surface charge is neutralized with acid, the casein micelles aggregate and any fatty organoleptic sensation is lost. The curd resulting from adjustment of the pH of skim milk to the isoelectric point of casein is the basis for several nonfat or lower-fat dairy products, including baker's cheese and cottage cheese.

The isoelectric precipitates from acidification of milk are readily redispersed by homogenization, and they can function as fat-mimetic particles, provided that recoalescence is prevented. The homogenized cottage cheese curds and fines described by Flick and Pisani (1970) remain dispersed in the creaming mixture because of coating with the other components of the mixture. In the case of the nonfat cottage cheese example, the other components in the creaming mixture include condensed skim milk and cultured buttermilk.

Normally it would be difficult to prevent reaggregation of homogenized cottage cheese curds at very low pH. However, cottage cheese curd or even concentrated casein micelles can be used as fat mimetics in low pH products, including such nondairy products as pourable or viscous salad dressings, as long as the particles are kept separate by a protective coating. Hydrocolloids are effective barrier dispersants, providing an anionic surface coating at pH values below the isoelectric point of the proteins. Many hydrocolloids are known to interact with milk proteins (see, e.g., Grinrod and Nickerson, 1968), and the interactions of κ-carrageenan with milk proteins have recently been described in detail (Xu et al., 1992). Anionic polysaccharides, such as carrageenan or carboxymethylcellulose, can be preblended with the cottage cheese curd or simply included in the formulation of the dressings.

Protein Microparticles from Heat Denaturation

Microparticulate morphology is a key contributor to the ability of proteins to function as fat mimetics. In this section, several examples of protein microparticles that form during conventional processing are described. Although none of these had been proposed as fat replacers, their potential for use as such is apparent, and the formation of these particles could be built right into the food process.

Whey proteins can spontaneously form microparticles without any special treatment other than heating. Jost et al. (1989) showed that heat-coagulated whey proteins form an open network structure composed of strands of spherical protein particles, with a diameter of about 0.1–0.2 μm (Fig. 14.2). This gel was produced simply by heating a 12% w/v dispersion of whey protein concentrate, containing 9.6% protein, for 30

FIG. 14.2 SEM micrograph of a whey protein gel produced from an aqueous dispersion of the protein. The bar corresponds to 1 μm. (*From Jost et al., 1989.*)

min at 90°C. Based on this microstructure, a low-shear process would be expected to easily disrupt the gel and separate the particles.

Ricotta cheese, a cheese made either by heat treatment exclusively from whey protein (Europe) or with casein and whey (U.S.), also demonstrates the particulate nature of whey proteins in a natural product (Kalab, 1990). The creamy or gritty organoleptic properties of ricotta, which is traditionally a low-fat product, depend to a great extent on the structure of particles such as those shown in Fig. 14.3. Based on this type of microstructure, one would expect homogenized ricotta to perform as a fat-mimetic similar to homogenized cottage cheese curd or possibly with superior fat-mimetic functionality.

Milk proteins are not the only proteins capable of spontaneously forming spherical aggregates. Ovalbumin also forms gels with a particulate substructure (Heertje and VanKleef, 1986). The structures that result from coagulating 20% solutions of the protein at 100°C are pH dependent. At pH 5, scanning electron microscopy (SEM) reveals a gel composed of spherical units 1 μm in diameter (Fig. 14.4). At higher pH

442   MILLER

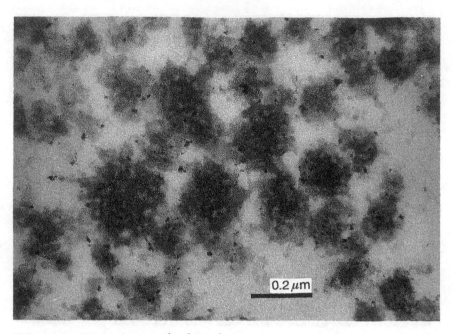

FIG. 14.3   TEM micrograph of North American ricotta cheese showing the
corpuscular ultrastructure of coagulated whey proteins. (*Courtesy of
M. Kalab, 1990. Reprinted with permission of John Wiley & Sons, Inc.*)

(pH 10), the protein forms a fibrous network without any apparent
particulate substructure. The data suggest that microparticulate morphology can be enhanced by carrying out the heat-coagulation process
at a pH where there are strong inter- and intramolecular interactions.
At the the pI of ovalbumin (near pH 5), intramolecular forces such as
hydrogen bonding and hydrophobic interactions strongly stabilize the
native conformation of the protein. As a result, heating causes only
partial disruption of conformation before aggregation occurs. But upon
heating at pH 10, the normally globular protein becomes unfolded and
extended due to electrostatic repulsion of the negatively charged chains.
Intermolecular forces are also low at this pH, so few junction zones
are formed, and a homogeneous gel network results. Tensile strength
measurements carried out on these gels show that the pH 5 gel breaks
at very low stress. The particles would be expected to be separated by a
relatively low-shear homogenization process.

Many other naturally occuring or spontaneously generated protein
microparticles could be cited. Some of these can be found in Barbut

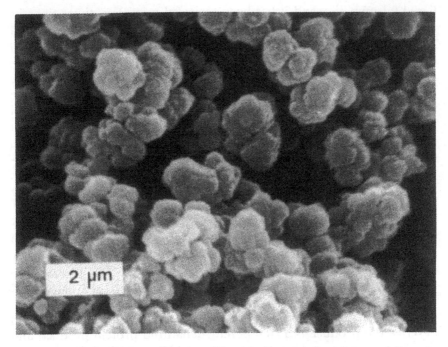

FIG. 14.4   SEM micrographs of ovalbumin gel (20 g/100 g), pH 5, following critical point drying. (*From Heertje and van Kleef, 1986.*)

(1994). The utility of these structures as fat mimetics has not been documented. If the particles are kept dispersed in the food product, exhibit no other adverse interactions with the other ingredients, and are stable to the conditions encountered during preparation of the product, then many of these microparticles should function well as fat-replacement ingredients.

Commercial Fat Mimetics
Based on Whey Protein Denaturation

Several commercial fat-mimetic ingredients based on whey proteins take advantage of heat denaturation to improve their functionality in lower-fat food products. These ingredients rely either on the natural tendency of whey proteins to form microparticle aggregates (described in the previous section) or on improved water-binding and emulsification capacity.

Simplesse® immediately comes to mind when one considers protein-based fat mimetics, and it has been reviewed numerous times since its introduction in 1987 [see Singer and Moser (1993) for a recent review]. Simplesse® is an all-natural fat substitute made by heat-coagulating protein under continuous-shear conditions (Singer and Dunn, 1990). Originally two versions were marketed: one from whey protein concentrate and the other from egg white, skim milk, sugar, and pectin. Both of these are made by a patented process that involves simultaneously heating and homogenizing the proteins (Singer et al., 1988). The process results in insoluble protein with microparticle morphology. The round aggregates of disulfide–cross-linked protein are approximately 0.1–3.0 μm in diameter. The whey protein–based version is the only one currently available. The milk and egg product will be discussed in the section regarding mimetics based on protein-protein interactions.

Another whey protein ingredient, Dairy-Lo™, was recently introduced by Ault Foods in Canada and is being marketed in the United States by the Pfizer Food Science Group (Anonymous, 1993). The whey protein in this ingredient is only partially denatured. The process subjects ultrafiltered whey protein concentrate, at pH 6.1, to a fairly mild heat treatment (80°C for 17 sec), which results in 60–80% denaturation (Asher et al., 1992). In their patent application, Asher et al. (1992) describe the product as a "tempered" denatured protein product, and the process as a temperate denaturation. The controlled denaturation process results in some unfolding and self-aggregation, but unlike the previously described product, this ingredient is not a microparticulated protein. As such, it retains many of the functionalities of whey proteins, such as water binding and emulsification, which are destroyed by more rigorous treatments. The loosely associated aggregates of whey protein are readily dispersed during mastication (Y. J. Asher, personal communication).

Since both of the aforementioned whey protein–based fat mimetics are structurally distinct, one cannot expect them to be functionally similar. Simplesse® works on the "ball-bearing" principle (see, e.g., Setser and Racette, 1992). The uniform size (<3.0 μm) and shape (spheroidal) allows the particles to roll smoothly and easily over one another, providing creaminess and lubricity normally associated with fat. This principle underlies the functionality of other microparticulate fat-mimetic ingredients, whether or not they are made of protein (Glicksman, 1991). Since microparticles have the structural features of emulsified fat droplets, Simplesse® would be expected to function best when replacing fat in oil-in-water emulsion products.

In contrast to Simplesse®, Dairy-Lo™ functions primarily as a water binder. This role is critical in many fat-free products in which the re-

TABLE 14.1 Commercial Fat Mimetics Based on Protein Denaturation

	Simplesse®[a]	Dairy-Lo™[b]
Protein source	Whey protein	Whey protein
Process	Simultaneous heat and shear	Mild heat treatment
Extent of denaturation	Complete?	Partial 60–80%
Particle size	0.1–3.0 μm	Not microparticulate
Functional mechanism	"Ball-bearing" principle	Enhanced water binding, emulsification

[a]*From Singer et al., 1988.*
[b]*From Asher et al., 1992.*

placement of fat with moisture is limited by the ability of the product to hold moisture. Other functionalities enhanced by the controlled denaturation include emulsification, air-cell stabilization, and prevention of iciness in frozen products (Anonymous, 1993). One of the main benefits cited by the inventors is the ability to use significant quantities of the ingredient in ice cream mix production (Asher et al., 1993). Apparently, whey protein concentrate had been demonstrated to have beneficial effects in lower-fat ice cream products, but its tendency to coagulate limited its usefulness. By pretempering Dairy-Lo™, coagulation of the mix during pasteurization is prevented. Table 14.1 contrasts some of the important differences between Simplesse® and Dairy-Lo™.

Other Microcoagulated Proteins

The formation of Simplesse® can also be thought of as a microcoagulation process, as opposed to a microfragmentation process. The difference between these processes is that in the former the formation of a continuous network is prevented, while in the latter it is encouraged. We will return to the microfragmentation process in our discussion of protein-polysaccharide complexes. Other processes for microcoagulation have been proposed. Some of these will be discussed in more detail below and are compared in Table 14.2. Recently a procedure was described for microcoagulation of whey protein by extrusion cooking at acid pH (Queguiner et al., 1992). It is proposed that the process restricts reaggregation at low pH; this should allow use of the ingredient as a fat replacer in acid foods, which would normally be incompatible with whey protein.

TABLE 14.2   Other Heat-Denatured Proteins Particles

	Queguiner[a]	Unilever[b]	Ziegler[c]
Protein source	Whey	Whey or plant storage protein	Egg white protein
Process	Extrusion cooking at low pH	Heat and low shear	Aqueous phase partitioning with gelatin
Particle size	Mean 11.5 μm (broad distribution)	Up to 30 μm	Mean 1–2 μm (narrow distribution)
Advantages	Stable at low pH	Simple process	Simple process; uniform, spherical particles.

[a]*From Queguiner et al., 1992,*
[b]*From Visser and Bakker, 1989; Brown et al., 1990.*
[c]*From Ziegler, 1991, 1992.*

Over the past few years, Unilever has filed several patent applications for fat mimetics based on microcoagulation of proteins by various heat- or acid-precipitation processes. Visser and Bakker (1989) describe a process for heat-denaturing whey proteins with little or no shear on the system. The resulting nonaggregated microparticles are mostly within the range of 0.1–10 μm. A subsequent application describes a similar process in which the whey is enriched in α-lactalbumin (Hakkaart et al., 1990). Mai et al. (1990) describe a procedure involving isoelectric precipitation of a variety of globular proteins, including whey, with low shear sufficient to prevent formation of large aggregates. The preferred protein source is egg white. Another application (Brown et al., 1990) addresses heat coagulation of plant-storage proteins, including those from soybean, rapeseed, field bean, cottonseed, and sunflower seed. These proteins are subjected to a simultaneous heat-and-shear procedure that results in formation of denatured protein globules from 0.1 to 30 μm in diameter.

The products resulting from several of these protein-microcoagulation processes are referred to as edible plastic compositions or dispersions (Visser and Bakker, 1989). Edible plastic dispersions occur in products like yogurt, cream, quark, tablespread, and processed cheese. The common features of these products is that they have a gel or semi-gel continuous matrix in which emulsified particles are dispersed. The difference between conventional fat-containing products and fat-mi-metic plastic dispersions is that, in the latter, both phases are aqueous. The concept is clearly defined in a Unilever patent in which nonfat tablespread products are formed from two condensed phases, of which one is continuous and the other is a dispersed aggregate-forming gel (Cain et al., 1990). Either phase can consist of protein-based fat mimetics; whey protein is ideal for the dispersed phase, and gelatin for the continuous.

The development of lower-fat foods is clearly still in the "art" stage. But once we start defining foods as "plastic compositions" and the fat mimetics as functional parts of composite materials, then we can begin to apply polymer theory and become more scientific about lower-fat food product development. This approach was taken to describe the rheology and microstructure of mixed gelatin–egg white gels (Ziegler and Rizvi, 1989; Ziegler, 1991).

By applying a mathematical model known as the Takayanagi equations to rheological data from gels of different concentrations of gelatin and egg white, Ziegler and Rizvi (1989) were able to hypothesize a phase-inversion point at which the gel changed from one supported primarily by egg white to one supported by gelatin. An interpenetrating polymer network was predicted at intermediate concentrations. The rheological interpretation was confirmed by microscopy (Ziegler, 1991). Figure 14.5 clearly shows that the aqueous phase was partitioned between the gelatin gel and the egg white gel. Microparticles of egg white gel formed spontaneously at higher concentrations of gelatin as a result of phase inversion. The microparticles were spherical, generally less than 2 μm diameter, and contained internal voids or regions of less dense protein gel (Ziegler, 1991, 1992).

The concept of aqueous phase partitioning becomes very important in fat-free products. In full-fat products, there are two distinct phases (lipid and aqueous), one of which is usually dispersed in the other. In products from which the fat has been removed, there can only be an aqueous phase. Soluble bulking agents or biopolymers can modify the properties of the aqueous phase, but it remains a single homogeneous phase. Fat-mimetic particles can help to partition the aqueous phase by sequestering a portion of the water in a dispersed compartment. The proportion of water in the dispersed and continuous aqueous phases depends on the relative ability of the fat mimetics to compete with the components of the continuous phase for the available moisture. Ideally one would want the mimetic particle to reduce the amount of water in the continuous phase to that found in the full-fat product. By comparing

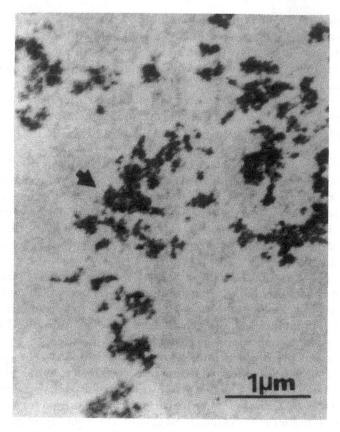

FIG. 14.5   TEM micrograph of a mixed gel containing 1.42% (w/w) type A gelatin (fine network) and 5.61% (w/w) egg white protein (coarse network). Arrow indicates region devoid of protein. (*From Ziegler, 1991.*)

phase volumes of gels, one may be better able to predict the functionality of fat-mimetic particles.

The physical behavior of fat in rennet-induced milk gels has been described as similar to fillers in synthetic polymer systems (Johnston and Murphy, 1984). In this study, the casein gel network is "plasticized" by heterogeneously distributed fat globules. Increasing the fat level affects the rheology of milk gels by increasing the initial rate of stress relaxation. In their study on rheological effects of emulsified milkfat on acid-induced milk gels, Xiong et al. (1991) conclude that fat globules may not be the inert fillers in the voids of the protein gel as had originally been hypothesized (Aguilera and Kinsella, 1989), but may actually behave as copolymers in the milk gel network. This concept is reinforced by a microstructural study of butterfat-containing whey protein gels (Yost and Kinsella, 1992), in which fat droplets are described as "active fillers."

By combining some of the mixed gel and phase-partitioning concepts (Ziegler and Rizvi, 1989; Ziegler, 1991) with the basic understanding of milk gels and the effects of emulsified butterfat (Johnston and Murphy, 1984; Xiong et al., 1991; Aguilera and Kinsella, 1989), it may be possible to create a fat-free compartmentalized aqueous system with the rheological properties of a full-fat product. Conceptually, this could be used for a variety of fat-free cheese and dairy products, using only the natural components found in conventional dairy products to create a fat-mimetic effect. The various possible structures of mixed and filled gels from dairy proteins (skim milk powder and/or whey protein concentrate) and butterfat (Aguilera and Kessler, 1989) suggest potential strategies for engineering gels with fat-mimetic properties.

Hydrophobic Protein Particles

One microparticulated protein-based fat mimetic has been developed in recent years that is uniquely different. It is called LITA® and was invented at Opta Food Ingredients, Inc. (Stark and Gross, 1992). LITA® is a water-dispersible microparticulated zein preparation. Zein is a naturally hydrophobic protein which is soluble in alcohol but insoluble in water. When an ethanolic solution of zein is introduced into an aqueous solution of polysaccharide, a stable colloidal precipitate results (Iyengar and Gross, 1991). Most of the colloidal particles are within the size range of 0.3–3.0 μm, although the particle size distributions can range from 0.1 to 8.0 μm without exhibiting either a powdery or gritty texture (Cook et al., 1991).

One of the most interesting features of LITA® is its microstructure. The scanning electron micrograph in Fig. 14.6 shows that LITA® consists of perfect spheres having uniform size, very smooth surface, and very dense packing of the protein within the spheres. Unlike the other heat- or acid-coagulated fat-mimetic ingredients, LITA® does not function by virtue of its water-controlling capacity. The particles contain 87% protein and very little water (about 7%, Cook et al., 1991). With particles this dense and nondeformable, any surface imperfections could contribute to unacceptable mouthfeel properties. They need to be perfect "ball bearings" to create the illusion of fat.

The surface of the zein microparticle would be extremely hydrophobic, and if left uncoated the particles would tend to aggregate in aqueous dispersions. Preferred aggregate-blocking agents include various poly-

FIG. 14.6   SEM micrograph of microparticles of LITA®. The bar corresponds to 1 μm. (*From Iyengar and Gross, 1991.*)

saccharides, such as gum arabic or carboxymethylcellulose. These are included in the aqueous medium during the formation of the spheres. The size, shape, and hydrophobic nature of LITA® make this an intriguing fat-mimetic material.

Mimetics Based on Multiple Protein Interactions

One of the original versions of Simplesse® was made by blending ultrafiltered egg white and condensed skim milk under conditions that result in the coprecipitation of the egg proteins and whey proteins onto a core nucleus of a casein micelle (Singer and Dunn, 1990). The process can be thought of as analogous to making snowballs. The core of the snowball is a densely packed region, in this case a casein micelle. As the snowball is rolled, it acquires a thicker and thicker coating but still retains its spherical shape. In the case of Simplesse®, the casein core is rolled in a patented fluid processor apparatus, a type of scraped surface heat exchanger, at a temperature sufficient to allow denatured egg and whey proteins to build up as a shell on the casein core (Singer et al., 1989).

The pH during the Simplesse® process is close to neutrality (pH 6–7). Denaturation at this pH would be expected to result in substantial cross-linking of the egg and whey proteins. This cross-linking would help to stabilize and form the microparticles. Transmission electron micrographs (TEM) of the egg and milk version of Simplesse® were published by Singer and Dunn (1990). In their micrographs, one can clearly see a core of material surrounded by a very thick layer of aggregated protein (Fig. 14.7). The authors took this one step farther and identified the egg albumen portion using immunogold labeling. By this technique, it is apparent that the outer layer is egg albumen that surrounds the casein micelle nuclei.

There is a striking similarity between the structures of egg/milk Simplesse® and the core-and-lining particles described by Harwalkar and Kalab (1988). Such particles were obtained by heat-coagulating casein micelles dispersions in the presence of β-lactoglobulin at 90°C at pH 5.2–5.5 (Fig. 14.8). The formation of the core-and-lining structure is highly dependent on pH and requires the presence of the milk salt system. If the conditions are not correct, a fused mass of whey protein and casein results. In addition, although many of the core-and-lining structures in Fig. 14.8 appear as distinct particles, it is apparent that they are part of a gel network, which is a significant distinction between this material and Simplesse®. Since particle aggregation would be detrimental to the fat-mimetic effect, the core-and-lining structures were never suggested as fat replacers. The Simplesse® process was specifically designed to result in nonaggregated particles (Singer et al., 1988).

Mimetics Based on Protein-Polysaccharide Interactions

The well-known interactions of milk proteins with polysaccharides and the use of hydrocolloids as barrier dispersants by complexing with fat-mimetic particles have been described. Complexes of proteins and polysaccharides may also be used to arrange proteins into various structures. There are three possible results from mixing solutions of proteins and polysaccharides, which were described by Tolstoguzov (1986): (1) a single homogeneous solution, (2) a liquid two-phase system in which the aqueous phase is partitioned between the thermodynamically incompatible protein and polysaccharide components, and (3) an insoluble electrostatic complex of protein with anionic polysaccharide. For a more detailed discussion of protein-polysaccharide interactions see Ledward (1994). Trailblazer, a microfragmented fat mimetic invented at Kraft

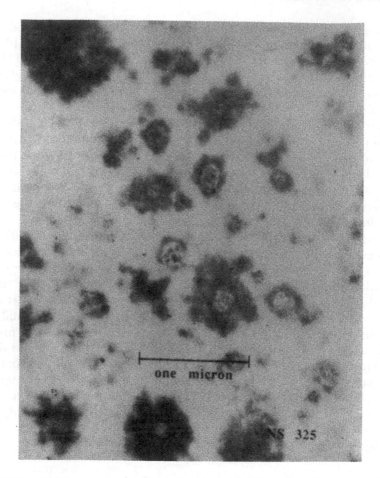

FIG. 14.7   TEM micrograph of protein microparticles formed from egg white protein and casein micelles by the Simplesse® microparticulation process. (*From Singer and Dunn, 1990, with permission of the authors and John Wiley & Sons, Inc.*)

General Foods, is based on possibility 3, but with some aspects of 2 as well (Chen et al., 1992).

The origins of Trailblazer can be found in a Kraft patent for spontaneous generation of protein/xanthan fibers (Chen and Soucie, 1985). These fibers closely resembled chicken breast when compacted and were developed as meat analogs (Soucie and Chen, 1986). We found that when

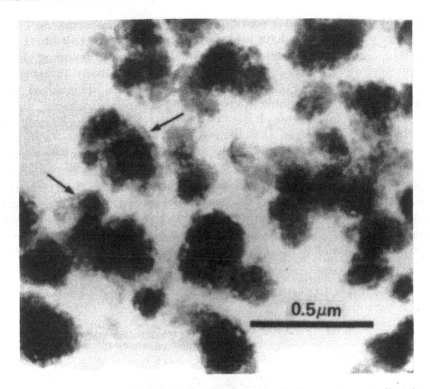

FIG. 14.8 TEM micrograph of heat-induced gels for casein micelles dispersed in a milk dialyzate in the presence of added β-lactoglobulin at pH 5.5 showing the core-and-lining structures (arrows). (*From Harwalkar and Kalab, 1988.*)

these fibers were fragmented, they formed smooth, creamy particles, which could be used to replace fat in many of our products, including various oil products (e.g., salad dressings) and dairy products (Chen et al., 1992). The FDA determined that Trailblazer is GRAS, based on evidence demonstrating that it is a mixture of ingredients already listed as GRAS (Shank, 1992).

The formation of Trailblazer results from the electrostatic interaction of protein and polysaccharide (Chen and Soucie, 1985). Anionic polysaccharides interact strongly with proteins at pH values below the isoelectric point of the protein (Tolstoguzov, 1986). Examples of polysaccharides capable of complex formation include both those with carboxyl groups (e.g., pectin and alginate) and those with sulfate groups (e.g., carrageenan). Many common proteins participate in complex formation. Only some of the complexes have a fibrous morphology. The best fibers are formed from xanthan with various globular proteins, including those from milk, egg, and soy sources (Chen and Soucie, 1985). In order to "set" the fibers and thereby prevent dissociation when pH is raised, it is best to use proteins capable of cross-linking upon denaturation.

Numerous compositions fall under the description of Trailblazer, but the typical version is made from a mixture of xanthan gum with egg white and whey protein (Chen et al., 1992). When the pH is dropped to just below the isoelectric point of the proteins, the proteins line up along the xanthan template and form a spontaneous fiber due to charge-charge interactions. Upon heating, the proteins cross-link by disulfide bond formation and form a stable fibrous network. The fibers are then fragmented by subjecting them to any one of a number of high-shear processes. The Trailblazer process is illustrated in Fig. 14.9. We believe that

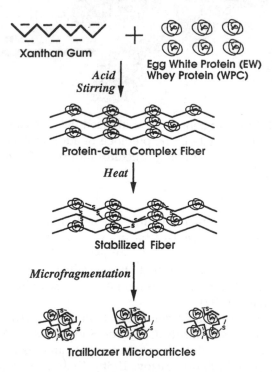

FIG. 14.9 Basis of Trailblazer technology showing steps required for the manufacture of Trailblazer microparticles. See text for further discussion of the process.

the functionality of Trailblazer in food products is derived from the space-occupying and water-binding characteristics of the protein-gum complex. The complex thus takes on functional characteristics different from the individual components when used separately.

The structure of Trailblazer differs from the other microparticulates described previously. Unlike the others, Trailblazer is formed by a microfragmentation process. This concept can be best illustrated by following the microstructure of Trailblazer through its manufacturing process. Scanning electron microscopy clearly shows the fibrous nature of initial protein-gum complex (Fig. 14.10). The fibers are in the form of thick ropelike bundles, which tighten upon cooking and denaturing of the proteins to form thinner, more compact strands (Fig. 14.11). Following the microfragmentation process, the resulting microparticles are irregularly shaped, with none greater than about 10 μm in their longest dimension (Fig. 14.12). Trailblazer microparticles retain an elongated axis and their fibrous morphology. There is considerable variability in the length-

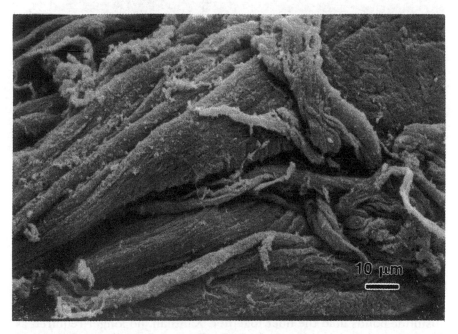

14.10 SEM micrograph of an egg white, whey protein, xanthan version of Trailblazer prior to heat setting the fibers. (*Micrograph by David Pechak, Kraft General Foods, Inc.*)

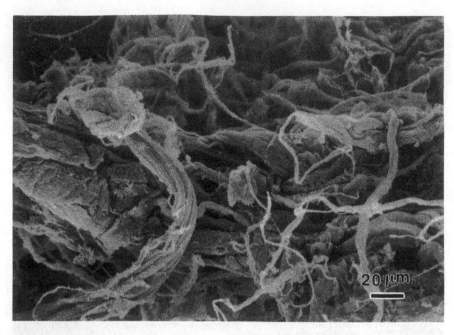

FIG. 14.11   SEM micrograph following heat denaturation of the same egg white, whey protein, xanthan version of Trailblazer shown in Fig. 14.10. (*Micrograph by David Pechak, Kraft General Foods, Inc.*)

to-width ratio among these particles, which is a function of the extent of processing during microfragmentation. The majority of particles are less than 5 μm in length based on a number average. Amorphous strands cover the surface of the particles. Although scanning microscopy does not allow discrimination between chemical components, we propose that these strands contain xanthan, which imparts a negative surface charge to the particles.

It was mentioned at the beginning of this section that Trailblazer has characteristics of both an electrostatic complex and a liquid two-phase system. Electrostatic interactions are most important during the initial stages of fiber formation, at which time xanthan functions as a template for protein alignment. Once the fibers are stabilized by heating, covalent disulfide bonds hold the fiber together. Xanthan, no longer required as a template, remains trapped within the fibers. At low pH, xanthan is still tightly complexed with the fibers. However, as the pH is raised above the isoelectric point of the proteins, xanthan no longer complexes with

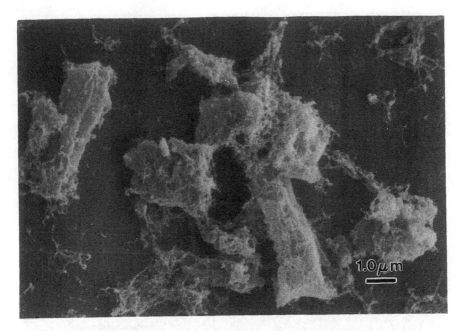

FIG. 14.12 SEM micrograph following the final microfragmentation processing of an egg white, milk protein, xanthan version of Trailblazer. (*Micrograph by David Pechak, Kraft General Foods, Inc.*)

the protein fibers. Some of the xanthan dissociates from the microparticles, while some of it coats the surface, providing an electrostatic barrier to protect against aggregation.

Most of the xanthan remains trapped in the particle gel interstices. This state may best be described as a Type II filled gel (Tolstoguzov, 1986), which has been microfragmented. Tolstoguzov classifies filled gels on the basis of the phase state of the system. The distinction between Type I gels (single-phase, soluble filler) and Type II gels (two-phase, gelled filler) also depends on the state of the dispersed phase. In Trailblazer particles under nonacidic conditions, the xanthan is not associated with the protein matrix and is in its typical gel state. Transmission electron microscopy was used to examine internal structure at high resolution (Fig. 14.13). The fibrous nature of a single Trailblazer particle is shown in a longitudinally sectioned profile. At this resolution, the particles appear as an open protein network with porous channels most likely containing trapped xanthan gel.

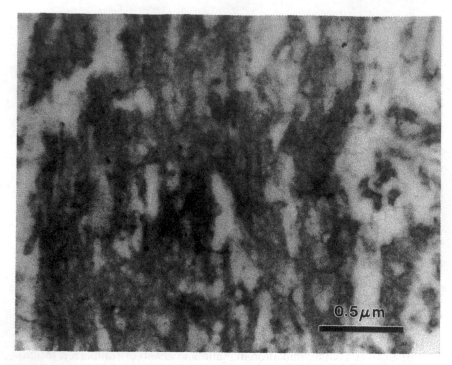

FIG. 14.13   TEM micrograph of a single microparticle of an egg white, whey protein, xanthan version of Trailblazer. (*Micrograph by Roger Unger, formerly of Kraft General Foods, Inc.*)

REQUIREMENTS FOR A FAT-REPLACEMENT SYSTEM

If you were to set out to design a fat-replacement particle, what are the physical properties you should target? If we accept the concept that a fat-mimetic material must physically take the place of fat in a product, and if we target the mimetic for replacement of fat in typical oil-in-water emulsion products, then we must structure the particles like oil droplets. Emulsified fat is characterized by particle size, shape, and surface properties. Some of the physical properties of fat droplets are compared with those of various fat mimetics in Table 14.3.

Particle Size

It is difficult to specify a precise particle size target for fat mimetics. The simplest target would be to match the range of oil droplets in food

TABLE 14.3   Requirements for Fat-Mimetic Particles

	Size	Shape	Surface	Water content
Emulsified Fat	0.1->10 μm	Spherical	Depends on emulsifier	None
Simplesse®[a]	0.1–3.0 μm	Globular	Whey protein (S-100) or pectin (S-300)	High
LITA®[b]	0.1–8 μm	Perfect spheres	Various hydrocolloids	Very low
Trailblazer	<11 μm	Fibrous	Xanthan	Very high

[a]*From Singer et al., 1988, 1991; and various Simplesse ingredient bulletins.*
[b]*From Cook et al., 1991.*

products, typically from 0.1 to 10 μm. It is certainly possible to form emulsions with larger droplet size, but stability of these emulsions is poor (see, e.g., Dickinson and Stainsby, 1982).

The upper limit must be defined by whether the particles can be detected as discrete entities in the mouth. The limit for detectability is generally accepted to be about 40 μm, but this value is not at all well defined, nor is it documented in the scientific literature. The particle size that can be detected is probably much less than 40 μm and possibly less than 20 μm. Our own experience with chalky and grainy dairy products supports this hypothesis. It may not even be possible to determine the size range with any precision because detectability depends on many other factors. Some of these factors are related to the properties of the particle and some to the other components of the product. For example, large particles (>40 μm) may be undetectable when suspended in a viscous hydrocolloid solution, but would feel chalky and gritty in most dairy products.

Each of the various protein-based fat-mimetic ingredients described in the previous sections of this paper has its own unique particle size range. In particle size studies carried out with Trailblazer in which various size particles between 3 and 14 μm were prepared by controlled particle-size reduction and placed into model fat-reduced viscous dressings, we concluded that somewhere between 7 and 11 μm, mouthfeel changed from creamy to chalky. Therefore, our target for Trailblazer was <7 μm. The ideal range for Simplesse® is between 0.1 and 3 μm; with particles >3 μm being perceived as gritty and those less than 0.1 μm as watery (Singer and Dunn, 1990). LITA® can range in size from

0.1 to 8 μm with no apparent chalkiness (Cook et al., 1991). Unilever claims particles up to 30 μm, but this must be interpreted cautiously in the absence of supporting documentation (Brown et al., 1990).

For best colloidal stability and limited detectability of the particle in the mouth, <10 μm is a good target for fat-mimetic particle size. It is important to keep in mind that there is probably no "magic" particle size range for fat-mimetic particles. While there may be an ideal range for a given material, this depends on other characteristics of the particle.

Shape and Deformability

It has been claimed that spheroidal geometry is a necessary feature of fat-substitute materials (e.g., Singer and Dunn 1990). Since emulsified fat droplets in products assume this shape, the assumption seems to be a valid one. In micrographs of Simplesse®, the microparticles appear as spheroidal, although at higher magnifications (Fig. 14.7), Simplesse® can at best be described as roughly globular aggregates (Singer and Dunn, 1990). They are not nearly as perfectly spherical as the LITA® zein particles shown in Fig. 14.6 (Iyengar and Gross, 1991).

Is a spherical shape really required for a fat-mimetic effect? Trailblazer particles are as far from spheroidal as one can get, being a completely fibrous and entangled structure. The answer is probably the same as the answer about particle size: that the requirement for spheroidal geometry is ingredient dependent. A fibrous deformable particle (e.g., Trailblazer) may not be as constrained by geometry as a rigid particle (e.g., LITA®) and may be able to slide over one another as easily as ball bearings. Conversely, nondeformable particles may require a more spherical shape. Particles that are intermediate in deformability (e.g., Simplesse®) may function well with a roughly globular shape.

Surface Properties

Surface properties of particles are important factors in their colloidal stability. All of the forces required to keep an emulsion stable are also needed to maintain a proper dispersion of microparticulate fat mimetics (see, e.g., Dickinson and Stainsby, 1982). The ideal surface of a particulate fat mimetic is unreactive with the other components in the system and repellent to other particles. A surface with these properties would allow the particles to slide over each other, producing a lubricating effect. In addition, an unreactive surface would preclude gel formation by particle interaction, which would destroy the perception of creaminess.

The desired surface properties can be achieved by using a barrier dispersant, usually a highly hydroscopic, negatively charged polysaccharide. Xanthan gum, pectin, and carboxymethyl cellulose are all accept-able polysaccharides for this role. With Trailblazer, the same xanthan component that had served as a template during fiber formation plays the part of protective colloid in the finished particles. The milk-and-egg protein version of Simplesse® uses pectin (Singer et al., 1991). Opta Food Ingredients has used neutral, sulfated, or carboxylated polysaccha-rides for LITA® (Cook et al., 1991).

Water Binding

The water-binding capacity of particulate fat-mimetic materials may be related to functionality, but the precise relationship has not been defined. Going from Trailblazer to Simplesse® to LITA®, we see a large variation in the ability to hold water. Trailblazer is much like a sponge, with an open capillary network that can bind and control a considerable amount of water. Furthermore, moisture is tightly held in its pores by hydrated xanthan trapped within the protein network. Suspensions of Trailblazer at about 10% easily stabilize all of the available water and have a pastelike consistency. A hydrated version of Simplesse®, which is no longer available, was sold at 42% solids with 22.7% protein, and had the consistency of heavy cream (Anonymous, 1992). LITA® imbibes very little water, being a hydrophobic particle, and thus does not contrib-ute substantially to the viscosity of a suspension (Cook et al., 1991).

However, there is probably an optimum hydrated volume for each microparticulate fat-mimetic ingredient beyond which the texture be-comes more gelatinous and less fatlike. It is important to have a high solids/low viscosity ratio. That is, unlike gums or starches that deliver high viscosity at very low solids levels, the fat-replacement systems need a higher solids level to provide water binding and mouthfeel characteris-tics. A short texture, perhaps reflected by a lack of thixotropy, is also an important criterion.

CONCLUSION

Fat contributes many critical aspects to product quality. It affects the taste, mouthfeel, and performance of a product. In creating a reduced-fat or fat-free product, the important functional properties of fat must be replaced if that product is to be successful. Since no single ingredient can replace the dozens of possible functional properties of fat in a prod-ucts, the concept of using a fat-replacement system has become popular in recent years. It is important to match the functional properties of the various ingredient options with those key to the product requirements in order to formulate a product-specific fat-mimetic system.

Proteins provide excellent building blocks for novel fat mimetics because of their complexity and diversity. I have described a few of the protein-based fat mimetics invented within recent years. The formation of each involves a different reaction mechanism, with the result that each is totally different from the others. Those chosen here exhibit specialized functionalities based on unique physical properties. By fully understanding the importance and interrelationships of the physical features of these protein-based ingredients to their functionality as fat mimetics, better ingredients can be designed for specific fat-mimetic needs in more healthful food products.

Most of the protein-based fat mimetics are in the form of microparticulates. These ingredients play a critical role in a fat-mimetic system by providing the dispersed properties of fat and oil necessary in typical oil-in-water emulsion products. In this role they physically occupy the space previously occupied by emulsified fat droplets. Beyond this simple role, the success or failure of a fat mimetic depends on its interactions with the remainder of the food matrix, as well as with other microparticles. Particle size, shape, and surface properties are all important to functionally replicate dispersed fat. Control of moisture is another important role for the protein-based fat mimetics, which can be performed well by either soluble or microparticulate ingredients.

ACKNOWLEDGMENTS

The author thanks Greg Cheng, Paul Wrezel, and Robert Martin for editorial advice and feedback on the manuscript. I would also like to thank Cheryl Brewer for the artwork and David Pechak and Roger Unger for the micrographs of Trailblazer.

REFERENCES

Aguilera, J. M. and Kessler, H. G. 1989. Properties of mixed and filled-type dairy gels. *J. Food Sci.* 54: 1213.

Aguilera, J. M. and Kinsella, J. E. 1989. Compression stress of dairy gels and microstructural interpretation. *J. Food Sci.* 54: 1224.

Anonymous. 1992. *Simplesse 100 Ingredient Bulletin.* The NutraSweet Company, Deerfield, IL.

Anonymous. 1993. *Unfold the Secret to Great-Tasting, Lowfat Foods.* Dairy-Lo™ technical brochure. Pfizer Food Science Group, New York.

Asher, Y. J., Mollard, M. A., Thomson, S., Maurice, T. J., and Caldwell, K. B. 1992. Whey protein product: Method for its production and use thereof in foods. International patent appl. WO 92/20239, Nov. 26.

Asher, Y. J., Mollard, M. A., Thomson, S., Maurice, T. J., and Caldwell, K. B. 1993. Whey and ice cream products and processes. International patent appl. WO 93/02567, Feb. 18.

Barbut, S. 1994. Protein gel ultrastructure and functionality. Ch. 13. In *Protein Functionality in Food Systems,* N. Hettiarachchy and G. Ziegler (Ed.). Marcel Dekker, Inc., New York.

Brown, C. R. T., Norton, I. T., and Wilding, P. 1990. Protein product. Eur. patent appl. 0 352 144 A1, Jan. 24.

Cain, F. W., Clark, A. H., Dunphy, P. J., Jones, M. G., Norton, I. T., and Ross-Murphy, S. B. 1990. Edible plastic dispersion. U.S. patent 4,956, 193, Sept. 11.

Chen, W. -S., Henry, G. A., Gaud, S. M., Miller, M. S., Kaiser, J. M., Balmaceda, E. A., Morgan, R. G., Baer, C. C., Borwankar, R. P., Hellgeth, L. C., Strandholm, J. J., Hasenhuettl, G. L., Kerwin, P. J., Chen, C. -C., Kratochvil, J. F., Lloyd, W. L., Eckhardt, G., De Vito, A. P., and Heth, A. A. 1992. Microfragmented ionic polysaccharide/protein complex dispersions. U.S. patent 5,104,674, Apr. 14.

Chen, W. -S. and Soucie, W. G. 1985. Edible fibrous serum milk protein/xanthan gum complexes. U.S. patent 4,559,233, Dec. 17.

Cook, R., Finocchiaro, E. T., Shulman, M., and Mallee, F. 1991. A microparticulated zein/polysaccharide composite with fat-like properties. Presented at the IBC Conference on Fat and Cholesterol Reduced Foods, Atlanta, GA, March 14–16.

Dickinson, E. and Stainsby, G. 1982. *Colloids in Food.* Applied Science Publishers, London and New York.

Farrell, H. M., Jr. 1988. Physical equilibria: Proteins. Ch. 1. In *Fundamentals of Dairy Chemistry,* 3rd ed., N. P. Wong, R. Jenness, M. Keeney, and E. H. Marth (Ed.), p. 461. Van Nostrand Reinhold, New York.

Flick, B. J. and Pisani, J. P. 1970. Cottage cheese creaming mixture. U.S. patent 3,506,456, Apr. 14.

Glicksman, M. 1991. Hydrocolloids and the search for the "oily grail." *Food Technol.* 45(10): 94.

Grinrod, J. and Nickerson, T. A. 1968. Effect of various gums on skimmilk and purified milk proteins. *J. Dairy Sci.* 51: 834.

Hakkaart, M. J., Kunst, A., and Leclercq, E. 1990. Edible compositions of denatured whey proteins. Eur. patent appl. 0 412 590 A1, July 9.

Harwalkar, V. R. and Kalab, M. 1988. The role of β-lactoglobulin in the development of the core-and-lining structure of casein particles in acid-heat-induced milk gels. *Food Microstructure* 7: 173.

Heertje, I. and Van Kleef, F. S. M. 1986. Observations on the microstructure and rheology of ovalbumin gels. *Food Microstructure* 7: 173.

Hynes, J. T. and Vakaleris, D. G. 1975. Preparation of a low fat cream cheese product. U.S. patent 3,929,892, Dec. 30.

Iyengar, R. and Gross, A. 1991. Fat substitutes. Ch. 11. In *Biotechnology and Food Ingredients*, I. Goldberg and R. Williams (Ed.), p. 287. Van Nostrand Reinhold, New York.

Johnston, D. E. and Murphy, R. J. 1984. Effects of fat content on properties of rennet-induced milk gels. *Milchwissenschaft* 39: 585.

Jost, R., Dannenberg, F., and Rosset, J. 1989. Heat set gels based on oil/water emulsions: An application of whey protein functionality. *Food Microstructure* 8: 23.

Kalab. M. 1990. Microparticulate protein in foods. *J. Am. Coll. Nutr.* 9: 374.

Keck, B. 1990. The Extra Light story: A lesson in cooperation. *Dairy Field Today* 173(3): 38.

Ledward, D. A. 1994. Protein-polysaccharide interactions. Ch. 8. In *Protein Functionality in Food Systems*, N. Hettiarachchy and G. Ziegler (Ed.). Marcel Dekker, Inc., New York.

Mai, J., Breitbart, D., and Fischer, C. D. 1990. Proteinaceous material. Eur. patent appl. 0 400 714 A2, Dec. 5.

Podolski, J. S. and Habib, M. 1992. Concentrated, substantially nonaggregated casein micelles as a fat/cream substitute and a method for reducing the fat content in food products. U.S. patent 5,143,741, Sept. 1.

Queguiner, C., Dumay, E., Salou-Cavalier, C., and Cheftel, J. C. 1992. Microcoagulation of a whey protein isolate by extrusion cooking at acid pH. *J. Food Sci.* 47: 610.

Riisom, T. 1991. Milk proteins as fat replacers. *Scand. Dairy Information* 5(4): 28.

Schirle-Keller, J. -P., Chang, H. H., and Reineccius, G. A. 1992. Interaction of flavor compounds with microparticulated proteins. *J. Food Sci.* 57: 1448.

Setser, C. S. and Racette, W. L. 1992. Macromolecule replacers in food products. *Crit. Rev. Food Sci. Nutr.* 32: 275.

Shank, F. R. 1992. Kraft Inc. Withdrawal of GRAS affirmation petition. *Fed. Reg.* 57(155): 35834.

Singer, N. S. and Dunn, J. M. 1990. Protein microparticulation: The principle and the process. *J. Am. Coll. Nutr.* 9: 388.

Singer, N. S. and Moser, R. H. 1993. Microparticulated proteins as fat substitutes. Ch. 9. In *Low Calorie Foods Handbook*, A. M. Altschul (Ed.), p. 171. Marcel Dekker, Inc., New York.

Singer, N. S., Speckman, J., and Weber, B. 1989. Fluid processor apparatus. U.S. patent 4,828,296, May 9.

Singer, N. S., Wilcox, R., Podolski, J. S., Chang, H. -H., Pookote, S., Dunn, J. M., and Hatchwell, L. 1991. Cream substitute ingredient and food products. U.S. patent 4,985,270, Jan. 15.

Singer, N. S., Yamamoto, S., and Latella, J. 1988. Protein product base. U.S. patent 4,734,287, Mar. 29.

Soucie, W. G. and Chen, W. -S. 1986. Edible xanthan gum-protein fibrous complexes. U.S. patent 4,563,360, Jan. 7.

Stark, L. E. and Gross, A. T. 1992. Hydrophobic protein microparticles and preparation thereof. U.S. patent 5,145,702, Sept. 8.

Tolstoguzov, V. B. 1986. Functional properties of protein-polysaccharide mixtures. Ch. 9. In *Functional Properties of Food Macromolecules*, J. R. Mitchell and D. A. Ledward (Ed.), p. 385. Elsevier Applied Science Publishers, London and New York.

Visser, J. and Bakker, M. 1989. Edible plastic compositions. Eur. patent appl. 0 347 237 A2, June 16.

Xiong, Y. L., Aguilera, J. M., and Kinsella, J. E. 1991. Emulsified milkfat effects on rheology of acid-induced milk gels. *J. Food Sci.* 56: 920.

Xu, S. Y., Stanley, D. W., Goff, H. D., Davidson, V. J., and Le Maguer, M. 1992. Hydrocolloid/milk gel formation and properties. *J. Food Sci.* 57: 96.

Yost, R. A. and Kinsella, J. E. 1992. Microstructure of whey protein isolate gels containing emulsified butterfat droplets. *J. Food Sci.* 57: 892.

Ziegler, G. R. 1991. Microstructure of mixed gelatin-egg white gels: Impact on rheology and application to microparticles. *Biotechnol. Prog.* 7: 283.

Ziegler, G. R. 1992. Process for producing microparticulated protein and the product thereof. U.S. patent 5,147,677, Sept. 15.

Ziegler, G. R. and Rizvi, S. S. H. 1989. Predicting the dynamic elastic modulus of mixed gelatin-egg white gels. *J. Food Sci.* 54: 430.

15
Edible Films and Coatings from Proteins

J. Antonio Torres

Oregon State University
Corvallis, Oregon

INTRODUCTION

An edible coating or film has been defined as a thin, continuous layer of edible material formed or placed, on or between, foods or food components. Its function is to provide a barrier to mass transfer, to serve as a carrier of food ingredients and additives, or to provide mechanical protection (Torres et al., 1985; Torres and Karel, 1985; Kester and Fennema, 1986; Guilbert, 1986; Hatzidimitriu et al., 1987; Torres, 1987; Guilbert and Biquet, 1989; Vojdani and Torres, 1989a,b, 1990; Krochta et al., 1988, 1990a,b; Gennadios and Weller, 1990; Rico-Peña and Torres, 1990a,b, 1991; Hagenmaier and Shaw, 1990, 1991a,b; Avena-Bustillos, 1992; Martin-Polo et al., 1992a,b; Mahmoud and Savello, 1992, 1993; Gennadios et al., 1993). Developments in edible-coating formula-

Technical Paper No. 10,277 from the Agricultural Experiment Station at Oregon State University.

tions with a wide range of gas and moisture permeability characteristics have extended their potential applications, and they could contribute to the stability of many processed foods (Table 15-1). Oxygen-barrier films are needed to prevent vitamin degradation, lipid oxidation, and other oxidative reactions. Flavor-exchange barriers are needed to retain the sensory individuality of food components. Light barrier films could protect pigments, flavors, and nutrients from photodegradation. However, in most cases, the functional property of interest in edible films and coatings is their resistance to the migration of moisture. In many processed foods, critical levels of water activity (a_w) must be maintained if the product, or a certain component of a multiphase food, is to retain quality and safety. Critical a_w values are important not only to prevent microbial outgrowth but also to prevent texture degradation and to minimize deteriorative chemical and enzymatic reactions (Kester and Fennema, 1986).

Only very recently has research been initiated to obtain quantitative information on the permeability of these "new" edible films. Permeability data for the fruit and produce coatings in commercial use for several decades are generally lacking (Hagenmaier and Shaw, 1990, 1991a,b, 1992). Independent of the permeability properties of the film, another critical difficulty is the need to produce coatings and films of even thickness and reproducible properties. For coatings, additional issues are surface adhesion and the need to keep them as thin as possible to facilitate consumer acceptance. In some cases, the coating could be flavored to make detection a consumer-acceptable experience.

Materials used to make edible coatings or films must have flavor and texture acceptable to consumers and possess flexibility so that they do not crack or fragment during product handling and storage. Additives, such as plasticizers, flavors, antimycotic agents, antioxidant agents, and light-energy absorbers can be added to improve their mechanical, organ-

TABLE 15.1   Suggested Applications for Edible Coatings and Films

Control of moisture migration/losses
Control of gas exchanges (O_2, CO_2, C_2H_4, etc.)
Control of oil and fat absorption and migration
Control of solute migration
Control of flavor and other volatile migration, exchange, and losses
Carrier of flavor, color, antimicrobial, and other food additives
Prevention or control of photo degration-oxidation
Improvement of mechanical handling properties of foods

oleptic, and protective properties over a wide range of storage conditions (Guilbert, 1986). Protein and polysaccharide polymer chains have the ability to associate by hydrogen bonding and, in some cases, electrostatic forces. Polysaccharides used in edible films and coatings include hydroxypropyl amylose and the cellulose derivatives methylcellulose, carboxymethylcellulose, hydroxypropylcellulose, and hydroxypropyl methylcellulose. Due to the hydrophilic nature of these polymers, only minimal moisture barrier properties can be expected (Kester and Fennema, 1986). Some other less commonly used polymers are plant and microbial polysaccharides, such as agar, carrageenan, alginate, pectin, dextran, gum ghatti, scleroglucan, pullulan, and curdlan (Vojdani and Torres, 1989b). Among the many proteins evaluated for films/coatings are collagen, gelatin, wheat gluten, soy protein, casein, whey protein, and zein, a corn protein fraction. In contrast to polysaccharides, proteins, and other synthetic polymers, hydrophobic substances such as waxes and triglycerides form thicker and more brittle films and coatings. Hydrophobic substances may need to be associated with other agents such as proteins to form improved films (Torres and Karel, 1985; Kester and Fennema, 1986; Hagenmaier and Shaw, 1990). Coatings and films from formulations containing lipids absorb less moisture and have much potential in the development of moisture, oxygen, and carbon dioxide barriers (Kester and Fennema, 1986; Guilbert, 1986). Hydrophobic substances for film formulations include beeswax, carnauba wax, paraffin wax, fatty acids, and acetylated glycerides (Krochta, 1992). Emulsion-based bilayer and multilayer films are more effective than pure protein and polysaccharide films because the characteristics of the whole film are enhanced by the individual contribution of each component (Kester and Fennema, 1986).

In general, increasing polymer chain length and polarity enhances film or coating cohesion (Kester and Fennema, 1986). A uniform distribution of polar groups along the polymer chain also enhances cohesion by increasing the likelihood of interchain hydrogen bonding and ionic interactions (Banker et al., 1966). Solvent systems for edible films or coatings are limited primarily to water, ethanol, or a combination of the two (Kester and Fennema, 1986). Plasticizers are added to films to reduce brittleness and increase flexibility through a decrease in film cohesion, but the plasticizer can also increase film permeability (Gontard et al., 1992; McHugh et al., 1993). Plasticizers must be compatible with the polymers and, if possible, be readily soluble in the solvent. Among the plasticizers used are mono-, di-, and oligosaccharides (e.g., glucose, high-fructose syrup, and honey), polyols (e.g., sorbitol, glycerol, glycerol derivatives, and polyethylene glycol), and lipids (e.g., fatty acids, monoglycerides and ester derivatives, phospholipids, and surfactants) (Guilbert, 1986). These additives alter the permeability properties to a different extent. For example, at equal weight concentrations, whey protein films plasticized with sorbitol exhibited significantly lower water vapor permeabilities than similar films containing glycerol, PEG 200, or PEG 400 (McHugh et al., 1993).

Edible Coatings and Agricultural Commodities

A key factor limiting the profitability of fresh fruit and vegetable production is their short shelf life. Major losses, both in quantity and quality, occur between harvest and consumption, with postharvest losses estimated to be 5–25% in developed countries and 20–50% in developing countries (Kader, 1985). Refrigeration is sometimes inadequate to retard fruit ripening and prevent other undesirable quality changes. In addition, refrigeration for prolonged periods of time may induce physiological damage (Smith et al., 1987). Early work by Kidd and West (1927) led to the development of storage under controlled oxygen and carbon dioxide atmospheres. A disadvantage of both refrigeration and controlled atmosphere storage is that removal from storage exposes the fruit or produce to ambient conditions, which initiates immediately the process of quality loss. The term "quality" is somewhat subjective, but firmness, flavor, and color are generally recognized to be important attributes (Knee, 1972, 1982; Banks, 1948b).

An effective form of fresh fruit preservation is the use of edible coatings to enhance or replace the natural coatings lost during fruit postharvest operations, particularly washing (Dalal et al., 1971; Smith et al., 1987; Shetty et al., 1989; Hagenmaier and Shaw, 1992). Generally, coating costs are lower than other techniques used to preserve fresh fruits (Kester and Fennema, 1986; Guilbert, 1986).

Fruits and vegetables continue to respire after harvest. Respiration is the overall process by which stored organic materials are broken down into simple end products with a release of energy. Produce and fruits stored at ambient conditions dehydrate, and these water losses diminish their fresh quality (Erbil and Muftugil, 1986). Water loss in fruits occurs via the dermal system, which includes the cuticle, epidermal cells, stomata, leutices, and trichomes. The cuticle, a layer that covers the fruit epidermis, is composed of surface waxes, cutin embedded in wax, and a mixed layer of cutin, wax, and polysaccharide (Banks, 1984a,b; Kader, 1985; Solomos, 1987). The thickness of the cuticle is an important factor in fruits because fruits containing a heavy cuticle generally lose less moisture (Claypool, 1940).

Artificial edible barriers may reduce water loss, lower internal oxygen and increase internal carbon dioxide and ethylene concentrations, reduce respiration rate, and retard ripening processes (Edmond et al., 1964; Pascat, 1986; Guilbert, 1986; Avena-Bustillos, 1992; Hagenmaier and Shaw, 1992). The extent to which these factors may be manipulated are coating and fruit specific. It is interesting to note that the literature on fruit-coating permeability to moisture and gases is rather limited, and thus prediction of a coating performance for a given fruit and storage condition [temperature, relative humidity (RH) and gas composition] is difficult.

Packaging films have been developed to create gas concentrations surrounding produce and fruits that approximate the beneficial effects of controlled atmosphere storage. Edible coatings as artificial barriers to gaseous diffusion could be used for the same purpose while reducing the risk of mechanical and microbial damage. These artificial barriers would not only reduce respiration, but also increase the carbon dioxide content of the internal coated-product atmosphere (Smock, 1935; Edmond et al., 1964; Kester and Fennema, 1986; Chu, 1986; Smith et al., 1987; Edmond et al., 1991).

Edible Coatings and Processed Foods

Moisture movement within food components and between food and the environment limits the storage stability of many foods, generating often complex packaging requirements. An example is the familiar cheese and cracker in many consumer lunchboxes, where these two components need to be packed separately to avoid hardening of the cheese and staling of the cracker (Pennisi, 1992). These complex packaging requirements are compounded by public concern for the environment and the valid, or invalid, association by consumers of food-packaging waste with overflowing landfills. "Reduce-Recycle-Reuse" technologies will need to be promoted and encouraged by the food industry to change this consumer perception. Aluminum is recycled the most because of energy savings possible when aluminum is obtained from recycled material instead of mineral ores. The process is considered generally safe because the high temperature of molten aluminum destroys bacteria and other toxic contaminants. The low viscosity of the molten metal facilitates the separation of suspended solids by simple technologies such as flotation and precipitation. Paper recycling has expanded, but improvements to recovery technologies are necessary to reduce fiber damage, particularly fiber length. Waxed paper and cardboard is in little demand except from thermoelectric plants where they are used as fuel. Plastic recycling is technologically difficult because it involves the separation of the multiple layers used to produce safe and effective packaging materials (Russell, 1989). Glass is a recyclable material, but there are few intrinsic economic advantages in using recycled glass because of the cost of transportation and separating containers by glass type.

An example of an ideal package based on edible film/coating technology would be food envelopes that could be tossed into a hot water sauce pan and dissolved into soup. These envelopes would contain ingredients protected by coatings against flavor exchange, moisture, and oxygen uptake and would be sold in simple and easy-to-recycle containers. This approach will require extensive research to relate information on molecular structure and properties to functional film and coating properties. Research has shown that both type of ingredients and process condition are important. For example, fluorescence microscopy studies have shown that starch–fatty acid films perform best as moisture barriers when the particle size distribution in the emulsion used to form the film is carefully controlled (Pennisi, 1992). Chitosan films combined with fatty acids were more effective moisture barriers when 12-C fatty acid was used. Computer simulations showed that chitosan molecules assumed helical configurations when combined with fatty acids. The hydrophobic end of the fatty acid would occupy the center of the helix closing a gap through which water could move through the film. Fatty acids with chains longer than 12-C may force a wider spacing between chitosan molecules and thus increase film porosity (Pennisi, 1992).

PERMEABILITY MEASUREMENTS

The development of edible films/coatings to reduce moisture transfer, oxidation, or respiration in foods involves measurement of permeabilities of films to water vapor, oxygen, and carbon dioxide. These measurement techniques must take into account assumptions made in calculations, instrument detection limits, and control of experimental conditions with particular emphasis on temperature and relative humidity.

Theoretical Considerations

In the absence of cracks or pinholes, the primary mechanism for gases and vapor transfer through a film or coating is diffusion. Diffusion depends on the molecular size and shape of the permeant, with smaller size particles having higher diffusion values. In general, linear molecules diffuse faster than branched or cyclic molecules. Small differences in shape may cause important changes in permeability.

Diffusion equations are used to evaluate mass transfer through films and coatings (Crank, 1956; Mannheim and Passy, 1985; Pascat, 1986; Loncin, 1986; Krochta, 1992):

$$F_{x,i} = -D_{x,i}\frac{\delta C_i}{\delta x} \tag{1}$$

where $F_{x,i}$ is the flow of permeant i in the direction x, $D_{x,i}$ is an apparent diffusion rate constant, and the partial derivative, $\delta C_i/\delta x$, is the permeant concentration gradient across the film or coating. The apparent diffusion rate constant is assumed to be independent of location within the film. Under steady-state and constant diffusion rate constant, it is possible to derive the following expression:

$$F_{x,i} = D_{x,i}\left(\frac{C_{i,1} - C_{i,2}}{L}\right) \tag{2}$$

where $C_{i,j}$ (j = 1 or 2) are the concentrations of the diffusant at the film surfaces, while L is the film or coating thickness. When measuring vapor and gas permeability, it is further assumed that an equilibrium exists between the film surface and the surrounding atmosphere (films) or bulk conditions in the food (coatings) and that heat of sorption is equal to that of desorption. Hysteresis with respect to sorption equilibrium, a fairly common phenomenon for hydrophilic polymers, is also neglected (Schwartzberg, 1986). The simplest expression and the one most often used assumes a linear relation between the coating/film and the surrounding water vapor or gas pressure:

$$C_{i,j} = S_i p_{i,j} \tag{3}$$

where S_i represents the solubility of permeant i in the film or coating, and $p_{i,j}$ (j = 1 or 2) is the permeant concentration in the environment on both sides of the film or coating. This leads to a modified Eq. (2):

$$N_{x,i} = D_{x,i}\left(\frac{S_i p_{i,1} - S_i p_{i,2}}{L}\right) \tag{4}$$

which leads to the following expression:

$$N_{x,i} = K_i\left(\frac{p_{i,1} - p_{i,2}}{L}\right) \tag{5}$$

where $K_i = D_i S_i$ is the permeability coefficient for permeant i through the film or coating.

The solubility coefficient dependence with water vapor or gas pressure can be determined from the sorption isotherm for the film or coating and may deviate significantly from the linear relationship assumed above. Also, the diffusion constant may increase with water vapor or gas pressure due to plasticization of the film by absorbed molecules (Cairns et al., 1974). A further assumption in the above equations is negligible resistance to mass transfer in the surrounding environment as compared to mass transfer in the coating/films. Expressions have been developed to analyze the case when comparable mass transfer resistances exist on both sides of the film (Krochta, 1992). Therefore, permeability is not a universal film property and values should be reported with a careful description of testing conditions, particularly relative humidity and temperature.

The concept of permeance has been used often in the literature on the permeability of edible coatings and films and can be defined as follows (Hagenmaier and Shaw, 1992):

$$K_i = P_i l \tag{6}$$

where P_i is the permeance of a coating with permeability K_i and thickness l. Permeance is considered a coating or film performance evaluation value and it is not a material property. It can be defined as the rate of water vapor or gas transmission through a unit area and induced by a unit of vapor or gas pressure difference separating the film or coating being tested (ASTM, 1992). The following expression can be used to determine the permeance of multiple layers (Hagenmaier and Shaw, 1992):

$$\frac{1}{P_i} = \sum_{j=1}^{n} \frac{1}{P_i^j} \tag{7}$$

Quantitative Evaluation of Coatings and Films

The advantages of a quantitative approach to food coating development was demonstrated by Hagenmaier and Shaw (1992) with the following example. Assuming a 170-g orange with a 150-cm^2 surface area, a respiration rate at room temperature of 11 mL O_2 (STP) kg^{-1} hr^{-1}, and an internal O_2 concentration of 18%, the following peel permeance ($P_{O_2}^{peel}$) value was estimated:

$$P_{O_2}^{peel} = \frac{11\left(\frac{mL\ (STP)}{kg\ hr}\right)\left(\frac{kg}{10^3 g}\right)170\ (g)\left(\frac{24hr}{day}\right)}{150\ (cm^2)\left(\frac{1m^2}{10^4\ cm^2}\right)(0.209 - 0.18)\ (atm)} \tag{8}$$

$$P_{0_2}^{peel} = 103,000\left(\frac{mL\ (STP)}{m^2\ day\ atm}\right) \tag{9}$$

where 20.9% is the atmospheric O_2 concentration. If this fruit were to be coated using a wax with $P_{0_2}^{coating} = 40,000$ mL (STP) m^{-2} day^{-1} atm^{-1}, the permeance of the coated peel could be determined as follows:

$$\frac{1}{P_{0_2}^{coated\ peel}} = \frac{1}{P_{0_2}^{peel}} + \frac{1}{P_{0_2}^{coating}} \tag{10}$$

$$P_{0_2}^{coated\ peel} = 29,000\left(\frac{mL\ (STP)}{m^2\ day\ atm}\right) \tag{11}$$

This calculation procedure can also be used to determine the permeability of coatings and films that are not self-supporting which is determined using a permeable film as a support (e.g., Hagenmaier and Shaw, 1991a).

Given the tendency of the edible materials used for coatings/films to interact strongly with water vapor, D_i and S_i, and consequently K_i and P_i, are values affected by testing conditions. The permeability of hydrophilic films at high RH is severalfold higher than values at low RH. This has been attributed to an increased hydration of the film matrix (Kamper and Fennema, 1984a,b, 1985; Hagenmaier and Shaw, 1991a). Therefore, permeability values should be measured near conditions of interest, and it is surprising that most edible film and coating permeability values reported in the literature have been measured using large (0–100%) RH gradients. More relevant evaluations would have been obtained if the RH gradient used in tests were smaller and were based on expected storage RH, package internal RH, and food water activity (a_w).

Moisture Permeability. A variety of procedures have been reported to determine the moisture permeability of edible coatings and films. The following sections will highlight differences and probable sources of errors rather than exact permeability-determination procedures.

ASTM E96-80 Test: The most frequently used methods to determine water vapor transmission rate (WVTR) are discontinuous (ASTM, 1992; Debeaufort et al., 1993). A container with water, saturated solution, or desiccant is covered with the test film and then placed in a constant RH chamber. The container is removed a few times from the chamber to determine by weight the amount of moisture crossing the film in a certain amount of time. At equilibrium, the rate of weight loss becomes a constant, which is then used in calculations of permeance (P_{water}) and permeability (K_{water}):

$$WVTR = \frac{\Delta W}{\Delta t \, A} \tag{12}$$

$$P_{water} = \frac{WVTR}{(p_1 - p_2)} \tag{13}$$

$$K_{water} = P_{water} l \tag{14}$$

where $\Delta w/\Delta t$ is the amount of moisture transferred per unit time, l is the film thickness, A is the area of film exposed to moisture transfer, p_i (i = 1 or 2) are the vapor pressures on either side of the film, and K_{water} is the moisture permeability of the film (Martin-Polo et al., 1992a). The recommended metric units for permeability are (amol m)/(m² sec Pa) where 1 amol = 10^{-18} g-mol (Gennadios et al., 1993). Modifications to the gravimetric method have been used by several authors working on edible coatings (Kamper and Fennema, 1984a, b, 1985; Kester and Fennema, 1986, 1989a,b; Martin Polo et al., 1992a,b; Gontard et al., 1992; Debeaufort et al., 1993).

Stagnant Air–Correction Method: The mass transfer coefficient above the cup can be increased by forced convection to ensure equal vapor pressure in the bulk chamber and in the air layer in direct contact with the film. If the mass transfer coefficient inside the cell is comparable to that in the film, the following measurement correction has been shown to be necessary (Krochta, 1992):

$$F_{w,x} = -\frac{P}{RT}\left(\frac{dx_w}{dx}\right) + x_w F_{w,x} = constant \tag{15}$$

where x_w is the molar fraction of water in the air inside the cell at a pressure P and temperature T. Integration of Eq. (6) between conditions on the surface of the desiccant, solution or water inside the cell, and those directly under the film yields the following expression:

$$p_{w,1} = P - (P - p_{w,c})\, e^{\left(F_{w,x}\frac{RTh}{PD_w}\right)} \tag{16}$$

where $P_{w,1}$ is the corrected water vapor under the film, $p_{w,c}$ is the water vapor pressure over the water, saturated solution or desiccant (= 0) in the film-covered container, h is the distance between the surface of this container and the film, and D_w is the diffusion rate constant of water in still air.

Permatran-W™ Test System: Permeation from a container with a saturated solution or water into an air stream at 0%RH is used in the MO-CON Permatran-W™ system (Modern Controls, Minneapolis, MN). A disadvantage of the device is the limiting detection range of the infrared sensor measuring the amount of moisture permeating the film at equilibrium conditions (J. B. Fritsche, Modern Controls, personal communication). Therefore, only air at 0% can be used to flow over the film being tested, and films with large water vapor permeabilities cannot be measured with this device (Krochta, 1992).

Continuous Gravimetric Permeability Determination: Frequent weighing of the container covered with the test film in the ASTM E96-80 procedure underestimates WVTR and moisture permeability (Debeaufort et al., 1993). These authors proposed the use of a continuous weighing system to determine moisture permeability and a ventilator to maintain a large flow rate within the chamber to minimize the time that internal RH conditions were affected by opening the chamber. A further advantage of the continuous measurement system was its significantly lower experimental variability, as indicated in Table 15.2.

Oxygen Permeability. Oxygen permeability seems to be independent of the oxygen partial pressure; however, it is strongly affected by the hydration status of the film or coating. Oxygen permeability can be determined with the Mocon Ox-Tran 100™ permeability tester (Modern Controls) calibrated at 0% RH with reference films provided by the National Bureau of Standards (ASTM, 1981; Hagenmaier and Shaw, 1991a, 1992). Permeability measurements at conditions other than 0% RH can be accomplished by holding films prior to testing in desiccators at the desired RH. During testing, gas streams are passed through bubblers with saturated salt solutions to adjust them to the desired RH (Hagenmaier and Shaw, 1991a).

Using a cell similar to the one described by Davis and Huntington (1977), we placed film samples between two chambers to determine the effect of film moisture content on oxygen permeability (Rico-Peña and Torres, 1990b). A RH-controlled nitrogen flow through one chamber

TABLE 15.2 Variation Coefficients of Continuous Versus Discontinuous Gravimetric Determination of Moisture Permeability

Permeability range (g/m/sec/Pa)	σ_x/X avg	
	Continuous	Discontinuous
$>10^{-11}$	$<5\%$	$10\text{--}45\%$
$<10^{-11}$	$<45\%$	$>70\%$

Source: Adapted from Debeaufort et al., 1993.

and a similarly adjusted oxygen flow through the other were used to adjust the film hydration status prior to oxygen permeability determination. After 2 hr, the inlet and outlet valves were turned off in the mentioned order to ensure equal pressure (atmospheric pressure) in the two cell compartments. At various times thereafter, samples were taken with a syringe from the nitrogen-flushed compartment. The increase in oxygen concentration was monitored using an oxygen headspace analyzer (MOCON LC-700F, Modern Controls). An amount of nitrogen equal to the volume of gas removed was added back to minimize changes in total pressure during testing. Ideal gas behavior was assumed to correct oxygen concentration measurements because of sampling and the nitrogen added back to minimize pressure changes.

In general, films and coatings made from polymers containing hydroxy, ester, and other polar groups tend to have a lower O_2 permeability than polymers with hydrocarbon and other nonpolar groups. Also, the presence of polar groups increases the RH effect on O_2 permeability (Ashley, 1985; Hagenmaier and Shaw, 1992). Rico-Peña and Torres (1990b) found a strong O_2 permeability dependence upon RH when the film was hygroscopic.

Oxygen permeability is important in the coating of agricultural commodities. Low oxygen permeability leads to the development of off-flavors as it induces an anaerobic metabolism, which leads to elevated internal ethanol and acetaldehyde concentrations (Erbil and Muftugil, 1986; Nisperos-Carreido et al., 1990; Hagenmaier and Shaw, 1992).

Permeability Measurements for Other Noncondensable Gases. Carbon dioxide, C_2H_4, and other noncondensable gases can be measured simultaneously using mixtures of these gases on one side of the film and on the other an inert gas such as N_2 or helium. Both gas streams need to be adjusted to the test temperature and RH. Concentration increase in the inert gas stream is measured by a gas chromatograph usually attached to a thermal conductivity detector (Hagenmaier and Shaw, 1991a, 1992).

In coatings for agricultural products, an additional parameter of interest is the ratio of noncondensable gases, which has been shown to vary much less than the permeabilities themselves (Ashley, 1985). For example, the O_2-to-CO_2 permeability ratio of emulsified polyethylene waxes used on agricultural commodities was 0.21 for several commercial formulations tested under a variety of experimental conditions; the same ratio for low-density polyethylene film was 0.35 (Hagenmaier and Shaw, 1991a). This information can be combined with the respiratory quotient

(RQ) and the desired internal oxygen (O_2^i) and carbon dioxide (CO_2^i) concentrations for a given fruit to obtain the following expression:

$$\frac{K_{CO_2}}{K_{O_2}} = \frac{20.9 - O_2^i}{(CO_2^i - 0.03) \, RQ} \tag{17}$$

where 20.9 and 0.03 are the O_2 and CO_2 concentrations in air (Hagenmaier and Shaw, 1992).

Thickness Measurements. An often ignored issue in permeability studies is the measurement of the coating/film thickness. The thickness of the films can be measured directly (Kester and Fennema, 1989a,b; Greener and Fennema, 1989a,b; Rico-Peña and Torres, 1991; Martin-Polo, 1992a; Gennadios et al., 1993; Debeaufort et al., 1993) or calculated from the net weight of the coating and the density of the material (Guilbert, 1986; Hagenmaier and Shaw, 1991b).

Debeaufort et al. (1993) compared micrometer screw (Palmer, 10 μm), electronic gauge (Sodexim S.A., 0.1–1 μm), and scanning electronic microscopy (SEM JEOL JSM 25-CF) measurements and found that if the two sides of the film were smooth, measured thicknesses were reproducible regardless of the method used. In their experimental tests (Table 15.3) this was the case for cellophane, pure methylcellulose, and laminated methylcellulose and paraffin wax films. Micrometer measurements yielded a larger standard deviation because of the instrument precision limitations. If the film surfaces were irregular, micrometer readings were about twice the value obtained with the electronic gauge, which were similar to SEM measurements. Electronic gauge measurements were considered superior for its simplicity, precision, and agreement with SEM measurements (Debeaufort et al., 1993).

TABLE 15.3 A Comparison of Film Thickness Measurement Methods

| Film tested | Contact measurements | | |
	Micrometer	Electronic	SEM
Cellophane	25 ± 10	19.8 ± 0.6	20 ± 4
Methylcellulose	20 ± 10	16.8 ± 0.4	17 ± 1
w/emulsified paraffin wax	130 ± 30	71.6 ± 6.2	68
w/laminated paraffin wax	90 ± 15	90.1 ± 0.6	87 ± 15

Source: Adapted from Debeaufort et al., 1993.

Temperature Effects on Permeability: Permeation Activation Energy. A film or coating is generally formed by a network of macromolecular chains and interstices. The thermal motion of the chains or their terminal groups yields cavities, which may be filled and crossed by the diffusing substances. Diffusion depends not only on the number and dimensions of the cavities, but also on the energy of activation (E_a) which is needed for the diffusing molecules to cross the polymeric membrane. This phenomenon is described by an Arrhenius model for a diffusant i as follows (Rogers, 1985):

$$K_i = K_{o,i} e^{\frac{-E_{a,i}}{RT}} \tag{18}$$

The activation energies for moisture and gas permeability can be determined from Arrhenius plots of $\ln K_i$ vs. inverse absolute temperature (T) (Hagenmaier and Shaw, 1991a). The values obtained are generally in the 0–15 kcal/g-mol area (Vojdani and Torres, 1990; Taoukis, 1989; Hagenmaier and Shaw, 1991a; Gennadios et al., 1993). For example, the average E_a for oxygen permeating emulsified polyethylene waxes used as fruit coatings is 4.7 kcal/g-mol, which means that permeability increases about 30% for a 10°C increase in temperature (Hagenmaier and Shaw, 1991a). Plasticizers added to coatings reduce polymer interchain attraction forces and thus can lower E_a values (Kester and Fennema, 1986).

Determination of Film Moisture Isotherms. Water vapor sorption of edible films and coatings may result in swelling and conformational changes of the structural polymer (proteins or polysaccharides). Therefore, a determination and interpretation of the water sorption properties of edible films is necessary to develop effective edible films and coatings (Gennadios and Weller, 1991). Film samples cut into small pieces and brought to zero moisture by freeze-drying or other suitable procedures can be exposed to constant RH and temperature. Constant RH environments can be generated using the saturated solutions recommended by the COST-90 project (Wolf et al., 1984). The weight gain of the samples is recorded to the nearest 0.1 mg until changes in moisture content are less than 0.001 g water/g dry matter. Similar procedures can be used to determine the desorption isotherm (Labuza, 1984). Gennadios and Weller (1991) evaluated several equations to describe the absorption isotherm for corn zein, wheat gluten, and wheat gluten/soy protein isolate films and found that the three-parameter (M_o, C, K) "G.A.B." equation (Bizot, 1984) to predict film moisture content (M) as a function of its water activity (a_w):

$$M = \frac{M_o C Ka_w}{(1 - Ka_w)(1 - Ka_w + CKa_w)} \qquad (19)$$

showed the best fit for all three types of films (Table 15.4). The degree of fit was evaluated using a mean relative deviation modulus (DM):

$$DM = \frac{100}{n} \sum_{i=1}^{n} \frac{|M_{exp,i} - M_{pred,i}|}{M_{pred,i}} \qquad (20)$$

where M_{pred} and M_{exp} are moisture content predicted with the G.A.B equation and measured experimentally, respectively. A DM < 5 indicates an extremely good fit, 5 < DM < 10 corresponds to a reasonably good fit, while DM > 10 indicates a poor fit (Gennadios and Weller, 1991).

Film-Orientation Effects. The moisture permeability properties of proteins and polysaccharides films with added hydrophobic materials (e.g., fatty acids) exhibit significant film-orientation effects. Films prepared by emulsification of these components are nonisotropic because of separation and creaming during film dehydration. For example, methylcellulose (MC) or hydroxypropyl methylcellulose (HPMC) suspensions containing lauric, palmitic, stearic, or arachidic acids form films when cast on glass plates at 80–85°C. The plates were left at room temperature

TABLE 15.4 Degree of Fit to Film Moisture Isotherms

| Film | G.A.B.[a] | | | | DM[b] | DM[c] | DM[d] |
	DM[e]	M_o	C	K			
Zein	8.0	0.124	0.346	0.890	85.3	20.1	48.1
Wheat gluten (WG)	4.7	0.105	0.990	0.945	33.1	10.1	29.3
WG/Soy protein isolate	0.112	3.44	0.985	0.896	36.8	13.8	33.8

DM, Deviation modulus.
[a]*From Bizot, 1984.*
[b]*From Smith, 1947.*
[c]*From Oswin, 1946.*
[d]*From Halsey, 1948.*

to promote fatty acid orientation within the wet film and then placed in a convection oven at 80–85°C for 15 min. SEM photomicrographs of such films after drying and cooling, showed fatty acid crystals on the film surface (Fig. 15.1) as well as discrete lipid particles within the polysaccharide matrix (Fig. 15.2a,c; Vojdani and Torres, 1989a). We prepared even more heterogeneous films by casting HPMC and fatty acid emulsions on top of a pure preformed HPMC film. We observed three layers: essentially pure HPMC, HPMC with discrete lipid particles, and a surface layer of pure lipids (Fig. 15.2b,d; Vojdani and Torres, 1989a). The same emulsion instability during film formation was observed in whey protein films in spite of the moderate ability of whey proteins to act as emulsifiers (McHugh and Krochta, 1993a). As in the polysaccharide films, discrete lipid particles remained within the protein matrix and a layer of lipids also formed on the film surface which appeared dull. Moisture-permeability tests with the dull side facing the high-relative-humidity chamber resulted in significantly lower values compared to films tested when oriented in the opposite direction (McHugh and Krochta, 1993a). This demonstrates the importance of reporting orientation when evaluating the moisture permeability of films containing proteins, polysaccharides, and lipids.

EDIBLE PROTEIN COATINGS AND FILMS

The utilization of proteins as edible film-forming agents has not been studied as extensively as polysaccharides (Kester and Fennema, 1986). More recent work has determined the properties and fabrication procedures for collagen, soybean, corn, wheat, and milk protein films (Krochta et al., 1988, 1990a,b; Aydt et al., 1991; Avena-Bustillos, 1992; Gontard et al., 1992; Mahmoud and Savello, 1992, 1993; Gennadios et al., 1993). Although edible films must withstand the normal use conditions encountered during application, food handling, and storage, the following discussion will focus on their permeability properties. Several research groups have provided extensive information on procedures for the evaluation of mechanical properties (e.g., Aydt et al., 1991; Gontard et al., 1992). An integrated approach combining permeability and mechanical test evaluations was suggested by McHugh and Krochta (1993b).

Collagen-Based Films

The most commonly utilized protein film is collagen, which is frequently utilized as a casing for processed meats. Gelatin, a protein de-

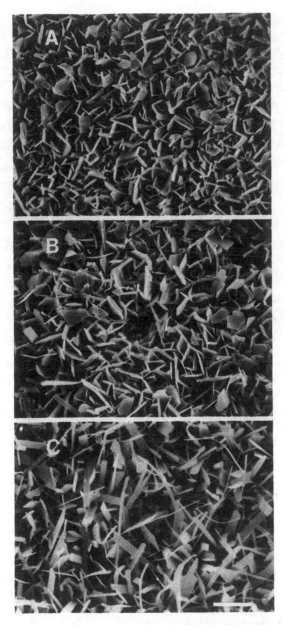

FIG. 15.1   Scanning electron microscopy photomicrographs of the surface of hydroxypropyl methylcellulose films containing (A) palmitic acid, (B) stearic acid, (C) or arachidic acid in a 3:1 weight ratio. (*From Vojdani and Torres, 1989a.*)

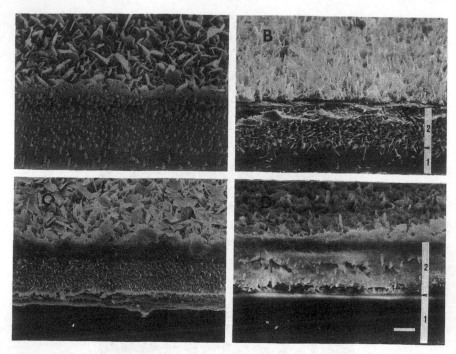

FIG. 15.2 Scanning electron microscopy photomicrographs of cross sections of edible films prepared from methylcellulose (MC) or hydroxypropyl methylcellulose (HPMC) and fatty acid emulsions in a 3:1 weight ratio. Films cast as (A) a single layer of MC and palmitic acid or (C) stearic acid or prepared as (B) a double layer of pure HPMC overlayed with HPMC and arachidic acid or (D) stearic acid. (*From Vojdani and Torres, 1989a.*)

rived from collagen, forms thermally reversible gels when warm aqueous suspensions of the polypeptide are cooled. The formation mechanism involves ionic cross-links between amino and carboxyl groups of amino acid side chains with a minor contribution of hydrogen bonds (Glicksman, 1982; Kester and Fennema, 1986). The moisture-barrier properties of dried gelatin, casein, serum albumin, and ovalbumin films can be enhanced by cross-linking these proteins with lactic or tannic acid. This effect was associated with a decreased chain mobility achieved by cross-linking. However, treatments with Ca ions were found to be ineffective (Kester and Fennema, 1986).

Wheat Gluten Films

Another protein studied as an edible film-forming agent is wheat gluten (WG), which requires a film-forming solution containing alcohol at either alkaline (pH > 10.5) or acidic (pH < 5) conditions (Anker et al., 1972; Aydt et al., 1991; Gontard et al., 1992). The alkaline-prepared WG films were rejected because of color and chewy texture when the casting solution pH was adjusted with sodium hydroxide. When the pH was adjusted with ammonia, the films were rejected for odor. WG films prepared using a volatile acid such as acetic acid were preferred for their visual and sensory properties (Gontard et al., 1992). A comprehensive response surface study was conducted by these authors to determine the effect of pH and the concentration of ethanol and gluten in the WG film–forming solution. The presence of two main protein classes in wheat gluten separated by solubility differences was reflected in the opacity of the films formed. At high ethanol concentrations (45–70 mL/ 100 mL solution), the heterogeneous film aspect correlated with the presence of the low molecular weight and alcohol-soluble gliadins. These appeared as transparent areas. The insolubilization or partial precipitation of the large molecular weight and acetic acid–soluble glutenins appeared as opaque areas. Conditions for the most transparent films seems to require a 32.5–45 mL ethanol/100 mL solution and a pH in the 2–4 range.

A low water solubility is important when films are in contact with water during processing and storage. The opposite is desirable when the intent is to design a package with premeasured dry food amounts to be dissolved in water or in hot food (Guilbert and Biquet, 1989). The lowest solubility for WG films, 40% after 24 hr in water at 20°C, was predicted by surface response analysis for film-forming solutions in the region defined by 20–40 mL ethanol/100 mL solution and pH 2–5 (Gontard et al., 1992). If high water solubility is desired, it would be preferable to use a nonvolatile acid (e.g., lactic acid) to adjust film-forming solution pH. Lactic acid remains in the film and acts as a plasticizer to facilitate the film-disintegration process. Solutions with 10% lactic acid and no glycerol disintegrated completely in water after only a few minutes at 20°C (Gontard et al., 1992).

Extensive covalent and noncovalent bonding between the gliadins and the glutenins are involved in WG film formation. Intermolecular and intramolecular disulfide bonds are most probably also involved. The extended random-coiled glutenin may provide the WG film matrix, whereas the smaller globular gliadins may be packed into this matrix generating a rather compact structure (Gennadios et al., 1993). The WG film heterogeneity and the opacity observed by Gontard et al. (1992) explained the high water vapor permeability of films from solutions at high ethanol concentration and pH > 4. On the other hand, low pH solutions promoted protein unfolding and exposure of hydrophilic groups. The latter facilitates the migration of moisture through the films and would explain the high moisture permeability values observed for the films prepared from low pH solutions. A reduction of permeability was observed when ethanol concentrations were increased to enhance the exposure of less hydrophilic groups and solution acidity was reduced to promote protein solubility. However, surface response analysis indicated that even lower values could be expected at neutral pH and low ethanol concentrations (Gontard et al., 1992).

Gennadios et al. (1993) observed that oxygen molecules could not readily permeate through the highly cross-linked gluten structure when dry. Wheat gluten films with very low oxygen permeability [2.4–7.6 (amol m)/(m^2 s Pa) at 0% RH and 37.8°C] and performing well as CO_2 barriers [^{1}0–4.6 (amol m)/(m^2 s Pa), at 0% RH and 37.8°C] were prepared by Aydt et al. (1991). These determinations were consistent with mean oxygen permeability values increasing from 0.9 at 7°C to 6.1 (amol m)/(m^2 s Pa) at 35°C as reported by Gennadios et al. (1993). These two research groups did not determine the effect of film-casting solution modifications, and it is probable that formulations for films with lower O_2 and CO_2 permeability could be identified by further testing.

Temperature increases permeability due to the enhanced motion of the polymer segments and the increased energy level of permeating molecules (Rogers, 1985; Gennadios et al., 1993). The applicability of the Arrhenius model to describe this behavior was confirmed for oxygen permeability through wheat gluten (WG) and wheat gluten/soy protein isolate (WG/SPI) films (Gennadios et al., 1993). WG films were prepared by combining 15 g wheat gluten, 72 mL 95% ethanol and 6 g glycerol with 12 mL 6N ammonium hydroxide and 48 mL distilled water while warming and stirring the mixture on a hot plate. WG/SPI film forming solutions were prepared in the same manner by substituting 4.5 g of wheat gluten with an equal amount of soy protein isolate. Soy protein isolate further increased cross-linking between protein molecules as reflected by oxygen permeability values of the WG/SPI films 30% lower than WG films.

The effect of temperature on oxygen permeability fitted well the Arrhenius model as demonstrated by a model mean deviation modulus (DM) under 4 estimated as follows (Gennadios et al., 1993):

$$DM = \left(\frac{100}{n}\right) \sum_{j=1}^{n} \frac{|K_{exp, j} - K_{pred, j}|}{K_{pred}} \qquad (21)$$

where $K_{exp, j}$ and $K_{pred, j}$ are the experimental and predicted oxygen permeability values and n is the number of experimental values. Arrhenius plots for WG and WG/SPI films gave coefficient of determination

(R^2) better than 0.995. The breaks observed sometimes in Arrhenius plots, and reflecting a structural transition of polymers from a crystalline to an amorphous state (Karel, 1975), were not observed in WG and WG/SPI films. This suggests that in the range tested, 7–35°C, no structural changes occurred in the protein matrix. Calculated values for the energy of activation (E_a) were 11.9 ± 0.5 and 10.8 ± 0.4 kcal/g-mol and were consistent with values obtained for permeation processes (Ashley, 1985; Taoukis, 1989; Vojdani and Torres, 1990; Rico-Peña and Torres, 1990b).

The effect of moisture content on the oxygen permeability of WG and WG/SPI films was not determined. The plasticizing and/or swelling effect of moisture would be expected to increase this property, as was demonstrated by Rico-Peña and Torres (1990b) for an edible film produced from methylcellulose and palmitic acid (3:1 weight ratio). This effect of moisture on the permeability of oxygen and other noncondensable gases could be solved by preparing edible laminated films similar to the ones developed for moisture-sensitive plastic barriers such as ethylene-vinylalcohol (EVA) and polyamide (PA) (Gennadios et al., 1993).

Corn Zein–Based Films

The aqueous alcohol-soluble protein extracted from corn gluten has been reported to form films with relatively good water barrier properties (Guilbert, 1986). Aydt et al. (1991) used a commercial zein solution containing ethanol and glycerin to produce cast films dried at 23 ± 1°C for 24 hr. Film thickness ranged from 75 to 100 μm and had generally poor mechanical properties; however, plasticizers suggested by Andres (1984) and others (e.g., Torres, 1987) were not included in the film-forming solution. Water vapor permeability (50–100% RH gradient, 26°C) of these films was similar to values for cellophane tested at the same conditions. These films had low oxygen [~0–160 (amol m)/(m^2 s Pa) at 0% RH and 37.8°C] and relatively low carbon dioxide [~ 66–630 (amol m)/(m^2 s Pa) at 0% RH and 22.8°C] permeability values. The wide variation in permeability values obtained by these authors was not explained but reflects the difficulties in using manual laboratory procedures to prepare edible films with consistent properties.

More reproducible oxygen permeability determinations were made by Gennadios et al. (1993). Mean values ranged from 1.8 at 7°C to 11.2 (amol m)/(m^2 s Pa) at 35°C and 0% RH. This temperature effect corresponded to an $E_a = 11.1 ± 0.7$ kcal/g-mol, a value very similar to the one obtained for oxygen permeating wheat gluten and wheat gluten/soy protein isolate films described above. This suggests that E_a is a good

TABLE 15.5   Energy of Activation (E_a) for K-sorbate Permeation through Polysaccharide-based Films

Film	E_a (kcal/g-mol)
Films embedded in 50% v/v aqueous glycerol	
Chitosan	6.98
Hydroxypropyl methylcellulose (HPMC)	6.62
Methycellulose (MC)	7.71
HPMC + MC (3:1 weight ratio)	7.27
Average	7.15 ± 0.46
Films embedded in water	
Chitosan	3.95
Methylcellulose (MC) films with fatty acids (MC:FA = 3:1, w/w) embedded in 50% (v/v) glycerol	
MC:lauric acid	7.71
MC:palmitic acid	7.84
MC:stearic acid	7.49
MC:arachidic acid	7.63
Average	7.67 ± 0.15

Source: Adapted from Vojdani and Torres, 1989b, 1990.

description of the interaction of the permeant molecule and the network that this molecule needs to cross. A similar behavior was observed for the permeation of potassium sorbate through various polysaccharide-based films (Table 15.5; Vojdani and Torres, 1989b, 1990). E_a values for K-sorbate permeation were similar for all polysaccharide if the solvent embedding the film was the same (50% v/v glycerol) in spite of nearly a 100-fold difference in permeability values achieved by changing the fatty acid (lauric, palmitic, stearic, or arachidic acid) and the polysaccharide (chitosan, methylcellulose, or hydroxypropyl methylcellulose). The E_a changed dramatically when the solution embedding the chitosan film was changed to water, which suggests that the solvent affected the molecular conformation of the film structural polymer. The major role of fatty acids seems to have been a reduction in the amount of solvent in the film matrix.

Milk Protein—Based Films

Krochta et al. (1990a,b) have evaluated edible coatings made from milk protein and vegetable oil derivatives and found them to reduce moisture loss substantially. Films formulated with these two food components had different permeability values depending on the protein/lipid ratio (Avena-Bustillos et al., 1990).

Calcium caseinate and acetylated monoglycerides were used to formulate edible coatings to extend the shelf life of fresh zucchini (*Cucurbita pepo* var. *melopepo*) (Avena-Bustillos, 1992). Evaluation of zucchini moisture losses showed a significant effect of coating application method with dipping reducing moisture loss 20% more than brushing. Microbial stability problems of the coating emulsion and of the coated fruits were controlled by the addition of 0.1% potassium sorbate. Note that although this agent is widely used in processed foods, it is not approved for use on fresh fruits and vegetables marketed in the United States (Hall, 1988; Avena-Bustillos, 1992). Response surface analysis was used to identify optimum formulations reducing moisture losses while decreasing the total solids concentration and the amount of acetylated monoglycerides relative to caseinates. The formulation identified contained 0.8% calcium caseinate and 0.7% acetylated monoglycerides. Respiration rate was also studied and shown to be affected only during the initial storage days. Ethylene production was not affected by the coatings tested. These caseinate-acetylated monoglycerides coatings were compared with the commercial coating Semperfresh™ (Inotek International Corp., Mentor, OH) at the 0.5 and 1.0% levels, which were found ineffective in reducing moisture loss from zucchini under the conditions tested.

Considerable interest exists for the development of edible coatings for freshly cut fruits and vegetables. Effective edible films could prevent product dehydration, modify internal product atmosphere to retard senescence, act as a barrier against microbial invasion, and serve as a carrier for additives such as antioxidants, flavorings, colorants, and even nutrients. Pure protein films have been found to form films with poor moisture- and oxygen-barrier properties (Guilbert, 1986). Protein-lipid emulsions have been used to prepare more effective films (Ukai et al., 1976; Krochta et al., 1988; McHugh and Krochta, 1993b). The oxygen permeabilities of whey and other protein films are lower than values for high-density polyethylene and compared favorably to EVA films. The moisture sensitivity of their oxygen permeability will need to be solved as indicated by an increase from 0.7 to 43.3 (mL μm)/(m^2 d kPa) for whey protein films containing sorbitol (3.5:1 weight ratio) when testing RH was increased from 40 to 70% RH (Table 15.6; McHugh and Krochta, 1993b).

A limitation of these films for application on cut fruit surfaces is that they are readily soluble in water. For example, casein films dissolve in water in less than 2 min. However, this was not the case when films were cast and then soaked for 3 min in sodium acetate or calcium ascorbate

TABLE 15.6   Oxygen Permeability (PO$_2$) at 23°C of Protein-based Films Compared to Synthetic Films

Film	RH	PO$_2$	Ref.
WPI:G = 5.7:1	50	18.5	McHugh and Krochta, 1993b
WPI:G = 2.3:1	50	76.1	McHugh and Krochta, 1993b
WPI:S = 2.3:1	50	4.3	McHugh and Krochta, 1993b
WPI:S = 1:1	50	8.3	McHugh and Krochta, 1993b
WPI:S = 3.5:1	40	0.7	McHugh and Krochta, 1993b
WPI:S = 3.5:1	70	43.3	McHugh and Krochta, 1993b
WG:G = 2.5:1	0	3.8	Gennadios et al., 1991
CZ:G = 5:1	0	7.7	Gennadios et al., 1991
Collagen	63	2.3	Lieberman and Guilbert, 1973
Collagen	93	89	Lieberman and Guilbert, 1973
LDPE	50	1870	Salame, 1986
HDPE	50	427	Salame, 1986
Cellophane	0	252	Taylor, 1986
EVOH	0	0.1	Salame, 1986
EVOH	95	12	Salame, 1986
Polyester 1470	0	17.3	National Bureau of Standards

Film ingredients are indicated as weight ratios. WPI, whey protein isolate; G, glycerol; S, sorbitol; WG, wheat gluten; CZ, corn zein; LDPE, low density polyethylene; HDPE, high density polyethylene; EVOH, ethylene vinyl alcohol (70% vinyl alcohol).
Source: Adapted from McHugh and Krochta, 1983b.

buffer solutions. This treatment prevented films from dissolving but had little effect on moisture transmission. However, soaking the buffer-treated films in distilled water to remove excess buffer reduced the film moisture transmission by about 50% (Table 15.7; Krochta et al., 1988). When the lipid amount was doubled, the permeability of the film was also reduced (Table 15.7), an observation consistent with other studies (e.g., Vojdani and Torres, 1990). All of the films tested showed increased water transmission rates during the initial testing hours, indicating that moisture uptake by the films increased their permeability. The greatest increase with time occurred in films that were only buffer-treated. Buffer-only–treated films increased their permeability so much that in 24 hr, moisture transfer measured had exceeded the amount of moisture transferred through the untreated films. This observation was interpreted by Krochta et al. (1988) as a confirmation that excess buffer promoted the movement of water through the films. Tests with calcium ascorbate–treated films showed no difference from those treated with sodium

Table 15.7 Water Transferred (WT) Across Casein Films as Compared
to Permeability Cell Measurements with No Film (100%-0 RH Gradient)

Film solution composition	Thickness (μm)	WT	Water solubility
No film	n/a	1.00	n/a
10% casein–5% glycerol	88	0.93	<2 min
PLUS: acetate buffer, 4.6 pH, 3 min	72	0.87	>24 hr
10% casein–10% Myvacet 5-07	129	0.48	<2 min
PLUS: ascorbate buffer, 4.6 pH, 3 min	n/d	0.49	>24 hr
PLUS: dist. water, 2 min	119	0.26	>24 hr
10% casein–20% Myvacet 5-07	184	0.20	<2 min
PLUS: ascorbate buffer, 4.6 pH, 3 min	217	0.20	>24 hr
PLUS: dist. water, 2 min	201	0.12	>24 hr
10% casein–10% carnauba wax	156	0.51	<2 min
PLUS: ascorbate buffer - dist. water	146	0.25	>24 hr

WT, a relative index of water transferred through the films in 6 hr (WT = 1 for no film); n/a, not applicable.
Source: Adapted from Krochta et al., 1988.

ascorbate or sodium acetate buffers. This illustrated the importance of adjusting the film pH to its isoelectric point (4.6 pH) and that Ca ions did not appear to be critical (Krochta et al., 1988). The latter observation was confirmed in films treated with calcium chloride solutions. Films made from emulsions with various waxes were not superior to those prepared with acetylated monoglycerides (Table 15.7). Krochta et al. (1990b) suggested that films with ascorbate buffer have the potential of providing antioxidant properties to coated foods. Further work on caseinate-based films offers interesting possibilities because of the ionic character of many of the amino acids present in casein.

The mechanical properties, moisture permeability, solubility, and hydrolyzability of whey protein films prepared using transglutaminase as a cross-linking agent were evaluated by Mahmoud and Savello (1992, 1993). A more economical technique to prepare such films is heating whey protein past the degradation temperature for β-lactoglobulin, which is the major component in whey (McHugh et al., 1993). The recommended heat treatment, 90°C for 30 min of 10% w/w whey solutions, promoted the formation of intermolecular disulfide bonds by thiol-disulfide interchange and thiol oxidation reactions (McHugh et al., 1993). Whey concentrations under 8% were insufficient to form acceptable films, while above 10% the solutions began to gel, making difficult the vacuum removal of air bubbles from the solution. The removal of entrapped air reduces the size (to ¹200 nm) and number of pores as shown in Fig. 15.3 for 10% w/w whey protein films. Above 12%, the whey gels formed were so strong that they could not be spread into films (McHugh et al., 1993). The effect on film formation of whey solution pH adjusted before and after heat denaturation was also investigated by these authors (Table 15.8). At pH 5, films could not be formed because

TABLE 15.8 Effect of Whey Protein Solution pH on Moisture Permeability (100%-0 RH Gradient)

Whey solution pH	Film thickness (μm)	RH[a] inside cup (%)	WVP[a] (g-mm/kPa-hr-m^2)
pH adjusted before heating:			
6	0.131	76.0	3.2[b]
7	0.122	77.7	2.7[c]
8	0.113	76.5	2.7[c]
9	0.125	77.6	2.8[b,c]
pH adjusted after heating:			
6	0.165	73.8	4.5[a]
7	0.122	77.7	2.7[c]
8	0.124	76.7	2.9[b,c]
9	0.126	76.9	2.9[b,c]
Ionic strength variation[c]:			
7	0.122	77.7	2.7[a]
7	0.131	77.9	2.9[b]

[a]The relative humidity at the inner surface of the film was adjusted for stagnant air effects and used to obtain corrected WVP values.
[b]WVP values with the same superior letters are not statistically different using Fisher's PSID multiple comparison test ($p < 0.05$).
[c]The second set of values correspond to films prepared from a whey solution with added HCl and NaOH to yield neutral pH but at a higher ionic strength than the control (first set of values).
Source: Adapted from McHugh et al., 1993.

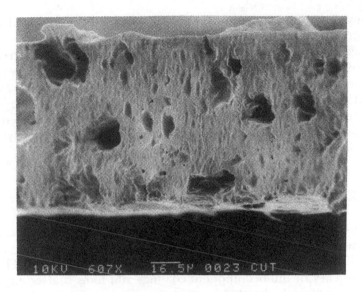

FIG. 15.3 Effect of air removal from a film casting mixture containing sorbitol and whey protein (preheated for 30 min at 90°C as a 10% w/w solution). Scanning electron micrographies for 62.5% protein/37.5% sorbitol dried films (a) with and (b) without vacuum application. (*From McHugh et al., 1983.*)

of whey protein precipitation. At pH 10, whey solutions formed strong gels and could not be used to prepare films. In the pH range 7–9, no statistically significant effects on WVP were observed. The increase in WVP of films prepared from whey solutions adjusted to pH 6 before or after heating may be an ionic strength effect, which would also explain why adding NaOH and HCl while retaining pH 7 produced more permeable films (Table 15.8). This would suggest that films prepared from proteins processed by dialysis or ultrafiltration to lower electrolyte concentrations could reduce moisture permeability.

FUTURE RESEARCH AND DEVELOPMENT NEEDS

Protein-based films possess limited water vapor barrier properties. Water vapor permeability decreases with ingredients added to increase hydrophobicity (Debeaufort et al., 1993; Guilbert and Biquet, 1989). The barrier efficiency of multicomponent coatings and films is affected by the distribution of the hydrophobic component on or within the film forming matrix (Kamper and Fennema, 1984, 1985; Martin-Polo et al., 1992a). The exact nature of the hydrophobic component and the protein used as the structural matrix for the film needs to be further analyzed. Film preparation techniques need to be interpreted in terms of their effect on this protein matrix–hydrophobic ingredients interaction.

Lipids, frequently used to increase the hydrophobicity of edible films, have different physical states depending upon their chemical composition, conformational structure, and molecular size (Martin-Polo et al., 1992b). Kamper and Fennema (1984), in their work on polysaccharide-based films, demonstrated that an increase in the degree of saturation increased the moisture-barrier properties of the film. Films prepared with the emulsion technique were more moisture resistant when paraffin and beeswaxes were used instead of corn oil. These authors suggested that this difference was related to water vapor solubility in the lipid component and on differences in the molecular organization of the lipid. Differences in oxygen permeability could be explained in terms of the previous factors and the relative polarity of the support where these lipids are incorporated.

The role of paraffin waxes and oils in controlling moisture permeability in edible films has been recently examined by Martin-Polo et al. (1992a,b). In their work, three cellulose derivatives (methylcellulose, filter paper, and cellophane) were used to prepare films by dipping, casting, and coating techniques. Disks of cellophane or filter paper were dipped for 30 sec in molten paraffin wax (95°C) or in paraffin oil at room temperature and then drained for 1 min. Paraffin wax and oil are mixtures of alkanes with different chain length. The major components in paraffin oil, $C_{16}H_{34}$ and $C_{14}H_{30}$, and in paraffin wax, $C_{26}H_{54}$ and $C_{40}H_{82}$, are responsible for the liquid-solid difference in physical state (Martin-Polo et al., 1992a). These two paraffins have similar hydrophobicities, which means that solubility of water in the coatings thus prepared would be similar and negligible (Martin-Polo et al., 1992a). Coated disks were kept for 18 hr at 70°C to remove excess wax and ensure a constant deposition of paraffin (0.045 ± 0.006 kg m^{-2}). The thickness of the coated filter paper was 190 ± 10 μm and 80 ± 10 μm for the coated cellophane. Methylcellulose (MC, 4.25 g), paraffin wax (4.25 g), and polyethylene glycol 400 (1.3 mL) were dissolved at 75°C in 75 mL of a 1:2 (v/v) mixture of distilled water and ethanol and used to cast films following the techniques described by Kester and Fennema (1989b). Cast emulsion films with a 1:0.025 methylcellulose to paraffin wax ratio and heated to 70°C were coated with molten paraffin wax preheated to 95°C. For all tested films, a significant decrease (1.3–62 times) in moisture permeability was observed when the paraffin oil or waxes were used (Table 15.9; Martin-Polo et al., 1992a). The most effective moisture

TABLE 15.9 Characterization of Films Prepared with Paraffin Oils and Waxes

Film tested	Thickness (μm)	Contact angle[a] (°)	$K \times 10^{11}$ (g/m/sec/Pa)
Emulsion and emulsion-coated films:			
Methylcellulose (MC)	70 ± 10	28 ± 1	10.8 ± 1.6
MC/paraffin oil	130 ± 20	51 ± 2	9.1 ± 1.0
MC/paraffin wax	200 ± 20	62 ± 3	9.6 ± 2.0
MC/paraffin wax + paraffin wax	140 ± 20	102 ± 3	0.2 ± 0.2
Dip-coated films:			
filter paper (FP)	180 ± 10	—[b]	35.1 ± 3.6
FP + paraffin oil	190 ± 10	71 ± 5	26.5 ± 4.6
FP + paraffin wax	190 ± 10	116 ± 2	29.0 ± 2.5
cellophane (CP)	30 ± 10	8 ± 1	5.7 ± 0.4
CP + paraffin oil	80 ± 10	n.m.[c]	9.3 ± 1.7
CP + paraffin wax	90 ± 10	104 ± 3	0.3 ± 0.04

[a]Contact angle for water determined at 20°C with the aid of a microscope equipped with a goniometric piece.
[b]Water was absorbed on filter paper.
[c]Not measured.
Source: Adapted from Martin-Polo et al., 1992a.

barriers were the wax-coated cast films and the wax-dipped cellophane films (Table 15.9). This property could be explained by a continuous layer of paraffin wax covering these films as observed by SEM (Fig. 15.4). The high permeability values for the MC-paraffin emulsion films are in some disagreement with the work by Kamper and Fennema (1984a), who found that emulsified films were more efficient than laminated ones when fatty acids were used in the methylcellulose emulsion. Martin-Polo et al. (1992a) speculated that the fatty acids solidify, forming spangles that are less efficient in providing a continuous hydrophobic surface. Further evidence of the role of the hydrophobic layer structure has been obtained by analyzing films composed of cellophane and a hydrophobic coating containing different proportions of a solid phase, paraffin wax or n-octacosane ($C_{28}H_{58}$), and a liquid one, paraffin oil or n-hexadecane ($C_{16}H_{34}$) (Martin-Polo et al., 1992b). In both cases a significant increase in moisture permeability was observed when the mixture contained less than 25% w/w of the solid phase. SEM and x-ray diffraction (XRD) analysis suggested that the efficiency of 100% paraffin wax–coated films were related to a parallel orientation of the wax crystals to the base sheet of the cellophane. XRD analysis showed that the presence of a liquid phase (paraffin oil) did not influence the orientation of the wax crystals. The significant increase of moisture permeability when the paraffin wax proportion was less than 25% was interpreted as an indication that the quantity of crystals was so low that it did not cover the cellophane surface.

The barrier properties of n-alkane mixtures were less (~10 times) than those for paraffin waxes. When n-alkane coatings containing less than 25% octacosane were analyzed by XRD, a unique orientation for the aliphatic chains perpendicular to the base sheet of the cellophane was observed. Other arrangements were observed at higher solid n-alkane proportions. For 100% octacosane, clusters of needles were observed by SEM, while stacking of separated flakes were apparent at 10% (Fig. 15.5). In going from 50 to 100% octacosane, the moisture permeability increased from 0.5×10^{-11} to 6.9×10^{-11} g/m/sec/Pa. This abnormal behavior was interpreted on the basis of void regions in the crystalline structure, which were filled with liquid n-hexadecane at lower octacosane concentrations. This would produce a greater resistance to the passage of the water molecules than in the absence of a liquid hydrophobic phase. The absence of a sufficient number of crystalline structures at octacosane concentrations below 25% would explain the increase in moisture permeability in this region (Martin-Polo et al., 1992b).

Large differences were noted between the effectiveness of paraffin waxes and oils and the substrates on which they were applied (Martin-Polo et al., 1992a). As expected, the highest permeability values were observed for films with the smallest contact angles for the more polar supports (Table 15.9). Permeability values estimated for the paraffin wax layer in the coated cast films and in the dipped cellophane films were 0.18×10^{-11} and 0.06×10^{-11} g/m/sec/Pa, respectively. This difference was interpreted as an indication that wax applications on a nonporous substrate (cellophane) was more effective than over the cast MC-wax film. In porous filter paper, i.e., a model for some food surfaces, the paraffin oil or wax fills the pores, but the moisture transfer may still occur through the paper fibers free of the paraffins. The value for the paraffin oil on the cellophane film was estimated to be 15×10^{-11} g/m/sec/Pa, which suggests a much lower effectiveness for liquid paraffin wax (Martin-Polo et al., 1992a). Finally, in many food applications, the coating may interact and probably extract some food components that alter the effectiveness of the coating.

CONCLUSION

The use of edible barriers, coatings or films, on foods is not a new technique. Coatings were used in China as early as the twelfth century to prolong the storage life of oranges and lemons (Hardenburg, 1967; Kester and Fennema, 1986). Food coating processes using wax and gelatin were patented in the 1800s (Guilbert, 1986). In the 1930s hot-melt paraffin waxes became commercially available for the coating of citrus fruits. In the early 1950s, carnauba wax oil-in-water emulsions, were developed for coating fresh fruits and vegetables (Kaplan, 1986). Shellac wax has also been used since the beginning of this century as an edible coating and has been approved by the Food and Drug Administration (FDA) for some foods since 1939 (Hardenburg, 1967; Fisher, 1981). However, only very recently have studies been initiated to interpret the role of the ingredients and processes used to generate edible barriers (e.g., Martin-Polo et al., 1992a,b).

The considerable efforts in developing edible films and coating from proteins and polysaccharides have not yet produced many commercial applications. This can be traced to the success of the polymer industry in supplying the food industry with packaging materials having a wide range of functional properties and economic advantages. But considering the number of advantages of edible film and coating technologies over these traditional polymeric materials as well as the rapidly expanding research efforts on alternative materials, it is reasonable to anticipate that the future in food packaging belongs to edible films (Gennadios and Weller, 1990).

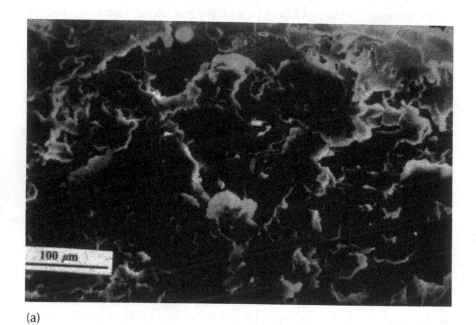

(a)

FIG. 15.4   Scanning electron micrographs of edible films cast from methyl-cellulose and paraffin wax emulsions. Electron beam directed (a) normal to the surface, or (b) at 45°. (*From Martin-Polo et al., 1992a.*)

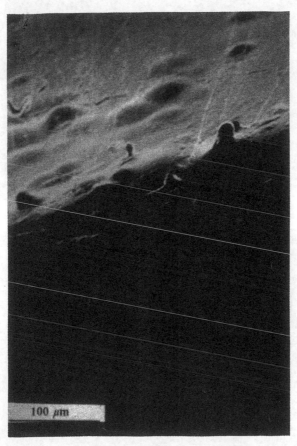

100 μm

(b)

(a)

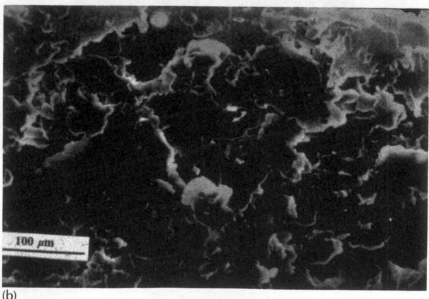

(b)

FIG. 15.5   Scanning electron micrographs of coated cellophane films with 100% n-octacosane and (b) 10% *n*-octacosane in *n*-hexadecane. (*From Martin-Polo et al., 1992b.*)

REFERENCES

Andres, C. 1984. Natural edible coating has excellent moisture and grease barrier properties. *Food Proc.* **45**(13): 48–49.

Anker, C. A., Foster, G. A., and Loader, M. A. 1972. Method of preparing gluten containing films and coatings. U.S. Patent 3,653,925.

Ashley, R. J. 1985. Permeability and plastic packaging. Ch. 7. In *Polymer Permeability*, J. Comyn (Ed.), p. 269. Elsevier Applied Science Publishers Ltd., London.

ASTM. 1981. Standard test method for oxygen and gas transmission rate through plastic film and sheeting using a coulometric sensor, Designation D 3985-81. In *Annual Book of ASTM Standards*, ASTM, Philadelphia, PA.

ASTM. 1992. Standard test methods for water vapor transmission of materials. Designation E-96-90. In *Annual Book of ASTM Standards*, ASTM, Philadelphia, PA.

Avena-Bustillos, R. J. 1992. Ph.D. thesis, University of California, Davis.

Avena-Bustillos, R. J., Krochta, J. M., Buhlert, J., and Davila-Ortiz, G. 1990. Water vapor permeability of casein-acetylated monoglyceride edible films. Presented at the IFT Annual Meeting, paper no. 639, June 16–20, Anaheim, CA.

Aydt, T. P., Weller, C. L., and Testin, R. F. 1991. Mechanical and barrier properties of edible corn and wheat protein films. *Trans. ASAE* **34**: 207–211.

Banker, G. S., Gore, A. Y., and Swarbrick, J. 1966. Water vapour transmission properties of free polymer films. *J. Pharm. Pharmac.* **18**: 457–466.

Banks, N. H. 1984a. Some effects of TAL-Pro-Long coating on ripening bananas. *J. Exp. Bot.* **35**(150): 127–137.

Banks, N. H. 1984b. Studies of the banana fruit surface in relation to the effect of TAL-Pro-Long coating in gas exchange. *Sci. Hort.* **24**: 279–286.

Bizot, H. 1984. Using the 'G.A.B.' model to construct sorption isotherms. In *Physical Properties of Foods*, R. Jowitt, F. Escher, B. Hallström, H. F. T. Meffert, W. E. L. Spiess, and G. Vos (Ed.), pp. 27–41. Applied Science Publishers, London.

Cairns, J. A., Oswin, C. R., and Paine, F. A. 1974. *Packaging for Climatic Protection*, pp. 4–48. The Institute of Packaging, Newnes-Butterworths, London.

Chu, C. L. 1986. Post-storage application of TAL Pro-long™ on apples from controlled atmosphere storage. *HortScience* **21**: 267–268.

Claypool, L. L. 1940. The waxing of deciduous fruits. *Am. Soc. Hort. Sci.* **37**: 443–447.

Crank, J. 1956. *The Mathematics of Diffusion*, p. 43. Oxford University Press, London.

Dalal, V. B., Eipeson, W. E., and Singh, N. S. 1971. Wax emulsion for fresh fruits and vegetables to extend their storage life. *Indian Food Packer* **25**(5): 9–15.

Davis, E. G. and Huntington, J. N. 1977. New cell for measuring the permeability of film materials. *CSIRO Food Res. Quart.* **37**: 55.

Debeaufort, F., Martin-Polo, M., and Voilley, A. 1993. Polarity homogeneity and structure affect water vapor permeability of model edible films. *J. Food Sci.* **58**: 426–429, 434.

Edmond, J. P., Senn, T. L., and Andrews, F. S. 1964. *Fundamentals of Horticulture*. McGraw-Hill Book Co., New York.

Edmond, J. P., Castaigne, F. Toupin, C. J., and Desilets, D. 1991. Mathematical modeling of gas exchange in modified atmosphere packaging. *ASAE* **34**(1): 239–245.

Erbil, H. Y. and Muftugil, N. 1986. Lengthening the postharvest life of peach by coating with hydrophobic emulsions. *J. Food Proc. Pres.* **10**: 269–279.

Fisher, K. D. 1981. Evaluation of the health aspects of shellac and shellac wax as food ingredients. Report FDA/BF-82/37. Federation of American Societies for Experimental Biology, Bethesda, MD.

Gennadios, A. and Weller, C. L. 1990. Edible films and coatings from wheat and corn proteins. *Food Technol.* **44**(10): 63–68.

Gennadios, A. and Weller, C. L. 1991. Moisture sorption of edible films. Paper No. 91-6521 presented at the ASAE Winter Meeting, December 17–20, Chicago, IL.

Gennadios, A., Weller, C. L., and Testin, R. F. 1991. Modifications of physical and barrier properties of edible wheat gluten-based films. Presented at the Conference of Food Engineering, March 10–12, Chicago, IL.

Gennadios, A., Weller, C. L., and Testin, R. F. 1993. Temperature effect on oxygen permeability of edible protein-based films. *J. Food Sci.* **58**: 212–214.

Glicksman, M. 1982. Functional properties. In *Food Hydrocolloids*, Vol. 1, M. Glicksman (Ed.), p. 47. CRC Press, Inc., Boca Raton, FL.

Gontard, N., Guilbert, S. G., and Cuq, J. L. 1992. Edible wheat gluten films: Influence of the main process variables on film properties using response surface methodology. *J. Food Sci.* **57**: 190–195, 199.

Greener, I. K. and Fennema, O. 1989a. Barrier properties and surfaces characteristics of edible bilayer films. *J. Food Sci.* **54**: 1393–1399.

Greener, I. K. and Fennema, O. 1989b. Evaluation of edible, bilayer films for use as moisture barriers for foods. *J. Food Sci.* **54**: 1400–1406.

Guilbert, S. G. 1986. Technology and application of edible protective films. In *Food Packaging and Preservation—Theory and Practice*, M. Mathlouthi (Ed.), pp. 371–394. Elsevier Applied Science Publishing Co., London, England.

Guilbert, S. G. and Biquet, B. 1989. Les films et enrobages comestibles. In *L'emballage des denrées alimentaires de grande consommation*, G. Bureau and J. J. Multon, (Ed.), pp. 320–359, Tech. et Doc. Lavoisier, Paris.

Hagenmaier, R. D. and Shaw, P. E. 1990. Moisture permeability of edible films with fatty acid and (hydroxypropyl)methylcellulose. *J. Agric. Food Chem.* **38**: 1799–1803.

Hagenmaier, R. D. and Shaw, P. E. 1991a. Permeability of coatings made with emulsified polyethylene wax. *J. Agric. Food Chem.* **39**: 1705–1708.

Hagenmaier, R. D. and Shaw, P. E. 1991b. Permeability of Shellac coatings to gases and water vapor. *J. Agric. Food Chem.* **39**: 825–829.

Hagenmaier, R. D. and Shaw, P. E. 1992. Gas permeability of fruit coating waxes. *J. Am. Soc. Hort. Sci.* **117**: 105–109.

Hall, D. J. 1988. Comparative activity of selected food preservatives as citrus postharvest fungicides. *Proc. Florida State Hort. Soc.* **101**: 184–187.

Halsey, G. 1948. Physical adsorption of non uniform surfaces. *J. Chem. Phys.* **16**: 931–937.

Hardenburg, R. E. 1967. *Wax and related coatings for horticultural products. A bibliography*. Agricultural Research Service Bulletin No. 51-15, United State Department of Agriculture, Washington, DC.

Hatzidimitriu, E., Guilbert, S. G., and Loukakis, G. 1987. Odor barriers properties of multi-layer packaging films at different relative humidities. *J. Food Sci.* **52**: 472–474.

Kader, A. A. 1985. Postharvest biology and technology: An overview. Ch. 2. In *Postharvest Technology of Horticultural Crops*. Division of Agriculture and Natural Resources, Cooperative Extension, University of California, Davis, CA.

Kamper, S. L. and Fennema, O. 1984b. Water vapor permeability of an
edible, fatty acid, bilayer film. *J. Food Sci.* **49**: 1482–1485.

Kamper, S. L. and Fennema, O. 1985. Use of an edible film to maintain
water vapor gradients in food. *J. Food Sci.* **50**: 382–384.

Kaplan, H. J. 1986. Washing, waxing and color-adding. In *Fresh Citrus
Fruits*, W. F. Wardowski, S. Nagi, and W. Grierson (Ed.), p. 379. AVI,
Wesport, CT.

Karel, M. 1975. Protective packaging of foods. Ch. 12. In *Principles of
Food Science. Part II: Physical Principles of Food Preservation*, M. Karel,
O. R. Fennema, and D. B. Lund (Ed.), p. 399. Marcel Dekker, Inc.,
New York.

Kester, J. J. and Fennema, O. 1986. Edible films and coatings: a review.
Food Technol. **40**(12): 47–59.

Kester, J. J. and Fennema, O. 1989a. An edible film of lipids and cellulose
ethers: performance in a model frozen-food system. *J. Food Sci.* **54**:
1390–1392.

Kester, J. J. and Fennema, O. R. 1989b. An edible film of lipids and
cellulose ethers: barrier properties to moisture vapor transmission
and structural evaluation. *J. Food Sci.* **54**: 1383–1389.

Kidd, F. and West, C. 1927. Gas storage of fruit. Food Investigation Spec.
Rept. 30, Dept. of Science and Industrial Research, Ditton Laboratory,
East Malling, Kent, U.K.

Knee, M. 1972. Anthocyanin, carotenoid and chlorophyll changes in the
peel of Cox's Orange Pippin apples during ripening on and off the
trees. *J. Exp. Bot.* **23**: 184–196.

Knee, M. 1982. Fruit Softening III. Requirements for oxygen and pH
effects. *J. Exp. Bot.* **33**: 1263–1269.

Krochta, J. M. 1992. Control of mass transfer in foods with edible coatings and films. Ch. 39. In *Advances in Food Engineering*, R. P. Singh
and M. A. Wirakartakusumah (Ed.), pp. 517–538. CRC Press, Inc.,
Boca Raton, FL.

Krochta, J. M., Hudson, J. S., and Avena-Bustillos, R. J. 1990a. Casein-
acetylated monoglyceride coatings for sliced apple products. Presented at the IFT Annual Meeting, paper no. 762, June 16–20, Anaheim, CA.

Krochta, J. M., Pavlath, A. E., and Goodman, N. 1990b. Edible films
from casein-lipid emulsions for lightly processed fruits and vegetables.
In *Engineering and Food, Proceedings of the Fifth International Congress on
Engineering and Food*, W. E. L. Spiess and H. Schubert, (Ed.), Vol. 2.
Elsevier Applied Science Publishing Co., London.

Krochta, J. M., Hudson, J. S., Camirand, W. M., and Pavlath, A. E. 1988. Edible films for lightly-processed fruits and vegetables. Paper No. 88-6523 presented at the ASAE Winter Meeting, December 13–16, Chicago, IL.

Labuza, T. P. 1984. *Moisture Sorption: Practical Aspects of Isotherm Measurement and Use.* American Association of Cereal Chemists, St. Paul, MN.

Lieberman, E. R. and Guilbert, S. G. 1973. Gas permeation of collagen films as affected by crosslinkage, moisture, and plasticizer content. *J. Polymer Sci.* **41**: 33–43.

Loncin, M. 1986. Mass transfer and permeability. In *Food Packaging and Preservation. Theory and Practice*, M. Mathlouthi (Ed.), pp. 1–6. Elsevier Applied Science Publishing Co., London.

Mahmoud, R. and Savello, P. A. 1992. Mechanical properties of and water vapor transferability through whey protein films. *J. Dairy Sci.* **75**: 942–946.

Mahmoud, R. and Savello, P. A. 1993. Solubility and hydrolyzability of films produced by transglutaminase catalytic crosslinking of whey protein. *J. Dairy Sci.* **76**: 29–35.

Manheim, C. and Passy, N. 1985. Choice of package for food with specific considerations of water activities. In *Properties of Water in Foods in Relation to Quality and Stability*, D. Simatos and J. L. Multon (Ed.), p. 375. Martinus Nijhoff Publishing, Dordrecht, The Netherlands.

Martin-Polo, M., Mauguin, C., and Voilley, A. 1992a. Hydrophobic films and their efficiency against moisture transfer. 1. Influence of film preparation techniques. *J. Agric. Food Chem.* **40**: 407–412.

Martin-Polo, M., Mauguin, C., and Voilley, A. 1992b. Hydrophobic films and their efficiency against moisture transfer. 2. Influence of physical state. *J. Agric. Food Chem.* **40**: 413–418.

McHugh, T. H. and Krochta, J. M. 1993a. Water vapor permeability properties of whey protein lipid emulsion films. *J. Am. Oil Chem. Soc.:* In review.

McHugh, T. H. and Krochta, J. M. 1993b. Sorbitol-versus glycerol-plasticized whey protein edible films: Integrated oxygen permeability and tensile property evaluation. *J. Agric. Food Chem.:* In review.

McHugh, T. H., Aujard J.-F., and Krochta, J. M. 1993. Plasticized whey protein edible films: water vapor permeability properties. *J. Food Sci.:* In review.

Nisperos-Carriedo, M. O., Shaw, P. E., and Baldwin, E. A. 1990. Changes in volatile flavor components of pineapple orange juice as influenced by the application of lipid and composite films. *J. Agric. Food Chem.* **38**: 1382–1387.

Oswin, C. R. 1946. The kinetics of package life. III Isotherm. *J. Chem. Ind.* (London) **65**: 419–421.

Pascat, B. 1986. Study of some factors affecting permeability. In *Food Packaging and Preservation. Theory and Practice*, M. Mathlouthi (Ed.), pp. 7–24. Elsevier Applied Science Publishing Co., London.

Pennisi, E. 1992. Sealed in edible film. Can consumable films pass muster with finicky eaters? *Sci. News* **141**(1): 12–13.

Rico-Peña, D. C. and Torres, J. A. 1990a. Edible methylcellulose-based film as moisture impermeable barriers in ice cream cones. *J. Food Sci.* **55**: 1468–1469.

Rico-Peña, D. R. and Torres, J. A. 1990b. Oxygen transmission rate of an edible methylcellulose-palmitic acid film. *J. Food Proc. Eng.* **13**: 125–133.

Rico-Peña, D. R. and Torres, J. A. 1991. Sorbic acid and K-sorbate permeability of an edible coating: water activity and pH effects. *J. Food Sci.* **56**: 497–499.

Rogers, C. E. 1985. Permeation of gases and vapours in polymers. In *Polymer Permeability*, J. Comyn (Ed.), p. 11. Elsevier Applied Science Publishers Ltd., London.

Russell, M. J. 1989. Barrier plastics. *Food Eng.* (May): 89–101.

Salame, M. 1986. Barrier polymers. In *Encyclopedia of Packaging Technology*, M. Bakker (Ed.), pp. 48–54. John Wiley & Sons, New York.

Schwartzberg, H. G. 1986. Modelling of gas and vapour transport through hydrophilic films. In *Food Packaging and Preservation. Theory and Practice*, M. Mathlouthi (Ed.), pp. 115–135. Elsevier Applied Science Publishing Co., London.

Shetty, K. K., Kochan, W. J., and Dwelle, R. B. 1989. Use of heat-shrinkable plastic film to extend shelf-life of 'Russet Burbank' potatoes. *HortScience* **24**: 643–646.

Smith S., Geeson, J., and Stow, J. 1987. Production of modified atmospheres in deciduous fruits by the use of films and coatings. *HortScience* **22**: 772–776.

Smith, S. E. 1947. The sorption of water by high polymers. *J. Am. Chem. Soc.* **69**: 646–651.

Smock, R. M. 1935. Certain effects of wax treatments on various varieties of apples and pears. *Am. Soc. Hort. Sci.* **33**: 284–289.

Solomos, T. 1987. Principles of gas exchange in bulky plants tissues. *HortScience* **22**: 766–771.

Taoukis, P. S. 1989. Temperature abuse indicators. In *Minimally Processed Refrigerated Foods. Proceedings of the 1989 IFT Short Course*, pp. 77–107, June 24–24, Chicago, IL. Institute of Food Technologists, Chicago, IL.

Taylor, C. C. 1986. Cellophane. In *Encyclopedia of Packaging Technology*, M. Bakker (Ed.), pp. 159–163. John Wiley & Sons, New York.

Torres, J. A. 1987. Microbial stabilization of intermediate moisture foods. Ch. 14. In *Water Activity: Theory and Applications*. L. B. Rockland, and L. R. Beuchat (Ed.), pp. 329–368. Marcel Dekker, Inc., New York.

Torres, J. A., and Karel, M., 1985. Microbial stabilization of intermediate moisture food surfaces. III. Effect of surface pH control and surface preservative concentration on microbial stability of an intermediate moisture cheese analog. *J. Food Proc. Pres.* 9(2): 107–119.

Torres, J. A., Motoki, M., and Karel, M. 1985. Microbial stabilization of intermediate moisture food surfaces. I. Control of surface preservative concentration. *J. Food Proc. Pres.* 9(2): 75–92.

Ukai, N., Ishibashi, S., Tsutsumi, T., and Marakami, K. 1976. Preservation of agricultural products. U.S. Patent No. 3,997,674.

Vojdani, F. and Torres, J. A. 1989a. Potassium sorbate permeability of methylcellulose and hydroxypropyl methylcellulose multi-layer films. *J. Food Proc. Pres.* 13: 417–430.

Vojdani, F. and Torres, J. A. 1989b. Potassium sorbate permeability of polysaccharide films: Chitosan, methylcellulose and hydroxypropyl methylcellulose. *J. Food Proc. Eng.* 12: 33–48.

Vojdani, F. and Torres, J. A. 1990. Potassium sorbate permeability of methylcellulose and hydroxypropyl methylcellulose films: Effect of fatty acids. *J. Food Sci.* 55: 841–846.

Wolf, W., Spiess, W. E. L., Jung, G., Weisser, H., Bizot, H., and Duckworth, R. B. 1984. The water-vapour sorption isotherms of microcrystalline cellulose (MCC) and of purified starch. Results of a collaborative study. *J. Food Eng.* 3: 51–73.

Index

Milton Keynes UK
Ingram Content Group UK Ltd.
UKHW020007071024
449327UK00031B/2682